Microchannel Phase Change Transport Phenomena

T0339634

Microchannel Phase Change Transport Phenomena

Edited by

Sujoy K Saha

Department of Mechanical Engineering,
Indian Institute of Engineering Science
and Technology, Shibpur, Howrah,
West Bengal, India

AMSTERDAM • BOSTON • HEIDELBERG • LONDON
NEW YORK • OXFORD • PARIS • SAN DIEGO
SAN FRANCISCO • SINGAPORE • SYDNEY • TOKYO
Butterworth-Heinemann is an imprint of Elsevier

Butterworth-Heinemann is an imprint of Elsevier
The Boulevard, Langford Lane, Kidlington, Oxford OX5 1GB, UK
225 Wyman Street, Waltham, MA 02451, USA

ISBN: 978-0-12-804318-9

British Library Cataloguing-in-Publication Data
A catalogue record for this book is available from the British Library

Library of Congress Cataloging-in-Publication Data
A catalog record for this book is available from the Library of Congress

For information on all Butterworth-Heinemann publications
visit our website at http://store.elsevier.com/

Contents

List of Contributors

Gian P. Celata Energy Technology Department, ENEA Casaccia Research Centre, S. M. Galeria, Rome, Italy

Lixin Cheng Department of Engineering, Aarhus University, Aarhus, Denmark

Jaqueline D. Da Silva Heat Transfer Research Group, Department of Mechanical Engineering, Escola de Engenharia de São Carlos (EESC), University of São Paulo (USP), São Carlos, São Paulo, Brazil

A.K. Das Department of Mechanical and Industrial Engineering, Indian Institute of Technology Roorkee, Roorkee, Uttarakhand, India

P.K. Das Department of Mechanical Engineering, Indian Institute of Technology Kharagpur, Kharagpur, West Bengal, India

Durga P. Ghosh Department of Mechanical Engineering, Indian Institute of Technology Patna, Patna, Bihar, India

Diptimoy Mohanty Department of Mechanical Engineering, Indian Institute of Technology Bombay, Mumbai, Maharashtra, India

Rishi Raj Department of Mechanical Engineering, Indian Institute of Technology Patna, Patna, Bihar, India

Gherhardt Ribatski Heat Transfer Research Group, Department of Mechanical Engineering, Escola de Engenharia de São Carlos (EESC), University of São Paulo (USP), São Carlos, São Paulo, Brazil

Sandip K. Saha Department of Mechanical Engineering, Indian Institute of Technology Bombay, Mumbai, Maharashtra, India

Sujoy K. Saha Department of Mechanical Engineering, Indian Institute of Engineering Science and Technology, Shibpur, Howrah, West Bengal, India

Foreword by G.F. Hewitt

Phase change phenomena in boiling and condensation are extremely complex even in channels of normal commercial size. When the channel is reduced to smaller sizes (as in microchannels), the structure of the flow may change dramatically and the models used for channels of larger size may no longer be appropriate. Heat transfer in microchannels has been studied extensively in recent years because of the importance of heat removal from electronic devices whose scale diminishes and whose power increases as time goes on. There has been extensive support for such work (and generally decreasing support for larger-scale studies), and this has led the heat transfer community to focus increasingly on the microchannel case. Even for adiabatic two-phase flow, large differences are observed from the usual correlations of flow regimes, void fraction, and pressure drop as apply in large-diameter channels. It is important to distinguish between regimens in which surface tension has a dominant role (e.g., bubble flow, slug flow) from those (e.g., annular flow) where the influence of surface tension is less. It is generally true that studies of flow regimens for larger channels can rarely be applied and extended to microchannels. For condensation heat transfer, the data for microchannels are overpredicted by relationships developed for larger channels. For evaporative (boiling) heat transfer, the regimens in microchannels also differ from those observed for larger channels, and this leads to characteristic differences in heat transfer behavior. This book is devoted to providing a tutorial information source on phase change (boiling and condensation) phenomena in microchannels, and this is clearly a valuable aim in view of the rapid development of the subject.

G. F. Hewitt, FRS, FRAE
Imperial College, London, UK
Former President, International Centre
for Heat and Mass Transfer

Foreword by Cees W.M. van der Geld

This is a time when hardback books are disappearing from university libraries and the gathering of knowledge becomes quick and flashy through Internet tools. But although hardback encyclopedias might be obsolete, there is still a need for thorough summaries that are well organized and well balanced and contain concentrated information in an accessible format. Such is the book you are reading now.

The books written by Landau and Lifshitz have always been some of my favorite learning books. Although they are overconcise, these books cover many topics in a thorough way. However, it is difficult to imagine that two authors would write a series of books like that in modern times. Two authors simply would not have the time anymore. That is also why the present book is written by several authors, each of them contributing in a particular field of expertise. The book is focused on flow with heat transfer in small-diameter tubes, but this topic is dealt with in an extensive way, summarizing most of the work of the past decades in this area. This makes the book a good read for both beginners and more experienced researchers in this area.

The topic has been very popular and was elaborated on all over the world, in famous laboratories in Israel, Switzerland, Germany, Brazil, China, and many other countries. It therefore stands to reason that the authors of this book also originate from various countries. We must admire them for their ability to conceive and realize a book with nine chapters without doubling certain aspects. The editor, Sujoy Kamar Saha, must have had a great hand in organizing in the endeavor.

This is a book that is bound to last for a long time in our libraries and on our laptops.

Cees van der Geld
Technische Universitiet Eindhoven
President, Assembly of World
Conferences on Experimental Heat Transfer,
Fluid Mechanics and Thermodynamics

Critical Review by Masahiro Kawaji

Studies on phase change transport phenomena in microchannels have grown in popularity since the early 1980s, due mainly to the needs for electronics cooling and for the development of micro heat exchangers, microreactors, and lab-on-a-chip devices. This monograph presents a timely review of past works on fluid flow and phase change heat transfer in microchannels. Many topics, including flow patterns, void fraction and pressure drop in two-phase flow, bubble growth and flow boiling phenomena, critical heat flux, flow instability, and condensation in microchannels, are covered in nine chapters. The authors of these chapters have collected and reviewed a large number of relevant publications, including some well-known studies applicable to conventional channels as a basis for understanding the unique features of transport phenomena in microchannels.

In each chapter, past publications are reviewed in a comprehensive manner. The methodologies and results presented in some publications are described in sufficient detail, so that the readers can gain a good understanding of the key findings without reading the original publications. Many tables are presented that summarize and compare a large number of past experiments including the channel geometry, working fluids, and experimental conditions covered. From these comparisons, one can get a good sense of the nature and type of experiments performed in the past. Many correlations of experimental results and analytical models published to date are also reviewed. They are compared with experimental results to show their predictive abilities. These comparisons give a useful indication of the applicability of the models and correlations to practical applications in electronics cooling and designing micro heat exchangers, microreactors, and microfluidics components. Thus, this book is highly recommended for both experienced researchers and new investigators, enabling them to perform microchannel research with a full understanding of the past accomplishments.

Masahiro Kawaji
Professor, Department of Mechanical Engineering
The City College of New York
The City University of New York
and
Professor Emeritus, Department of Chemical Engineering and Applied Chemistry
University of Toronto

Critical Review by Lounès Tadrist

It is with great pleasure that I review this collective work, edited by Professor Sujoy Kumar Saha, on transport phenomena in microchannels phase change. The two-phase microfluidic field has experienced sustained growth in recent decades, and research has been abundant due to the necessity for understanding the fundamental mechanisms and to design devices using this technique for heat transfer. Indeed, although phase change mechanisms have been studied for several decades for the purpose of applications to the sectors of energy and industry, phase change phenomena in a confined environment such as microchannels have known real development only during the past two decades.

This attraction by the scientific community was strongly motivated by the development of microtechnologies and nanotechnologies and miniaturization of devices such as heat exchangers. The intensification of trade has led the designers of exchangers to reduce the hydraulic diameters of the channels, which can reach submillimeter sizes.

This book presents the knowledge and recent developments on transport phenomena with phase change. It deals with hydrodynamic, thermal, and nucleation aspects of microchannels. The various topics covered in this book enable the reader to discover the various facets of the phenomena involved and their specificities when the geometric confinement becomes important in comparison with the scales of physical phenomena.

Each chapter deals with an important topic in this field of phase change in microchannels. It includes a presentation of the main phenomena encountered, methods of analysis, and key findings in the existing literature. Each chapter includes a comprehensive reference list, and the reader can find ample details in the original reports cited. After general introduction is provided in Chapter 1, Chapters 2 and 3 cover the topics of nucleation, growth of bubbles, flow patterns, void fraction, and film thicknesses. These are the main ingredients encountered in two-phase flow with change of phase. Chapter 4 is devoted to the presentation of the flow boiling heat transfer phenomena with the main correlations for heat transfer. Chapter 5 deals with pressure drop, and several models are presented and compared. Chapter 6 deals with the critical heat flux for flow boiling in microchannels. The part related to two-phase flow instabilities is detailed in Chapter 7. The condensation in microchannels is treated Chapter 8. The main phenomena are also described, together with a detailed review of the main studies and results on heat transfer and pressure drop in the literature. Conclusions are drawn in Chapter 9, which also gives direction for further research.

This is a very interesting book for graduate students who already have basic knowledge of hydrodynamics, thermodynamics, and heat transfer and for researchers and engineers who are interested in the design of microdevices where transport phenomena with phase change in microchannels occur. This is a topic for the future that has not yet been finished; it is being explored in terms of knowledge and will provide great benefits in many applications such as the heat transfer enhancement and thermal control in MEMS.

Lounès Tadrist, Professor
Aix-Marseille Université, CNRS, IUSTI 7343
Technopole Chateau Gombert
5, Rue Enrico Fermi
13453 Marseille Cedex 13
France

Editorial by Sujoy Kumar Saha

Cluster of microchannels are being used for high heat flux thermal management in electronic devices. Single-phase flow is not good for this purpose; phase-change (boiling and condensation) with the flow of small mass inventory of fluid through microchannels is the solution. However, as of now, the thermo-fluid dynamics of physical phenomena occurring in this type of flow, being very complex, are far from being understood clearly. The predictive models, correlations, and experimental findings cannot be taken with a satisfactory level of confidence universally. Intense research is being done all over the world to understand the physics of such flow and the thermo-hydraulic behavior of such systems. New literature is continuously available.

The publishing giant Butterworth-Heinemann (imprint of Elsevier), a leading international publisher of books and eBooks for science and technology, approached the editor of this volume. Therefore, an attempt has been made to collate the most recent related information at one place in this book. We hope the volume will be a timely and welcome addition to the literature for the community of researchers, professionals, and graduate students.

The editor thanks all the contributing authors of this book. They are the internationally acclaimed leading experts in the field. Many other experts could not participate in the program due to their commitments elsewhere; all of them did appreciate the idea, and we have missed the opportunity to work with them.

The Editor joins with all the authors to thank Prof. G. F. Hewitt and Prof. C. W. M. van der Geld profusely for their priceless "Foreword" to the book. We are also thankful to Prof. M. Kawaji and Prof. L. Tadrist for their critical review of the book. Thanks are also due to the Staff of the publishing house for efficient handling of the project with their professional excellence.

Sujoy Kumar Saha
IIEST Shibpur, India

Introduction

Sujoy K. Saha[1], Gian P. Celata[2]
[1]Department of Mechanical Engineering, Indian Institute of Engineering Science and Technology, Shibpur, Howrah, West Bengal, India; [2]Energy Technology Department, ENEA Casaccia Research Centre, S. M. Galeria, Rome, Italy

The concept of a heat exchanger and its continuous technological development is essentially need based. Efficient heat recovery for energy loss minimization, environmental protection, and the role of heat exchangers in improving the performance of heat engines are among the motivating factors, leading to the development of various configurations of industrial heat exchangers and compact heat exchangers. The development of compact heat exchangers is driven primarily from the transportation sector (automotive, aircraft, submarine, and spaceship), where space limitation and weight are prime factors. The present electronic era is witnessing tremendous development, and the integration of electronic technology with mechanical devices has improved instrumentation and control, leading to improved performance. Although technological advances in electronic industry have made it possible to add many features to one platform, this has resulted in increased component concentration at the chip level, which, in turn, causes greater heat generation from a smaller area and might lead to overheating of the components if the heat is not readily dissipated. Therefore, the cooling of electronic devices has emerged as a new challenge to researchers and scientists. The task of removing a large amount of dispersed heat from a constrained small space is often beyond the capability of conventional cooling techniques. New methods of heat removal at least one order larger than that of conventional ones are required. To solve this problem, in early 1981, Tuckerman and Pease [1] introduced microchannel heat sink technology by performing experiments on a silicon-based microchannel heat sink for electronic cooling. They predicted that single-phase forced convective cooling in microchannels could remove heat at a rate of 1000 W/cm^2. Unfortunately, tests with water yielded enormous pressure drops at high fluxes (about 1 bar at 181 W/cm^2). Nevertheless, after this work, many investigations have been conducted with the purpose of gaining further understanding of heat transfer and fluid flow within microchannel heat sinks. To differentiate microchannels from conventional heat exchangers, different criteria for classifications have been proposed. Mehendale et al. [2] proposed classification based simply on the dimensions (hydraulic diameter) of the channels, as given in Table 1.1.

On the other hand, classification provided by Kandlikar et al. [3] is based on flow consideration, as given in Table 1.2.

Although the classification has been developed mainly from gas flow considerations, they are recommended for single- and two-phase flow applications to provide

Microchannel Phase Change Transport Phenomena. http://dx.doi.org/10.1016/B978-0-12-804318-9.00001-7

Table 1.1 **Classification of Heat Exchangers [2]**

Class of Heat Exchanger	Range of Hydraulic Diameter (D_h)
Conventional	$D_h \geq 6$ mm
Compact	1 mm $\leq D_h \leq 6$ mm
Macro or mini	100 μm $\leq D_h \leq 1$ mm
Micro	1 μm $\leq D_h \leq 100$ μm

From Mehendale et al. [2].

Table 1.2 **Classification of Heat Exchangers [3]**

Class of Heat Exchanger	Range of Hydraulic Diameter (D_h)
Conventional	$D_h \geq 3$ mm
Macro or mini	200 μm $\leq D_h \leq 3$ mm
Micro	10 μm $\leq D_h \leq 200$ μm

From Kandlikar et al. [3].

uniformity in the channel classification. There is no doubt that electronics industry has created a market for miniature heat exchangers. The need for micro heat exchangers is a result of the miniaturization of the electronics, which leads to denser packaging of components. This has led to higher heat fluxes, and consequently, the cooling problem is being obviated by new manufacturing methods to fabricate complex geometries on a very small scale. Material science engineers and researchers are the key players since they have made possible new manufacturing methods for micro design and are striving for low production costs of microchannels. Intense research activities are going on related to the mini-/microchannel fabrication and thermohydraulic performance of fluids in micro-passages. Depending on the phase of the coolant that flows through the microchannels, the study of heat transfer and fluid flow in microchannels can be divided into two subsections: single-phase flow and two-phase flow.

If single-phase cooling is compared with two-phase cooling, it is observed that there is a linear increase in stream temperature with increasing heat; this linear temperature rise contributes to greater surface temperature gradients. Two-phase cooling offers several inherent advantages over single-phase cooling. It is possible to achieve uniformity in temperature in two-phase cooling even for high heat fluxes. Also, because two-phase cooling has a larger capacity to absorb heat, comparatively higher heat fluxes can be dissipated by circulating a smaller quantity of coolant, compared with single-phase cooling. In the present book, we sought to cover different aspects of microchannel phase change transport phenomena.

References

[1] D.B. Tuckerman, R.F. Pease, High performance heat sinking for VLSI, IEEE Electron. Device Lett. EDL-2 (1981) 126–129.

[2] S.S. Mehendale, A.M. Jacobi, R.K. Shah, Fluid flow and heat transfer at micro and meso scales with application to heat exchanger design, Appl. Mech. Rev. 53 (7) (2000) 175–193.

[3] S.G. Kandlikar, J. Shailesh, S. Tian, Effect of channel roughness on heat transfer and fluid flow characteristics at low Reynolds numbers in small diameter tubes, in: Proceedings of 35th National Heat Transfer Conference, ASME, Anaheim, CA, 2001. Paper 12134.

Onset of Nucleate Boiling, Void Fraction, and Liquid Film Thickness

2

Durga P. Ghosh[1], Rishi Raj[1], Diptimoy Mohanty[2], Sandip K. Saha[2]
[1]Department of Mechanical Engineering, Indian Institute of Technology Patna, Patna, Bihar, India; [2]Department of Mechanical Engineering, Indian Institute of Technology Bombay, Mumbai, Maharashtra, India

2.1 Onset of Nucleate Boiling

2.1.1 Introduction

Onset of nucleate boiling (ONB) refers to the transition of heat transfer mode from the single-phase liquid convection to a combination of convection and nucleate boiling. It is identified by the formation of vapor bubbles in a pool of liquid or during a flow. Nucleation is broadly divided into two categories: homogeneous nucleation and heterogeneous nucleation. The formation of a vapor bubble completely inside a superheated liquid mass is termed "homogeneous nucleation." Theoretically, the upper limit on the superheat for homogeneous nucleation within a liquid mass at constant pressure is very high and equal to the spinodal limit that results from thermodynamic consideration [1]. However, for most practical systems, nucleation is typically observed somewhere on the walls of the containment vessel or a solid–liquid interface elsewhere in the system. Such type of nucleation on the interface of a solid and liquid is termed "heterogeneous nucleation."

While the theoretical value of superheat required for facilitating the heterogeneous nucleation from an atomically smooth surface has been estimated to be very high, and often close to the homogeneous nucleation limit, most experiments however report a significantly lower value. This is usually explained by the fact that typical surfaces used in practical applications (including micro channels) are far from atomically smooth and the fluid is contaminated. As a result, the surface is not completely wet by the liquid and there is always some entrapment of vapor/gas in the cavities or around the contaminant. Hsu [2] explained that the trapped vapor/gas bubbles in a pool of liquid can grow only if the temperature of the surrounding fluid is greater than the saturation temperature of the vapor present inside the bubble (i.e., the superheat criteria are met). Since the vaporization at the liquid–vapor/gas interface in the cavity is relatively easier, the estimated values of superheat for the onset of nucleate boiling from Hsu's criteria are comparable to the typical superheats observed in experiments. Considering the fact that most practical engineering surfaces are either contaminated or contain irregularities,

Microchannel Phase Change Transport Phenomena. http://dx.doi.org/10.1016/B978-0-12-804318-9.00002-9

the discussion in this chapter will be limited to the practical case of heterogeneous nucleation from entrapped vapor/gas in the cavities.

While the original model of Hsu was developed for pool boiling, most of the models for flow boiling are also inspired by the vapor/gas entrapment theory of Hsu [2] due to its simplicity and the realistic prediction values. The effect of velocity during flow boiling in these models is captured by accounting for the resulting decrease in the thickness of the thermal boundary layer due to the flow. A recent surge in activities pertaining to flow in boiling in microchannels has seen an increased focus on additionally modeling the change in bubble dynamics and boundary layer thickness to capture the effect of confinement due to microchannel walls. Contrary to the conventional sized channels where the bubble growth is mostly affected by liquid subcooling, the effect of increased heating from side walls modifies the bubbles' dynamics due to the presence of superheated liquid around the bubble and the bubble growth is hindered by confining walls of the microchannels.

In this section, we present a detailed discussion on the various aspects of onset of nucleate boiling during flow boiling in microchannels. We start with a discussion of the classic model of Hsu [2] for pool boiling to explain the concept of vapor/gas entrapment. Subsequent studies that capture the effect of various other important parameters during flow boiling in microchannels, such as flow rate, microchannel geometry, subcooling, dissolved gas concentration, and contact angle, are discussed next. The chapter closes with a discussion on the state-of-the-art pertaining to the use of micro-/nano-structured surfaces and various coatings to tune nucleation for an effective increase in boiling heat transfer coefficients.

2.1.2 Nucleation during Pool Boiling

In this section, we discuss the semitheoretical model of nucleation proposed by Hsu [2]. It was proposed that for a vapor/gas bubble to grow in a pool of liquid (Fig. 2.1), the temperature of the liquid surrounding the bubble should be greater than the saturation temperature of the vapor inside:

$$T_l > T_b \tag{2.1}$$

Figure 2.1 Schematic of Hsu's model [2] for bubble survival.

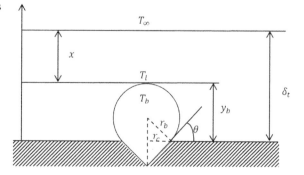

where T_b and T_l represent the temperature of vapor inside the bubble and liquid surrounding the bubble, respectively. The temperature profile in the liquid close to wall was modeled using the transient conduction in a semi-infinite medium as follows:

$$\frac{\partial(\Delta T)}{\partial t} = \alpha \left(\frac{\partial(\Delta T)}{\partial x^2} \right) \tag{2.2}$$

where $\Delta T = T - T_\infty$. The solution to Eq. (2.2) was obtained by considering the following boundary conditions:

$$\Delta T = 0 \text{ at } t = 0, \ \Delta T_w = T_w - T_\infty \text{ at } y = 0 \text{ and } \Delta T = 0 \text{ at } y = \delta_t$$

where δ_t and T_w are the thickness of the thermal boundary layer and the temperature of the superheated wall, respectively. The solution yields:

$$\varepsilon = \frac{\Delta T}{\Delta T_w} = \varphi + \frac{2}{\pi} \sum_{n=1}^{\infty} \frac{\cos n\pi}{n} \sin n\pi\varphi e^{-n^2\pi^2\tau} \tag{2.3}$$

where $\varphi = \frac{x}{\delta_t}$ and $\tau = \frac{\alpha t}{\delta_t}$. The bubble temperature was determined by using the Clausius–Clapeyron equation for superheat together with Young–Laplace equation to provide the following expression:

$$T_b - T_{sat} = +\frac{2\sigma T_{sat}}{h_{fg}\rho_v r_b} \tag{2.4}$$

where the Clausius–Clapeyron equation is given as $\frac{dp}{dT} = \frac{h_{fg}\rho_v}{T}$ and the Young–Laplace equation is given as $p_v - p_l = \frac{2\sigma}{r_b}$ [1], where p_v is pressure of the vapor inside the bubble, p_l is pressure of the liquid surrounding the bubble, r_b is bubble radius, $\Delta T_b = T_b - T_\infty$, $\Delta T_{sub} = T_{sat} - T_\infty$, σ is surface tension, T_{sat} is saturation temperature of liquid, h_{fg} is latent heat of vaporization, ρ_v is vapor density, and r_b is bubble radius. The bubble height, y_b, can be calculated from simple geometrical considerations, shown in Fig. 2.1 as,

$$y_b = (1 + \cos\theta)r_b \tag{2.5}$$

$$y_b = \delta_t - x \tag{2.6}$$

where θ is the contact angle for the given combination of solid surface and liquid. Equations (2.4)–(2.6) can be rearranged to derive a relationship as follows:

$$\Delta T_b = \Delta T_{sub} + \frac{2\sigma T_{sat}}{h_{fg}\rho_v \left(\frac{\delta_t - x}{C} \right)} \tag{2.7}$$

where $C = 1 + \cos \theta$. This equation can be rearranged as follows:

$$\varepsilon_b = \frac{\Delta T_b}{\Delta T_w} = \frac{\Delta T_{sub}}{\Delta T_w} + \frac{2\sigma T_{sat} C}{h_{fg} \rho_v \delta_t \Delta T_w (1 - \varphi)} \tag{2.8}$$

It is well known that additional cavities are gradually flooded by the surrounding liquid and no active nucleation sites may be found if the waiting period is long enough. Hence, for $\tau = \infty$, Eq. (2.3) reduces to $\varepsilon = \varphi$. Now, the limiting values can be determined by writing $\varepsilon_b = \varphi_b$. Equation (2.8) can be written as:

$$\varphi_b = \frac{\Delta T_{sub}}{\Delta T_w} + \frac{2\sigma T_{sat} C}{h_{fg} \rho_v \delta_t \Delta T_w (1 - \varphi_b)}$$

Solving for φ_b, we have,

$$\varphi_b = \frac{\left(1 + \frac{\Delta T_{sub}}{\Delta T_w}\right) \pm \sqrt{\left(1 + \frac{\Delta T_{sub}}{\Delta T_w}\right)^2 - \frac{8\sigma T_{sat} C}{h_{fg} \rho_v \Delta T_w \delta_t}}}{2} \tag{2.9}$$

Hsu assumed the bubble to be a truncated sphere as shown in Fig. 2.1. The relation between bubble radius r_b, cavity radius r_c, and bubble height y_b with the assumed contact angle of $53.1°$ was used to arrive at the following:

$$y_b = 2r_c \tag{2.10}$$

Hence, solving for φ_b and using Eqs (2.6) and (2.10), the final equation for active range of cavities was derived as follows:

$$\{r_{cmin}, r_{cmax}\} = \delta_t / 4 \left(\frac{\Delta T_w}{\Delta T_w + \Delta T_{sub}} \right)$$
$$\times \left[1 \pm \sqrt{1 - \frac{12.8\sigma T_{sat}\rho_v (\Delta T_w + \Delta T_{sub})}{\rho_v h_{fg} \delta_t \Delta T_w^2}} \right] \tag{2.11}$$

Based on Eq. (2.11), a graph showing the active range of cavities for different thermal boundary layer thicknesses was plotted for saturated water under atmospheric pressure (Fig. 2.2). Important conclusions derived from the plot can be summarized as:

1. A certain value of superheat is required before any cavity becomes active,
2. Although the superheat condition is satisfied, there exist only a finite range of cavities that can become active nucleation sites, and
3. Nucleation of bubbles becomes difficult and the range of active cavities shrinks when the thermal boundary layer thickness is reduced.

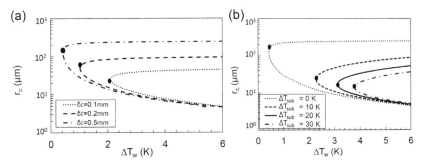

Figure 2.2 Active range of cavities (a) for different thermal boundary layer thickness during saturated boiling, and (b) for different levels of subcooling and a constant thermal boundary layer thickness of 0.5 mm [2] (black dots in the represent ONB).

While Hsu's criteria for pool boiling does not depend on the size of the heaters, experiments reported in the literature widely discuss the delay in nucleation on reducing the heater size (Raj et al. [3]). The inability to capture this effect can be attributed to the simplifying assumption made in Hsu's model where the heater was assumed to be infinite such that the thickness of the thermal boundary layer could be captured using the model for transient conduction in semi-infinite medium. However, practical experiments are performed on finite size heaters where the thickness of the thermal boundary layer is reduced due to the increased losses from the boundary/sides of the heater. The decrease in the thickness of the thermal boundary layer due to the decrease in the heater size would accordingly imply a delay in nucleation as per Fig. 2.2.

2.1.3 Effect of Various Parameters on ONB during Flow Boiling

In the previous section, we discussed Hsu's criteria, which provide the basic framework behind the nucleation of a bubble from a trapper vapor/gas in a cavity during pool boiling. However, flow boiling in microchannels is different in various ways, including the fact that the imposed flow alters the thickness of the thermal boundary layer and, unlike flat plate heaters mostly used during pool boiling, heating may additionally be facilitated from the sides of the microchannel, again influencing the boundary layer thickness. Moreover, bubble behavior during ONB largely depends on the conditions under which the experiments are carried out. Table 2.1 provides a summary of various correlations and models that have been developed to capture the effect of various parameters on ONB during flow boiling in microchannels.

2.1.3.1 Effect of Flow Rate

The thermal boundary layer thickness is altered due to the presence of flow during boiling in microchannels. If the flow rates are high, the thermal boundary layer thickness is reduced. As a result, the ONB is delayed and the range of active cavities at a given superheat is also found to shrink (Fig. 2.2). Further increase in flow rates may

Table 2.1 ONB Models for Different Parametric Conditions

Study	ONB Models Developed	Parametric Effects Considered	Remarks
Hsu [2]	$q_{onb} = k_f h_{fg} \rho_v (\Delta T_w)^2 / 12.8\sigma T_{sat}$ $\{r_{cmin}, r_{cmax}\} = \delta_t / 4 \left(\dfrac{\Delta T_w}{\Delta T_w + \Delta T_{sub}} \right)$ $\times \left[1 \pm \sqrt{ 1 - \dfrac{12.8\sigma T_{sat}\rho_v (\Delta T_w + \Delta T_{sub})}{\rho_v h_{fg} \delta_t \Delta T_w^2} } \right]$ (2.11)	• Thermal boundary layer thickness • Superheat • Subcooling	• For a bubble to grow, the minimum temperature of the liquid should be at least equal to the saturation temperature of the vapor inside it. • Assumed truncated shaped bubbles and the bubble top surface to be at a distance of 1.6 r_b away from the wall. • Solved the transient conduction problem. • While originally developed for pool boiling, often used for flow boiling as well. Foundation for the models developed later.
Bergles and Rohsenow [4]	$q_{onb} = 15.60 p^{1.156} (\Delta T_w)^{2.30/p^{0.0234}}$ q_{onb} is in Btu/h, lb/in^2, temperatures are in °F (2.12)	• Thermal boundary layer thickness • Superheat	• The bubble will grow if the temperature of the surrounding liquid at a distance equal to that of

Continued

Author	Notes	Parameters	Eq.	Equation	Remarks
	Temperature of liquid near the wall was calculated by assuming conduction heat transfer.				• bubble radius from the wall is at a temperature above the vapor temperature. • Solution was predicted graphically. • Bubbles of hemispherical shape was assumed.
Sato and Matsumura [5]		• Thermal boundary layer thickness • Velocity • Subcooling	(2.13)	$$q_{onb} = \frac{k h_{fg}}{8\sigma T_s (v_g - v_l)} \left(\frac{D^{0.2} v_v^{0.8}}{0.023 Pr^{0.4} k u^{0.8}} - \Delta T_{sub} \right)^2$$	• Postulated that bubble does not get thermal energy directly from the heating surface, but gains it indirectly from the surrounding liquid. • Assumed steady temperature profile near the heated surface and used equivalent thickness of superheated layer.
Davis and Anderson [6]		• Thermal boundary layer thickness • Superheat • Contact angle • Subcooling	(2.14)	$$q_{onb} = k_f h_{fg} \rho_v (\Delta T_w)^2 / 8\sigma (1 + \cos\theta) T_{sat}$$ $$\{r_{cmin}, r_{cmax}\} = \delta_t \sin\theta/2(1 + \cos\theta) \left(\frac{\Delta T_w}{\Delta T_w + \Delta T_{sub}} \right)$$ $$\times \left[1 \pm \sqrt{1 - \frac{8\sigma T_{sat}\rho_l(\Delta T_w + \Delta T_{sub})}{\rho_v h_{fg}\delta_t \Delta T_w^2}} \right]$$	• Strictly integrated Clasius-Clapeyron equation. • Derived a more precise equation for ONB. • Considered truncated bubble and introduced contact angle as variable.

Table 2.1 ONB Models for Different Parametric Conditions—cont'd

Study	ONB Models Developed	Parametric Effects Considered	Remarks
Kandlikar et al. [7]	$$\{r_{cmin}, r_{cmax}\} = \delta_t \sin\theta/2.2 \left(\frac{\Delta T_w}{\Delta T_w + \Delta T_{sub}}\right)$$ $$\times \left[1 \pm \sqrt{1 - \frac{9.2\sigma T_{sat}\rho_l(\Delta T_w + \Delta T_{sub})}{\rho_v h_{fg}\delta_t \Delta T_w^2}}\right]$$ (2.15) $$\frac{y_s}{r_b} = 1.10$$	• Thermal boundary layer thickness • Superheat • Velocity • Contact angle • Subcooling	• Derived the nucleation criteria by using the temperature in the liquid boundary layer at distance y_s as the temperature at the top of the bubble of radius r_b. • Bubble will grow if this temperature is above the saturation temperature inside the bubble. • Experimentally found that at increased flow rate thermal boundary layer is decreased and hence the active range of cavities also decreases.
Ghiaasiaan and Chedester [8]	$$q_{onb} = k_f h_{fg}\rho_v(\Delta T_w)^2/C\sigma T_{sat}$$ $$C = 22\xi^{-0.765}$$ $$\xi = \frac{\sigma_l - \sigma_w}{\rho_l u R^*}$$ $$R^* = \left[\frac{2\sigma k_f T_s}{q_{onb}''\rho_v h_{fg}}\right]^{(1/2)}$$ (2.16)	• Velocity • Superheat • ξ = ratio of thermocapillary force to aerodynamic force	• ONB occurs when aerodynamic force balances the suppressing effect of the thermocapillary force on the bubble. • Used Davis and Anderson's co-relation to develop a semi empirical model.

| Qu and Mudawar [9] | $$q_{onb} = \frac{(\Delta T_{sub})}{\left(\dfrac{\overline{T}_{w_w}}{k_s}\right) + \dfrac{A_t}{\rho_l c_p u N A_c}}$$ | (2.17) | • Velocity
• Superheat
• Contact angle
• Subcooling | • Dimensionless temperature field $\overline{T}$ is determined across the heat sink unit cell.
• T_{out} is determined from the energy balance equation by assigning a small value to q''.
$$\rho_l c_p u_{in} N A_c$$ $$(T_{out} - T_{in}) = q'' A_t$$
• Liquid temperature T is then found by substituting the values of T_{out} and q'' in $\overline{T}$.
• The value of q'' is increased until a bubble satisfies the superheat criteria.
• The q'' value at the condition is incipient boiling heat flux.
• A bottom wall bubble located close to the corner was first to satisfy super heat criteria. |
| Li and Cheng [10] | $$T_w - T_{sat} = \frac{T_{sat}\exp(C_g)}{h_{fg}\rho_v}\left(\frac{2\sigma}{r} - K_h C_g\right)$$ | (2.18) | • Superheat
• Velocity | • Classic nucleation kinetics theory was |

Table 2.1 ONB Models for Different Parametric Conditions—cont'd

Study	ONB Models Developed	Parametric Effects Considered	Remarks
Liu et al. [11]	$$\sqrt{T_w} - \sqrt{T_{sat}} \geq \sqrt{\frac{2\sigma C q_w}{h_{fg}\rho_v k_f}} \quad (2.19)$$ $$C = 1 + \cos\theta$$ $$\sqrt{T_\infty + \frac{qWL}{\rho_l C_p u (n w_c H_c)} + \left(\frac{\dfrac{a_s}{1+2\eta a_s}\dfrac{w_c+w_w}{H_c}}{(Nu_{fd}k_f)/D_h}\right)q} - \sqrt{T_{sat}} \quad (2.20)$$ $$= \sqrt{\frac{2\sigma Cq\left(\dfrac{a_s}{1+2\eta a_s}\dfrac{w_c+w_w}{H_c}\right)}{h_{fg}\rho_v k_f}}$$ $$\{r_{cmin}, r_{cmax}\} = r_c^* \pm \sqrt{\frac{\left(T_w + \dfrac{2\sigma Cq_w}{\rho_v h_{fg}k_f} - T_{sat}\right)^2 - 4\dfrac{2\sigma Cq_w}{\rho_v h_{fg}k_f}T_w\,\dfrac{\sin\theta}{1+\cos\theta}}{2\dfrac{q_w''}{k_f}}} \quad (2.21)$$	• Superheat • Velocity • Contact angle • Subcooling	• used to obtain fluid nucleation temperature. • Effect of dissolved gas incorporated in the model. • Dissolved gas decreases the bubble emergence temperature. • Bubble is assumed to be truncated sphere. • Bubble nucleus will grow if the temperature of the fluid at distance from the wall equal to bubble height is greater than super heat requirement. • For given wall heat flux, inlet velocity and fluid inlet temperature inequality (19) is used to determine if ONB will occur. • For varying heat flux, the threshold heat flux required to initiate

	$$r_c^* = \dfrac{\left(T_W + \dfrac{2\sigma C q_W}{\rho_v h_{fg} k_f} - T_{sat}\right)}{2\dfrac{q_w}{k_f}} \dfrac{\sin\theta}{1+\cos\theta}$$	(2.22)		ONB is obtained from Eq. (2.20)
Kandlikar [12]	$$q_{onb} = \dfrac{k_f \sin\theta_r}{1.1 r_c} \left[\Delta T_w - \dfrac{2\sigma T_{sat}\sin\theta_r}{h_{fg}\rho_v r_c}\right]$$	(2.23)	• Thermal boundary layer thickness • Superheat • Velocity • Contact angle	• Effect of bubble nucleus on thermal hydraulic field around it by means of numerical simulation • If cavity having critical radius is not available then nucleation will occur on appropriate size cavity that yields minimum nucleation wall super heat under a given set of conditions.

R^* = critical radius of cavity, $\overline{T}$ = dimensionless temperature, A_r = platform area of heat sink top surface, A_c = cross-sectional area of microchannel, C_g = solubility of dissolved gas, K_h = Henry's constant, θ_r = receding contact angle.

also result in complete suppression of nucleation. Some of the important studies on this effect are discussed next.

To obtain insights into the onset of nucleate boiling as well as its suppression under flow conditions, Kandlikar et al. [7] conducted experimental investigations on a rectangular channel of the dimension 3×40 mm using water as the coolant. The heater surface consisted of polished aluminum. A subcooling of $10\,^\circ C$ was maintained throughout the experiment. Along with visual observations using the high-speed camera, they also analyzed the steady state heat transfer data to develop a predictive tool (Eq. (2.15) in Table 2.1) for onset of nucleate boiling. The range of active cavities was found to shrink as the flow rate was increased, eventually resulting in complete suppression of nucleation.

Qu and Mudawar [9] conducted flow boiling experiments on copper microchannels of the dimensions 231 µm wide and 712 µm deep. Deionized water was used as coolant, the velocity was varied in the range of 0.13−1.44 m/s, and the inlet subcooling was maintained at 10, 40, and 70 K. Visual observations of the onset of nucleate boiling was performed. They used the combination of force balance of the bubble along with the superheat criteria to develop a model (Eq. (2.17) in Table 2.1) for predicting the heat flux required to trigger nucleation (q_{onb}). The incipience heat flux was reported to increase with an increase in the inlet velocity (Fig. 2.3).

Similarly, Li and Cheng [10] conducted numerical analysis for a flow in microchannel having the dimensions of 60×60 µm and a length (L) of 1 cm with water as the test fluid. Assuming fully developed flow, velocity profile for single-phase flow is given as:

$$u(y) = 6u_m \left[\frac{y}{H_c} - \left(\frac{y}{H_c} \right)^2 \right] \tag{2.24}$$

where u_m is mean flow velocity, H_c is microchannel height.

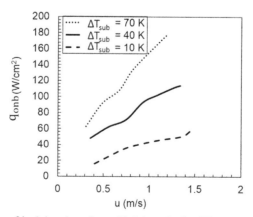

Figure 2.3 Variation of incipient heat flux with inlet velocity [9].

The energy equation of hydrodynamically fully developed flow is given as:

$$u \frac{\partial T}{\partial x} = \frac{k_f}{c_p \rho_l} \frac{\partial^2 T}{\partial y^2} \tag{2.25}$$

where u is given by Eq. (2.24) and the solution to Eq. (2.25) can be obtained by considering the following boundary condition:

$$-k_f \frac{\partial T}{\partial y} = q_w \text{ at } y = 0, \quad T_\infty = T_w \text{ at } y = H_c, \quad T = T_\infty \text{ at } x = 0,$$

$$-k_f \frac{\partial T}{\partial y} = 0 \text{ at } x = L$$

$$\tag{2.26}$$

It was not possible to get an analytical solution for Eqs (2.25) and (2.26); hence, to get the temperature distribution, they adopted numerical analysis for velocity and temperature profiles at mass flow rate of 0.0166 to 1.6 m/s. They found that for a constant heat flux of 300 W/cm^2 at low mean flow velocity of 0.0166 m/s, nucleation was observed in most parts of the channel, but when the mean flow velocity was increased to 1.6 m/s, the bubble nucleation in the microchannel was completely suppressed.

Liu et al. [11] conducted both experimental and theoretical analyses and found that incipient heat flux increases with increase in inlet velocity. They used deionized water as the working fluid and performed flow boiling experiments on 275-μm-wide and 636-μm-deep copper microchannels. Compared to Hsu's model (Eq. (2.4)), the temperature T_{sat} on the right side of the superheat equation was suggested to be replaced by T_b as follows:

$$T_b - T_{sat} = \frac{2\sigma C T_b}{h_{fg} \rho_v y_b} \tag{2.27}$$

A linear temperature profile around the bubble nucleus was assumed:

$$T_l = T_w - q_w y / k_f \tag{2.28}$$

where q_w is effective wall heat flux, k_f is thermal conductivity of fluid. For a bubble to nucleate, the necessary condition is $T_l = T_b$. So, from Eqs (2.27) and (2.28), we have:

$$T_w - \frac{q_w y_b}{k_f} = \frac{T_{sat}}{1 - \frac{2\sigma C}{h_{fg} \rho_v y_b}} \tag{2.29}$$

Equation (2.29) is a quadratic equation in y_b with the two solutions as follows:

$$y_b = \frac{\left(T_w + \frac{2\sigma C q_w}{\rho_v h_{fg} k_f} - T_{sat}\right)}{2\frac{q_w}{k_f}} \pm \frac{\sqrt{\left(T_w + \frac{2\sigma C q_w}{\rho_v h_{fg} k_f} - T_{sat}\right)^2 - 4\frac{2\sigma C q_w}{\rho_v h_{fg} k_f} T_w}}{2\frac{q_w}{k_f}} \tag{2.30}$$

From Fig. 2.1, we can see that,

$$r_c = y_b \frac{\sin\theta}{1 + \cos\theta}$$

Hence, active range of cavity is given as:

$$\{r_{cmin}, r_{cmax}\} = r_c^* \pm \frac{\sqrt{\left(T_w + \frac{2\sigma C q_w}{\rho_v h_{fg} k_f} - T_{sat}\right)^2 - 4\frac{2\sigma C q_w}{\rho_v h_{fg} k_f} T_w}}{2\frac{q_w}{k_f}} \frac{\sin\theta}{1 + \cos\theta} \quad (2.21)$$

where

$$r_c^* = \frac{\left(T_w + \frac{2\sigma C q_w}{\rho_v h_{fg} k_f} - T_{sat}\right)}{2\frac{q_w}{k_f}} \frac{\sin\theta}{1 + \cos\theta} \quad (2.22)$$

Energy balance principle was used to derive the exit fluid temperature if the fluid inlet temperature and velocity were specified:

$$T_e = T_\infty + \frac{qWL}{\rho_f c_p u n w_c H_c} \quad (2.31)$$

where T_e is exit fluid velocity, u is inlet velocity of the fluid, W is width of test section, L is length of the test section, ρ_f is density of the fluid, c_p is specific heat, n is number of microchannels, w_c is width of microchannel, H_c is height of microchannel, and q is applied heat flux.

Assuming uniform convective heat transfer along the surface and fully developed flow:

$$T_w = T_e + \frac{q_w D_h}{Nu_{fd} k_f} \quad (2.32)$$

where Nu_{fd} is the Nusselt number for fully developed flow in rectangular channel (Shah and London [13]) and is given as $Nu_{fd} = 8.235$ $\left(1 - \frac{1.883}{a_s} + \frac{3.767}{a_s^2} - \frac{5.814}{a_s^3} + \frac{5.361}{a_s^4} - \frac{2}{a_s^5}\right)$, D_h = hydraulic diameter.

Effective wall heat flux and applied heat flux are related as:

$$q_w = \frac{q a_s(w_c + w_w)}{(1 + 2\eta a_s) H_c} \quad (2.33)$$

where a_s is aspect ratio, w_w is microchannel fin thickness, an η is fin efficiency.

For both roots of Eq. (2.21) to be real:

$$\left(T_w + \frac{2\sigma Cq_w}{\rho_v h_{fg} k_f} - T_{sat}\right)^2 - 4\frac{2\sigma Cq_w}{\rho_v h_{fg} k_f} T_w = 0 \qquad (2.34)$$

By rearranging Eq. (2.34), we get Eq. (2.19) as reported in Table 2.1:

$$\sqrt{T_w} - \sqrt{T_{sat}} \geq \sqrt{\frac{2\sigma Cq_w}{h_{fg}\rho_v k_f}} \qquad (2.19)$$

For a given heat flux, fluid inlet velocity, and inlet temperature, the T_w required to initiate ONB is determined by Eq. (2.19).

Equation (2.19) can be rearranged by using Eqs (2.31)–(2.33) to obtain:

$$\sqrt{T_\infty + \frac{qWL}{\rho_l c_p u(nw_c H_c)} + \frac{\left(\frac{a_s}{1+2\eta a_s}\frac{w_c+w_w}{H_c}\right)q}{(Nu_{fd}k_f)/D_h}} - \sqrt{T_{sat}} = \sqrt{\frac{2\sigma Cq\left(\frac{a_s}{1+2\eta a_s}\frac{w_c+w_w}{H_c}\right)}{h_{fg}\rho_v k_f}} \qquad (2.20)$$

Hence, for a given value of fluid inlet temperature, fluid inlet velocity and microchannel size, heat flux required to trigger ONB can be evaluated from Eq. (2.20). Once the incipience heat flux for different fluid inlet velocities was determined, they were then substituted in Eq. (2.21) to determine active cavity range corresponding to each inlet velocity. It was observed that lower superheat was needed to initiate nucleation (ΔT_{onb}) when the flow velocity was lower. From Fig. 2.4, it can

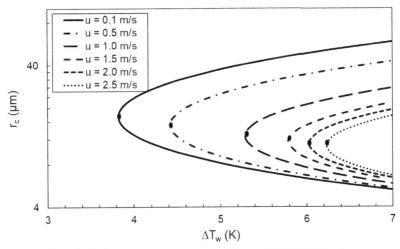

Figure 2.4 Effect of velocity on active cavity range (ΔT_{sub} of 15 K) [11] (black dots correspond to ONB).

be observed that nucleation of bubble becomes difficult as flow velocity is increased and thus active cavity range is reduced. It can also be inferred from Fig. 2.4 that even for a particular superheat, more nucleation sites are activated if velocity is reduced.

Bertsch et al. [14] performed experiments on copper microchannel of hydraulic diameter 1.09 and 0.54 mm and using R-134a and R-245fa as test fluids. They varied the mass fluxes in the range of 42−334 kg/m² s (i.e., for R-134a, velocity was varied in the range 0.034−0.27 m/s, and for R-245fa, velocity was varied in the range 0.031−0.24 m/s). It was concluded that mass flux has a strong effect on onset of nucleate boiling where increasing the mass flux required higher heat flux (q_{onb}) to initiate nucleation. This observation was true for both the test fluids.

2.1.3.2   Effect of Microchannel Size and Geometry

Microchannel size can have a definite effect on nucleation characteristics as the bubble size in transversal direction is limited by the channel walls. Also for ordinary sized channels, the bubble is primarily affected by presence of bulk liquid around the bubble, whereas in microchannels, it is also surrounded by heated walls. Hence, in the transversal direction, the temperature gradient is large and in regions close to the wall extensive evaporation occurs.

Hence, to understand the thermophysical process, two hypothetical concepts "evaporating space and fictitious boiling" were introduced [15]. Evaporating space is defined as the space required for evaporation. Fictitious boiling occurs when internal evaporation and bubble growth are not achieved even though nucleation conditions are reached. In this case, liquid was postulated to absorb larger heat compared with the liquid-sensible heat. Fictitious boiling was proposed to take place when not enough space is available for evaporation. But later, bubbles as small as 40 μm [16] were observed and the concept of fictitious boiling seems debatable.

Peng et al. [17] postulated that when microchannel size is very small, higher superheats are required to initiate nucleation (Fig. 2.5). They derived a dimensionless

Figure 2.5 Effect of diameter on heat flux [17].

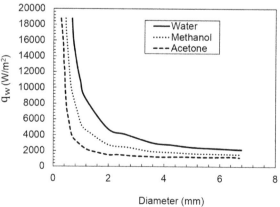

parameter N_b to describe nucleation characteristics, which is given by the following equation:

$$N_b = \frac{h_{fg}\alpha_v}{\pi(v_g - v_l)qD_h}$$
(2.35)

where α_v is thermal diffusivity of vapor phase and the condition for nucleation is given by $N_b \leq 1$.

Ghiaasiaan and Chedester [8] also predicted that boiling incipience heat flux is strongly affected by the channel diameter. The nature of this dependence was not clearly explained, and they suggested that more experimental verifications are required for channels with significantly smaller diameters.

Liu et al. [11] also conducted experimental investigations and numerical analysis to study the effect of microchannel size on nucleation. The effect of channel size can also be inferred from their model where the hydraulic diameter was incorporated as one of the parameter affecting the onset of nucleate boiling (Eqs (2.32) and (2.20) discussed previously). By using Eq. (2.20), the incipience heat flux for a given value of hydraulic diameter is determined and the active range of cavities for different hydraulic diameter can be obtained by substituting the obtained value of incipience heat flux in Eq. (2.21) and is shown in Fig. 2.6. It can be observed from the figure that the early nucleation can be achieved by decreasing the channel size.

Harirchian and Garimella [18] conducted an extensive investigation on microchannel dimensions on heat transfer characteristics using FC-77 as the test fluid. A subcooling of 5 °C was maintained. They used seven test pieces of different widths ranging from 100 to 5850 μm. They also observed the same dependency of the channel dimension on the onset of nucleate boiling. For example, Fig. 2.7 shows the effect of microchannel width on heat transfer for a constant mass flux of 700 kg/m^2 s. It can be seen that the ONB was observed at smaller superheat for channels with smaller widths. Moreover, when the width was increased beyond 400 μm, the boiling curves

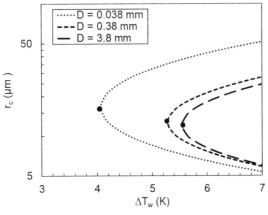

Figure 2.6 Effect of hydraulic diameter on nucleation (ΔT_{sub} of 15 K) [11] (black dots correspond to ONB).

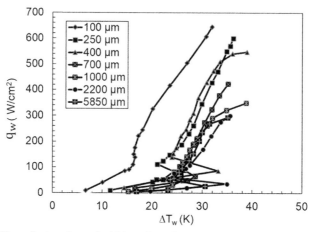

Figure 2.7 Effect of microchannel width on heat flux [18].

were observed to cluster together, suggesting that heat flux is independent of size for larger dimension microchannels.

2.1.3.3 Effect of Heat Flux

As discussed in a previous section, a minimum superheat is required for nucleation to take place. Until and unless the minimum superheat is reached, there will be no nucleation even if cavities of appropriate sizes are present. It is also clear from Hsu's model that activation of cavities is only possible if the superheat increases beyond a critical value.

Basu et al. [19] conducted experimental investigations and adopted both visual observations and temperature and heat flux measurements to identify the location of ONB along the length of the microchannel. Heat transfer coefficient variation along the length of the microchannel was plotted, and the location of a steep change in the slope of this curve was found to coincide with the visually observed locations of ONB (the arrow points to the location of ONB, Fig. 2.8). The heat transfer coefficient in the single-phase regime (6 W/cm^2, mass flux of 346 kg/m^2 s, and inlet subcooling of 26.5 °C) decreased monotonically as we move along the microchannel length. As the heat flux was increased, ONB was initiated and the location of ONB was observed to shift upstream. For example, for a heat flux of 14.7 W/cm^2, the nucleation initiated at about $z = 15$ cm, whereas for a heat flux of 19.7 W/cm^2, the nucleation was observed to initiate at about $z = 5$ cm.

Li and Cheng [10] performed a numerical study (as discussed in Section 2.1.3.1) to show that nucleation was completely suppressed at low heat fluxes. Figure 2.9 shows the effect of heat flux on surface temperature for a constant velocity of 0.167 m/s for saturated water. The arrow in the figure represents the homogeneous nucleation temperature. It is seen that nucleation was not observed for a lower heat flux of 250 W/cm^2. However, the homogeneous nucleation was observed on increasing the heat flux to larger values.

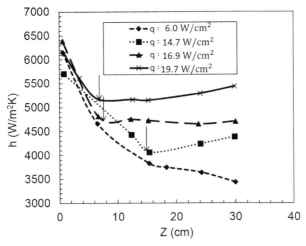

Figure 2.8 Effect of heat flux on nucleation [19].

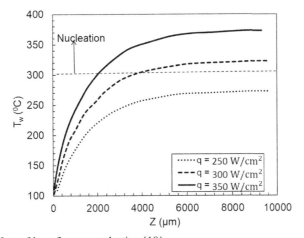

Figure 2.9 Effect of heat flux on nucleation [10].

2.1.3.4   Effect of Subcooling

Subcooling is one of the most important parameters governing the onset of nucleate boiling. This has been identified by many researchers, and we can see that most of the models presented in Table 2.1 have incorporated subcooling as one of the parameter [2,4–7,9,11]. Among different parameters, fluid inlet temperature has maximum influence on incipient heat flux [11]. From the model developed by Liu et al. [11], the active range of cavities for different subcooling can be determined by using Eqs (2.20) and (2.21). Active range of cavities for different subcooling based on this model is shown in Fig. 2.10, where we can see that there is a decrease in active range of cavity with increase of subcooling.

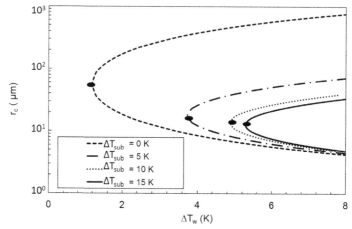

Figure 2.10 Effect of subcooling on bubble nucleation [11] (black dots correspond to ONB).

Qu and Mudawar [9] deduced from their experimental results (details of their experimental conditions are presented in Section 2.1.3.1) that as subcooling was reduced, it led a decrease in the incipient heat flux, and their experimental results (Fig. 2.3) were in agreement with the developed model.

A numerical analysis was carried out by Ghiaasiaan and Chedester [8] for microtubes with diameters in the range of 0.1−1 mm. They postulated that ONB occurs when aerodynamic force balances the suppressing effect of the thermocapillary force on the bubble. The semiempirical model developed by them is presented in Table 2.1 as Eq. (2.16). According to their numerical analysis, it was found that with the increase of subcooling, incipience heat flux increased monotonically.

Experimental investigations carried out by Liu et al. [11] (experimental setup detailed in Section 2.1.3.1) showed that as subcooling is increased, it leads to larger incipient heat flux. Figure 2.11 shows a curve between fluid inlet temperature and incipient heat flux, which clearly shows that with the increase of fluid inlet temperature, there is a decrease in the incipient heat flux.

2.1.3.5 Effect of Dissolved Gasses

Liquids contain dissolved gases, the amount of which may vary according to temperature and solubility. Henry's law states that the amount of gas that dissolves in a given type and volume of liquid is directly proportional to the partial pressure of that gas in equilibrium with that liquid at a constant temperature. The solubility of gases usually decreases with an increase of temperature as shown via an example in Fig. 2.12. Thus, the local solubility of the liquid is reduced near the superheated surfaces, thereby resulting in the release of dissolved gases in the form of bubbles. These bubbles trigger further nucleation (i.e., induce a two-phase flow) and have implications on the bubble and heat transfer behavior.

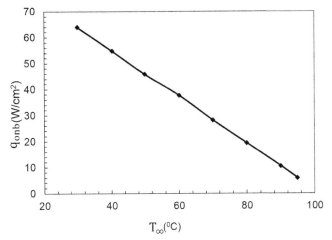

Figure 2.11 Effect of fluid inlet temperature on nucleation [11].

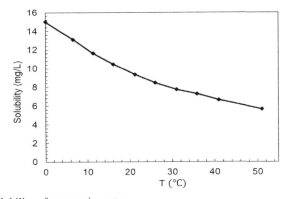

Figure 2.12 Solubility of oxygen in water.

Experimental investigations were carried out by Li and Peterson [20] on a micro-heater of geometry $100 \times 100 \ \mu m^2$, using deionized water as the test fluid. The nucleation temperature was predicted from the measured temperature data and a three-dimensional heat conduction model was adopted. Subcooling was about 75 for gassy water and 60 °C for degassed water. They also drew very similar conclusion that presence of dissolved gases resulted in a decrease in nucleation temperature. They observed that in the case of gassy water, the nucleation temperature from a smooth surface was around 135 °C, whereas for degassed water, the value was near 180 °C.

Experiments conducted by Müller-Steinhagen et al. [21] to study the influence of various gasses like He, N_2, Ar, CO_2, and C_3H_8 on flow boiling heat transfer in an annulus with a heated core revealed that there was an increase in heat transfer coefficients due to the desorption of gases.

Steinke and Kandlikar [22] conducted experimental investigations on six parallel microchannels of 207 μm as hydraulic diameter to determine the effect of dissolved gases on heat transfer during flow boiling through microchannels using water as the coolant. An apparatus capable of delivering water with different level of dissolved air was developed. They observed that for dissolved oxygen content of 8 ppm nucleation occurred for a surface temperature lower than the saturation temperature. The nucleation sites become active at such lower temperature due to desorption of air. Figure 2.13 shows that in the 8 ppm case, transition from single-phase to two-phase is faster than in 5.4 and 1.8 ppm cases.

Li and Cheng [10] developed a model that incorporated the effect of dissolved gas and found that the presence of dissolved gases decrease the nucleation temperature. They adopted Hsu's criteria of nucleation; that is, nucleation was assumed to initiate when the liquid temperature was greater than the vapor temperature inside the bubble. Classic kinetics of nucleation was used to compute the nucleation temperature. For homogeneous nucleation:

$$J_{homogeneous} = N_o \left(\frac{kT_w}{h} \right) exp \left[-\frac{16\pi\sigma^3}{3kT_w \left(p_v - p_l \right)^2} \right] \qquad (2.36)$$

where $J_{homogeneous}$ is homogeneous bubble nucleation density (1 cm^{-3} s^{-1}), N_o, is molecule number per unit volume, k, is Boltzmann constant, and h, is Prancle constant.
For heterogeneous nucleation:

$$J_{heterogeneous} = N_o^{2/3} \Psi \left(\frac{kT_w}{h} \right) exp \left[-\frac{16\pi\sigma^3\omega}{3kT_w \left(p_v - p_l \right)^2} \right] \qquad (2.37)$$

Figure 2.13 Effect of dissolved gases [22].

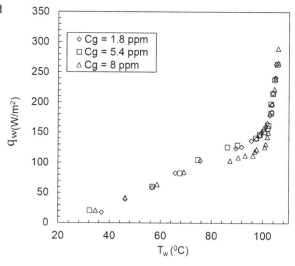

where $J_{heterogeneous}$ is heterogeneous bubble nucleation density ($1 \text{ cm}^{-2}\text{ s}^{-1}$), Ψ is surface available for heterogeneous nucleation per unit bulk volume of liquid phase, and ω is geometric correction factor for minimum work required to form critical nucleus.

$$\Psi = \frac{1}{2}(1 + \cos\theta) \tag{2.38}$$

$$\omega = \frac{1}{4}(1 + \cos\theta)^2(2 - \cos\theta) \tag{2.39}$$

From the Clausius−Clapeyron equation and Eqs (2.36) and (2.37), we obtain:

$$T_w - T_s = \frac{T_s}{h_{fg}\rho_v}\sqrt{\frac{16\pi\sigma^3\omega}{3kT_w\ln\left(\frac{N_o^{\gamma}kT_w\Psi}{Jh}\right)}} \tag{2.40}$$

for homogeneous nucleation $\gamma = \Psi = \omega = 1$, and for the heterogeneous equation γ is 2/3, ω and Ψ are given by Eqs (2.38) and (2.39). According to Ward et al. [23], total pressure inside the bubble is given as:

$$p_b = p_v + p_g \tag{2.41}$$

$$p_v = p_s(T_w)\exp\left(-C_g\right) \tag{2.42}$$

$$p_g = K_h C_g \tag{2.43}$$

where p_b is total pressure inside the bubble, and p_g is pressure of gas inside the bubble. Substituting Eqs (2.41) and (2.43) into Eq. (2.37),

$$T_w - T_s = \frac{T_s \exp\left(C_g\right)}{h_{fg}\rho_v}\left(\sqrt{\frac{16\pi\sigma^3\omega}{3kT_w\ln\left(\frac{N_o^{\gamma}kT_w\Psi}{Jh}\right)}} - K_h C_g\right) \tag{2.44}$$

From the Young−Laplace equation, we have,

$$r_c = \frac{2\sigma}{p_s(T_w) - p_l} \tag{2.45}$$

Nucleation temperature can also be calculated directly by using Eq. (2.45), Eqs (2.41)−(2.43), and the Clausius−Clayperon equation as:

$$T_w - T_{sat} = \frac{T_{sat}\exp\left(C_g\right)}{h_{fg}\rho_v}\left(\frac{2\sigma}{r} - K_h C_g\right) \tag{2.18}$$

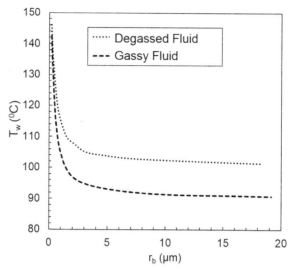

Figure 2.14 Effect of dissolved gases [10].

The required expression developed by them is given in Table 2.1 as Eq. (2.18). In their analysis, they concluded that for bubble size of greater than 1 μm, ONB can be triggered at a temperature lower than the saturation temperature if dissolved gases are present in water. This effect is clearly depicted in Fig. 2.14, where the steep change in slope refers to ONB.

Cioncolini et al. [24] carried out experimental investigations to study the effect of dissolved air in subcooled and saturated flow boiling of water at pressure range of 177−519 kPa for varying mass fluxes and heat fluxes. They used water saturated with air for experiments and compared the results with models derived for degassed liquids. They found that dissolved gases have negligible effect in flow boiling.

2.1.3.6 Effect of Contact Angle

All the practical surfaces are far from being atomically smooth; they contain many surface irregularities. If these surfaces are not completely wetted (i.e., the contact angle is high), these cavities may contain entrapped gases. Because vaporization at liquid−gas interface is relatively easier, it would require less superheat to initiate nucleation.

In the model developed by Hsu to predict the active range of cavity for onset of nucleate boiling, the author used a constant value of contact angle to derive Eq. (2.11). After that, many researchers initially proceeded according to Hsu's model but included the effect of contact angle in their models [5−8,10−12] later. Since then, many experimental and numerical analyses have been performed to predict the influence of contact angle on onset of nucleate boiling.

Li and Cheng [10] (details explained in Section 2.1.3.1 and Section 2.1.3.5) performed numerical analysis and, using their model, predicted that with an increase of contact angle, there is a decrease in the nucleation temperature. Figure 2.15 is

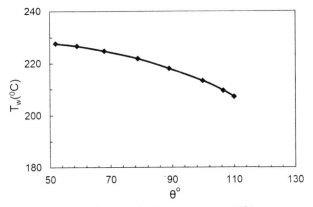

Figure 2.15 Effect of contact angle on nucleation temperature [10].

obtained by using Eq. (2.44) for saturated dissolved air in water for solubility of 11.5×10^{-3} mg/ml.

Similarly, Liu et al. [11] (details explained in Section 2.1.3.1) reported that a bubble with a smaller contact angle would require more heat to initiate nucleation but the effect was not significant. In their study, when contact angle was increased from 30° to 90°, the change in incipient heat flux was only 8.6%.

Experimental investigations carried out by Basu et al. [19] (details explained in Section 2.1.3.3) show that the nucleation temperature increases as there is a decrease in the contact angle. The static contact angle was measured by placing droplets at different locations on the copper surface and taking photographs using a CCD camera. Contact angle was measured from these photographs. This phenomenon can be observed in Fig. 2.16.

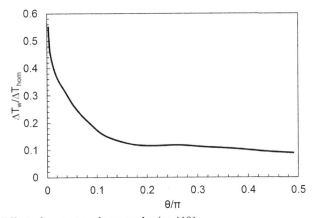

Figure 2.16 Effect of contact angle on nucleation [19].

2.1.4 Influence of ONB on Boiling Phenomenon

In general, onset of nucleate boiling marks the transition from single-phase boiling to two-phase boiling. In comparison to single-phase systems, two-phase systems are known to dissipate higher heat flux with little temperature difference due to the high latent heat of vaporization. However, ONB is also known to trigger instabilities in the boiling system, prompting researchers to develop a better understanding of the influence of ONB on boiling phenomenon.

Experimental investigation was carried out by Hapke et al. [25] on nickel-based cylindrical tubes having a 1.5-mm inner diameter and deionized water was used as the working fluid. The test fluid was degassed by heating it to a temperature of 85 °C. The mass flux was varied in the range 100 to 500 kg/m^2 s, and heat flux was varied in the range of 5—20 W/cm^2. They observed strong pressure as well as mass fluctuations during evaporation in minichannels.

Similarly, Kandlikar et al. [26] performed experimental analysis on six 1-mm-hydraulic diameter channels using water as the test fluid. They also reported pressure fluctuations with 1-Hz frequency.

Proceeding on similar terms, Huh et al. [27] conducted flow boiling experiments and observed pressure and mass flux fluctuations once two-phase flow was encountered. Unlike previous studies [25,26], these fluctuations had long periods and large amplitudes. Pressure and mass flux fluctuations observed in their experiments are shown in Fig. 2.17. In their experiments, fluctuation in wall temperatures was also observed.

Zhang et al. [28] observed transient pressure pulses in the range of 69—138 kPa at the onset of nucleate boiling, which can adversely affect the device as well as local boiling phenomena being hampered.

2.1.5 Recent Trends

Onset of nucleate boiling is critical as it signifies the transition from single-phase to multiphase regime. Effects of various parametric conditions have been studied extensively by many researchers, which have been discussed in previous sections. Recently, many new methods have been developed to improve the ONB characteristics so that more heat can be dissipated from the surface. On a heating surface with a smooth

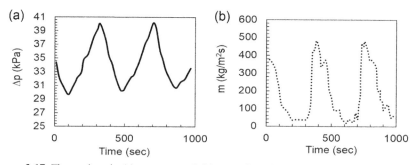

Figure 2.17 Fluctuations in (a) pressure and (b) mass flux [27].

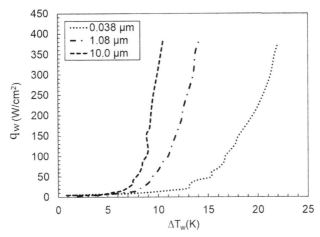

Figure 2.18 Effect of roughness on heat flux [29].

wall, rapid bubble growth follows the onset of nucleate boiling and instabilities are triggered almost instantaneously.

It has been shown by many researchers that the presence of surface defects can enhance boiling heat transfer as they provide more nucleation sites. Jones et al. [29] conducted pool boiling experiments to study the effects of surface roughness on heat transfer using water and FC-77 as test fluids under saturated conditions. The polished surfaces had roughness (R_a) in the range of 0.027−0.038 μm, while the rough surfaces had roughness in the range of 1.08−10 μm. Figure 2.18 shows boiling curves for water with different surface roughness values. From Fig. 2.18, it is clear that an increase in roughness results in a decrease in the value of superheat to initiate nucleation. This is due to the fact that more active cavity sizes are present in rougher surfaces.

McHale et al. [30] used two different coolants on copper surface enhanced with sintered copper powder carbon nanotubes to study pool boiling heat transfer. They used HFE-7300 and deionized water as working fluids. The results are summarized in Table 2.2.

Table 2.2 Results Obtained by McHale et al. [30]

	HFE-7300		Deionized Water	
Surface	ΔT_w (K)	r_b (μm)	ΔT_w (K)	r_b (μm)
Bare copper	22.5 ± 2.0	0.279 ± 0.026	9.7 ± 0.6	3.51 ± 0.22
Metallized CNT	12.6 ± 1.2	0.498 ± 0.049	4.8 ± 1.2	7.46 ± 1.89
Sintered copper	6.6 ± 2.3	1.07 ± 0.38	0.4 ± 0.1	88.6 ± 22.2
Sintered copper + metallized CNT	0.8 ± 0.2	8.27 ± 2.07	0.5 ± 0.2	79.2 ± 31.7

Table 2.3 Different Types of System Considered by Lu and Pan [31]

System	Description
Type 1	Diverging microchannel without ANS
Type 2	Diverging microchannel having 13 ANS located along the downstream half of the channel
Type 3	Diverging microchannel having 25 ANS located throughout the channel

From Table 2.2, it can be deduced that nucleation temperature can be reduced by carbon nanotubes on the surface. However, the large-scale sintered coating proves to be more efficient than the carbon nanotubes, as can be noted from Table 2.2.

Similarly, the presence of artificial nucleation sites (ANS) also enhances heat transfer characteristics in flow boiling in microchannels. Lu and Pan [31] carried out experimental investigations on microchannels with diverging cross section with and without artificial nucleation sites using water as test fluid for varying mass flux ($99-297$ kg/m^2 s) and subcooling ($28-48$ °C). Laser etching was done on the bottom surface of the channel, which served as ANS. Three different types of test sections were used, as listed in Table 2.3. From their study, it was found that for a given set of conditions, type 3 microchannel required less wall superheat to initiate nucleation.

Morshed et al. [32] grew nanowires on the bottom surface of a copper microchannel having a hydraulic diameter of 672 µm via electrochemical deposition process and carried out experimental analysis using deionized water as the test fluid. Experiments were conducted under varying conditions of mass flux and subcooling. For a mass flux of 45.95 kg/m^2 s and fluid inlet temperature of 22 °C, the superheat required to initiate nucleation was about 2.9 K, whereas for bare copper surface, superheat for same mass flux was around 6.6 K. Their findings report about 25% enhancement of heat transfer by copper nanowire integration.

Alam et al. [33] carried out experimental investigations on silicon microchannels to study the effect of surface roughness on flow boiling heat transfer. They used deionized water with an inlet temperature of 91 °C for two different mass fluxes of 390 and 650 kg/m^2 s. The surface roughness with values of $R_a = 0.6$, 1.0, and 1.6 µm was examined. Similar to previous studies, they also observed that with increase of roughness, there was an increase in heat transfer.

Bai et al. [34] conducted flow boiling experiments on microchannels with metallic porous coating using anhydrous ethanol as the working fluid. The inlet fluid temperature was maintained at 30 °C. Solid state sintering was used to fabricate porous coated microchannels. Copper particles with diameters of 30, 55, and 90 µm were used for this purpose. From their experiments, it was found that substantial enhancement of heat transfer can be achieved in porous coated microchannels. Also, it can be seen in Fig. 2.19 that for a mass of 182.8 kg/m^2 s, 55-µm-diameter particle produced

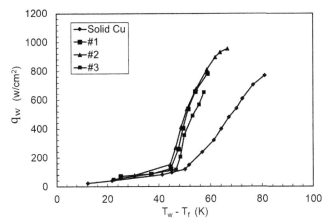

Figure 2.19 Variation of heat flux for porous coated and solid microchannels [34].

the most optimum results. This can be explained by the fact that pores with larger size are flooded by the subcooled liquid, thereby hindering the nucleation process.

Deng et al. [35] performed experimental analysis on porous reentrant microchannels having a hydraulic diameter of 786 μm and compared the results with solid copper microchannels with the same reentrant configuration. Deionized water was used as the working fluid. Experiments were carried out for three different inlet temperatures (33, 60, and 90 °C), and the mass flux was varied in the range 125–300 kg/m² s. Their results showed that ONB occurred at a lower super heat for the porous microchannel as more nucleation sites were activated due to entrapped gases in the porous medium.

Yang et al. [36] conducted experimental investigations with silicon nanowires grown on the inner walls of a silicon microchannel and compared the results with plain wall microchannels with and without inlet restrictors. Their results showed a significant improvement in heat transfer coefficient with the use of silicon nanowire integrated microchannels. For a mass flux of 389 kg/m² s and fluid inlet temperature of ∼22 °C, the heat transfer coefficient was enhanced by 248% and 326% compared with plain wall microchannels with and without inlet restrictors, respectively.

Kumar et al. [37] conducted flow boiling experiments to study the effect of carbon nanotubes on heat transfer. The test section had dimensions of 25 × 20 × 0.4 mm. Water was used as the test fluid and a subcooling of 10 K was maintained in their study. Experimental results were analyzed for three different mass fluxes of 283, 348, and 427 kg/m² s. From Fig. 2.20, we can see that for a mass flux of 283 kg/m² s, much higher heat flux is achieved in the case of carbon nanotubes compared with a bare copper surface. They reported critical heat flux was enhanced by 21.6% with the use of carbon nanotubes.

Law et al. [38] experimentally investigated flow boiling heat transfer in oblique finned microchannels. In oblique finned microchannels, oblique cuts were made at specific locations, producing oblique fins with secondary microchannels. The test section

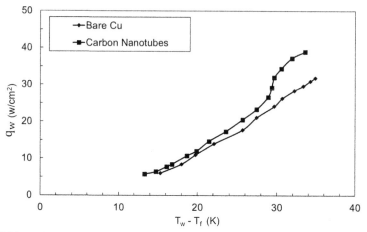

Figure 2.20 Variation of heat flux for bare copper and with integrated nanotubes [37].

consisted of 40 microchannels, each having a width of 0.3 mm, while the width of oblique cuts was 0.15 mm. FC-72 was used as the test fluid, and the inlet temperature was maintained at 29.5 °C. The mass flux was varied in the range of 175–350 kg/m² s. They reported minimal fluctuations in the inlet pressure, thereby indicating that flow boiling instabilities can be reduced to a great extent by the use of oblique finned microchannels.

The use of artificial nucleation sites is gaining much popularity as a method to enhance boiling heat transfer; hence, many investigations are being carried out by researchers on various aspects of this technique [39–43]. The use of artificial nucleation sites no doubt improves boiling characteristics; however, they are not very effective to minimize the effects of flow instabilities [44]. Hence, to minimize instabilities, use of pressure drop elements (PDEs) is implicated. PDEs are small orifice incorporated into the inlet to the channel and are found to be very effective in reducing the fluctuations.

Kuan and Kandlikar [45] carried out experiments to investigate the effect of PDEs on flow boiling. In their experiments, they compared the results between flow boiling without PDEs and flow boiling with 6.1% PDEs (i.e., a 6.1% microchannel opening). Their investigations revealed that the stabilities was improved with the use of PDEs, when high mass flow rates were employed.

Kandlikar et al. [44] performed experiments on 1054×197-µm parallel rectangular microchannels to study the effects of PDEs and fabricated nucleation sites on stabilization of flow boiling in microchannels. The results were reported for following five cases, which are tabulated in Table 2.4.

They observed severe instabilities for case (a) (i.e., with no PDEs or ANS). With the use of 51% PDEs, stability was improved; however, it failed to stop back flow completely. Similarly, use of ANS alone provided less resistance to back flow, leading to unstable flow conditions. They found that 4% PDEs with ANS provided a very stable flow. The results are summarized in Table 2.5.

Table 2.4 **Cases Investigated by Kandlikar et al. [44]**

Case	Description
(a)	No PDEs, no ANS
(b)	51% PDEs only
(c)	ANS only
(d)	51% PDEs with ANS
(e)	4% PDEs with ANS

Table 2.5 **Summary of Test Results [44]**

Case	Average Surface Temperature (°C)	Pressure Drop (kPa)	Pressure Fluctuation (±kPa)	Stability
Without PDEs or ANS	114.5	16.3	1.4	Unstable
With 51% PDEs	115.0	16.2	0.8	Partially stable
With ANS only	112.4	14.0	3.8	Unstable
With 51% PDEs and ANS	113.0	12.9	0.9	Partially stable
With 4% PDEs and ANS	111.5	39.4	0.3	Completely stable

Szczukiewicz et al. [46] conducted flow boiling experiments on microchannels of dimensions 100×100 μm with R245fa, R236fa, and R1234ze(E) as test fluids. From their experiments, it was concluded that two-phase flow flashed by inlet restrictors efficiently mitigated the flow instabilities that are most commonly encountered in flow boiling. In their study, inlet restrictors with an expansion ratio of 2 (50 μm wide) was used and a subcooling of 5.7 K was maintained. Eight different regimes were distinguished during their experiment as follows:

1. Single-phase flow with vapor bubbles at outlet manifold.
2. Single-phase flow followed by two-phase flow with back flow.
3. Unstable two-phase flow with back flow developing into jet flow.
4. Jet flow.
5. Single-phase flow followed by two-phase flow without back flow.
6. Two-phase flow with back flow triggered by bubbles formed in the flow loop before the test section.
7. Flashing two-phase flow with back flow.
8. Flashing two-phase flow without back flow.

Due to flashing effect, lower heat flux was required to initiate nucleation, and wall temperature overshoot at onset of nucleate boiling was avoided. Hence, regime 8 was the most desired regime.

Xu and Xu [47] carried out experimental investigations on microchannel having dimensions $7500 \times 100 \times 250$ μm. Pure water and nanofluid were used as the working fluids and the inlet temperature of the fluid was maintained at room temperature ~ 27 °C. Al_2O_3 particles of 40 nm were used to prepare the nanofluid. Because a very low concentration (0.2%) of nanoparticles was used, particle deposition over the heater surface was not observed. In contrast to flow boiling with pure water, higher heat transfer was observed with the use of nanoparticles as test fluid. This can be attributed to the fact that with the use of nanoparticles, smaller bubbles were formed, which inhibited formation of dry area over the heater surface.

2.2 Void Fraction in Microchannels

2.2.1 Introduction

Several studies have been conducted over the years to understand various aspects of two-phase flow in microchannel for designing efficient and reliable thermal management systems for commercial and defense applications. Void fraction is one of the important factors in two-phase flow; it is used to calculate several other parameters of interest such as the heat transfer coefficient, pressure drop, two-phase flow density, two-phase flow viscosity, and average velocities of the respective phases.

Void fraction is defined in multiple ways. Volumetric void fraction is defined as the ratio of vapor volume to the total volume for a fixed channel length. In the area-averaged void fraction, fraction of cross-sectional area of channel occupied by the vapor phase is calculated. The chordal void fraction is evaluated by dividing the length of the vapor phase by the channel length. The local void fraction is referred to the ratio of time spent by vapor phase at a particular point to the total time. The void fraction in a channel varies both along the length of the channel and with time.

2.2.2 Different Methods of Void Fraction Measurement

Several methods have been developed for measuring the void fraction, some of which are discussed in this section. Pujara et al. [48] summarized the void fraction measurement techniques as listed in Table 2.6.

2.2.2.1 Experimental Methods

Singh et al. [49] used image processing technique to study the heat transfer and different flow regime maps during the flow boiling of water in the microchannel of hydraulic diameter of 140 μm. A 2-inch, 275 ± 25-μm-thick, p-type, double-side polished silicon wafer was used to fabricate microchannels with dimensions of $173 \times 119 \times 20$ μm using CMOS technique. Surface roughness was less than 0.1 μm, and sealing was

Table 2.6 Different Void Fraction Measurement Techniques

Techniques	Applied To	Remarks
Quick closing valve	Measurement of volumetric void fraction	1. Intrusive method 2. Requires finite time for closing of the valve 3. Requires considerable time for bringing the system back to the steady state between successive experiment
Conduction probe	Local time-averaged and chordal void fraction measurement	1. Intrusive method 2. Limited to measure bubbly and slug flow regime
By pressure drop	Measurement of volumetric void fraction	1. Nonintrusive method 2. Friction pressure drop and acceleration pressure drop neglected, manometer line filled by single phase
Radiation absorption and scattering method	Chordal void fraction measurement	1. Nonintrusive method 2. Expensive, difficult to handle high energy radiation, presence of metal wall induces error
Laser beam method	Chordal void fraction measurement, flow pattern identification	1. Nonintrusive method 2. Expensive
Photographic method	Chordal, cross section void fraction measurement	1. Nonintrusive method 2. Subjective 3. Error due to operation performed on image
Impedance method	Volumetric void fraction, low cost, suitable for transition measurement	1. Nonintrusive method
Capacitance method	Measurement of volumetric void fraction	1. Nonintrusive 2. Relatively low cost 3. Not easy to calibrate the capacitance with void fraction due to signal dependency on the void fraction and flow patterns

achieved with a quartz plate. Deionized water was used as the working fluid. Chrome-gold thin-film stack was used as micro-heater material as it showed good linearity with respect to temperature along with high chemical and thermal stability.

An image processing algorithm was developed to estimate the void fraction and evaluate the percentage of different flow regimes and heat transfer coefficient as the function of position, heat flux, and mass flow rate. In image processing, images were first recorded using a camera. The vapor region was identified next, and void fraction was estimated as the ratio of area of vapor to total area of the microchannel in the image. The images correspond to the plane normal to the flow direction (and not the cross-sectional view) of the microchannel. Steps for image processing techniques are:

1. First, background image was read and boundaries were whitened for edge detection. Microchannel was extracted by cropping. Whitening was required to know whether bubbles were fully enclosed by edges and its area was well defined.
2. Sample image was read, and microchannel was extracted by cropping.
3. Background was removed to reduce noise. It was done by subtracting background image from the sample image.
4. Sample image was then converted to intensity image or grayscale from RGB file type. Histogram equalization was done before converting to a binary image.
5. Noise was removed by median filter.
6. Edges were identified and vapor region with closed contours was filled.
7. Small objects with less than 40 pixels were left out for calculation of total area of vapor.

They observed that void fraction increased along the downstream of flow in the microchannel. This was observed due to the fact that the fraction of liquid in the microchannel decreased due to evaporation, which made a monotonic increase in the void fraction. Also, it was observed that the void fraction increased with lowering the volumetric flow rate. This was because a larger liquid fraction was evaporated to vapor at a given heat flux as the volumetric flow rate was reduced, thereby increasing the void fraction.

Gijsenbergh and Puers [50] developed a capacitive fringing fields-based sensor to measure void fraction in silicon microchannels of 100 μm (width) by 500 μm (height) cross section. The test section consisted of 60 channels having dimensions of 500 μm (depth) × 100 μm (width) × 1000 μm (long) and made of silicon die of 650-μm thickness. The change in permittivity due to change in fluid content was detected via the sensors connected to the top and the bottom electrodes. The top electrode was located at the back of each channel of silicon microchannels, and the bottom electrode was mounted on a glass substrate of each channel and insulated with a layer of Pyrex glass as shown in Fig. 2.21. The capacitance network of the system is shown in Fig. 2.22, where the total capacitance was given by,

$$C_{total} = 2C_{Silicon\ wall} + \left[\frac{1}{C_{Silicon\ ceiling}} + \frac{1}{C_{ch}} \right]^{-1} \tag{2.46}$$

The individual capacitance, $C_{Silicon\ wall}$, $C_{Silicon\ ceiling}$, and C_{ch} depend on the electrode area, the layer thickness, the absolute permeability ε_{abs} and relative

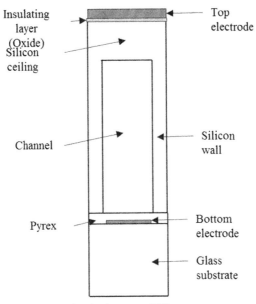

Figure 2.21 Schematic diagram of a single sensor [50].

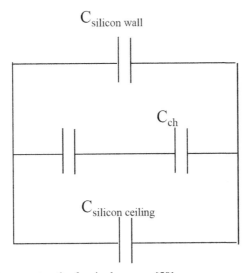

Figure 2.22 Capacitance network of a single sensor [50].

permeability ε_{rel}. As the geometry of the sensor is fixed, all capacitances except relative permeability ε_{rel} are known. The relative permeability ε_{rel}, which varies due to change in fluid content in channel, is estimated for annular flow using numerical simulation. The relative permeability ε_{rel} at two extremes, viz. channel completely filled with air and water, are evaluated and subsequently, the variation in ε_{rel} with different proportions of air—water is correlated, which is found to be quasilinear in nature.

The capacitance measured by the sensor is digitized using a 16-bit sigma-delta capacitance-to-digital converter (CDC). The digital signal is then converted into an analog voltage using a digital-to-analog converter (DAC). Their result shows a linear relationship with the measured voltage and void fraction.

Paranjape et al. [51] developed an electrical impendence-based sensor to measure void fraction in microchannels with a square cross section of 780×780 μm and length of 50.8 mm. Deionized water and air were used in their two-phase experiments. The faces of two electrodes made of 304 stainless steel were flush mounted to the opposite sidewalls of the microchannel made of acrylic sheeting as shown in Fig. 2.23. The electrodes were positioned at 25.4 mm from the inlet of microchannel. The cross section of electrodes was same as that of the microchannel. The electrical conductivity of water was enhanced by the addition of a small quantity of morpholine and ammonium-hydroxide (individual quantity of 1 mg per 1 l of water) and pH of about 7. The impedance-based void fraction (α_{imp}) was calibrated using a time-averaged void fraction (α_{time}) determined from the image processing of the videos obtained from the high-speed camera. The relationship between normalized admittance (G^*) and void fraction (α_{imp}) measured by impedance meter was obtained with the following expression,

$$\alpha_{imp} = 1 - G^* \tag{2.47}$$

where $G^* = \frac{G_m - G_1}{G_0 - G_1}$. G_m is instantaneous two-phase mixture admittance, G_0 is admittance with zero void fraction (i.e., only deionized water) and G_1 is admittance

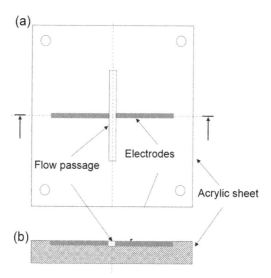

Figure 2.23 Schematic of Impedance meter test cell (a) top view and (b) sectional view [51].

with the void fraction equal to 1 (i.e., only air). The correlation between α_{imp} and α_{cal} was found after plotting α_{imp} versus α_{time} as,

$$\alpha_{cal} = -1.18\alpha_{imp}^3 + 1.57\alpha_{imp}^2 + 0.61\alpha_{imp} \tag{2.48}$$

This model was stated to be applicable to various flow regimes such as bubbly flow, cap-bubbly flow, slug flow, churn-turbulent flow, and long slug flow.

Fogg et al. [52] developed a new fluorescent technique to measure local liquid fraction in two-phase micro-flows. Photodiode and dye were used to obtain the high-speed transient isothermal flow with different optical methods. This technique could be used in boiling and condensation in future experiments when two diodes and dye would be used. In this study, the theory for both isothermal and boiling flows was developed, and the technique was applied to isothermal flows involving air and deionized water with fluorescein as the fluorescent dye.

Fluorescence is a luminescence phenomenon by which a molecule gets energized by absorbing a photon and thereafter emits another photon as it returns to its ground state. Emitted radiation is of longer wavelength than excitation radiation due to the fact that some of the energy is lost during collision with neighboring molecules. This is known as Stoke's shift. It was also known that not all absorbed radiation results in an emitted photon at the fluorescent wavelength. The ratio of emission to absorption is known as the quantum efficiency, φ, which is a function of temperature. Because the time the molecule stays in the excited state is typically short ($\sim 10^{-9}$ s), high-speed measurements of transient phenomena were possible.

Fluorescence was used separately to measure both temperature and void fraction. By using two dyes that emit at different wavelengths and could be measured with two optical detectors, such as photodiodes or CCD arrays, both the temperature and liquid fraction of the measurement volume could be calculated. In this study, the focus was on isothermal flows using a single dye. The fluorescein dye has absorption and emission peaks at 494 and 520 nm, respectively.

Assuming negligible scattering, reflection, and refraction, the power absorbed is,

$$P_a = IAT_g\left(1 - e^{(-\bar{a}_{\lambda i}D\beta)}\right) \tag{2.49}$$

$$\bar{a}_{\lambda i} = \alpha_{e,\lambda i} + ca_{\lambda i,dye} \tag{2.50}$$

where P_a is the power absorbed by the dye solution, I is the intensity of the excitation light on the microchannel, T_g is the transmittance of the glass cover, A is the area of the channel illuminated, α_e is the extinction coefficient of water, $\bar{a}$ is the molar extinction coefficient of the dye, D is the depth of the channel, β_l is the volume averaged liquid fraction, and c is the molar concentration of the dye.

If $\bar{a}_{\lambda i}D\beta_l \ll 1$, the optically thin assumption is valid and $e^{(-\bar{a}\lambda iD\beta)} \approx 1 - \bar{a}_{\lambda i}D\beta_l$ allowing the absorbed power by the dye to be expressed as,

$$P_{a,dye} = IAT_g c_{dye}a_{\lambda i,dye}D\beta_l \tag{2.51}$$

Fluorescent power emitted is given by,

$$P_{a,dye} = IAT_g c_{dye} a_{\lambda i, dye} D \beta_l \phi_{dye} \tag{2.52}$$

Assuming the fluid as optically thin for the emitted fluorescence, the power absorbed by the detector after transmission back through the two-phase mixture and the glass was,

$$P_d \approx IAT_g^2 c_{dye} a_{\lambda i, dye} D \beta_l \phi_{dye} = \Gamma \phi \beta_l \tag{2.53}$$

Γ is a constant $= IAT_g c_{dye} D$
ϕ depends on temperature $= \phi_{dye} a_{\lambda i, dye}$

Normalization of data was required for removing the dependence of intensity and dye concentration.

$$\widehat{P} = \frac{P(T, \beta)}{P(T_{ref}, \beta)} = \frac{\phi(T)\beta}{\phi(T_{ref})} \tag{2.54}$$

The temperature dependence drops out and the normalized power is the liquid fraction, $\widehat{P} = \beta_l$.

For heated flow, two dyes were used. Liquid fraction for each dye was given by,

$$\beta = \widehat{P}_1 \frac{\phi_1(T_{ref})}{\phi_1(T)} = \widehat{P}_2 \frac{\phi_2(T_{ref})}{\phi_2(T)} \tag{2.55}$$

One can then obtain,

$$\frac{\phi_1(T)}{\phi_2(T)} = \frac{\widehat{P}_2}{\widehat{P}_1} \frac{\phi_1(T_{ref})}{\phi_2(T_{ref})} \tag{2.56}$$

$\phi_1(T_{ref})$ and $\phi_2(T_{ref})$ were obtained from the calibration curves of dye 1 and dye 2 respectively. This provided a value for $\frac{\phi_1(T)}{\phi_2(T)}$, which allowed the temperature of the liquid in the measurement volume to be determined from the ratio of the calibration curves for the two dyes. Once the temperature was calculated, liquid fraction could be evaluated from Eq. (2.55).

Figure 2.24 represents the schematic of the optical equipment that was used for isothermal static and transient measurement. In this setup, emission from 200 W mercury arc lamp was passed through a filter and then to a dichroic mirror, which only reflected light in the range of 460−500 nm to the measurement volume in the channel. Emitted light passed back through the dichroic and then through the filter, which only allowed passage of wavelengths above 510 nm. Optical paths of upright microscope were used to split the light to the CCD camera and a two-eyepiece microscope. Photodiode was fixed to one of the eyepieces. A 0.33× demagnification lens was used along

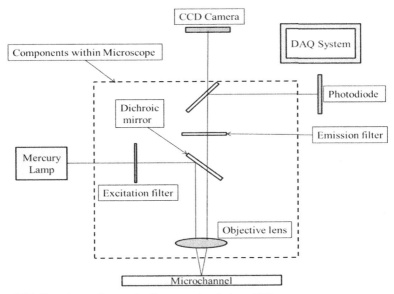

Figure 2.24 Experimental setup for fluorescent technique [52].

with camera to ensure that same size image was captured by both CCD arrays and photodiode.

A static void was generated in a single microchannel and an array of parallel microchannels in order to compare the void measured with the photodiode to a CCD image of that void. A syringe pump was used to flow air and fluorescein solution into the channels. Once void was present in the measurement volume, the void was captured by both the photodiode and the CCD camera.

The measured void was calculated as,

$$\beta = \frac{I_m - I_o}{I_f - I_o} \tag{2.57}$$

where β is volume-averaged liquid fraction, I_m is the measured intensity for the void in question, I_o is the measured intensity when the channels were completely empty ($\beta_l = 0$), and I_f is the measured intensity when the channels are completely full ($\beta_l = 1$).

The measured liquid fraction was verified using the images. The binary representation of the vapor distribution within the measurement volumes was obtained by thresholding the intensities from the CCD array. From the binary representation, the number of liquid pixels for a given image was compared with the number of liquid pixels for the image when the channels were completely filled to obtain the liquid fraction. However, the binary representation of the image neglected the information from the liquid films above and below many of the vapor structures, which restricted an accurate measurement of the liquid fraction from a single threshold value.

This study demonstrated the use of a simple and inexpensive fluorescent technique for steady-state and transient measurements in isothermal two-phase microchannel flow. As predicted by the one-dimensional analysis, the dependence of the emitted intensity on liquid fraction was found to be linear. Analysis indicated that a single dye was required for isothermal measurements while two dyes were required for boiling flows.

Ide et al. [53] developed an optical measurement system to measure void fraction, plug lengths, and velocities of the two-phase system of nitrogen gas and deionized water in a circular microchannel of 100-μm diameter. Two-phase mixture of nitrogen gas and deionized water was obtained in a T-junction attached to the microchannel. A tube with length of 2.0 m and an inner diameter of 250 μm, through which each phase was flowing, was connected to the T-junction. Figure 2.25 shows the optical measurement system, which consisted of four optical fibers, each with the core diameter of 250 μm, placed at a distance of 36 mm from the T-junction. The optical fibers were set in a plastic plate with an interval of 0.25 mm and were connected to an infrared LED light source. The light passed through the transparent microchannel was guided by the optical fibers was detected by infrared photo-detectors and converted to an electric signal in the form of voltage. They studied two cases where a flow control valve was attached to upstream (case 1) and to downstream (case 2) of the gas-side tube. Using optical fibers and infrared photo-detectors, the void fraction in microchannel was measured by detecting the intensity of the signal level. The signal level in terms of voltage was correlated to different flow regimes. They found an almost linear relationship within a variation of ±10% between time-averaged void fraction data from optical fiber measurements (α) and video images (α_v) was obtained. They also showed that time-averaged void fraction for case 1 varies with homogeneous void fraction in a nonlinear manner according to the correlation developed by Kawahara et al. [69] and is given by,

$$\alpha = \frac{C_1 \beta^{0.5}}{1 - C_2 \beta^{0.5}} \tag{2.58}$$

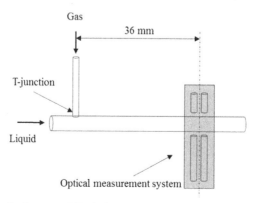

Figure 2.25 Schematic diagram of Optical measurement system [53].

And in case 2, it follows the Armand [96] equation as,

$$\alpha = 0.83\beta \tag{2.59}$$

In case 1, authors attributed the presence of compressible gas volume at the inlet of T-junction to causing intermittent gas supply to the microchannel that results in large slip liquid and gas phases. This compressible gas volume is absent in case 2, and thus a steady flow of gas occurs in the microchannel, causing a smaller gas plug.

Recently, Zhou et al. [54] used the capacitively coupled contactless conductivity Detection (C^4D) technique to measure void fraction measurement of gas—liquid two-phase flow in millimeter-scale horizontal pipe with inner diameters of 2.8, 3.9, 5.3, and 7.0 mm under adiabatic condition. A T-junction was used to create a mixture of water and nitrogen, which was used as the working fluid. The C^4D sensor consisted of an insulating pipe, one metal excitation electrode, one metal pick-up electrode, AC source, an inductor, a grounded shield of the pick-up electrode, and a data acquisition unit as shown in Fig. 2.26. The corresponding equivalent circuit is shown in Fig. 2.27.

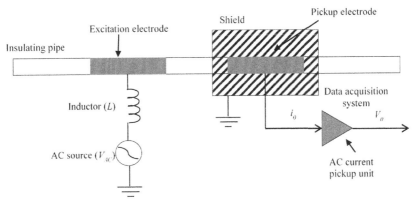

Figure 2.26 Schematic diagram of C^4D sensor [54].

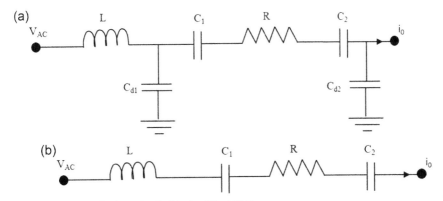

Figure 2.27 (a) Equivalent circuit (b) simplified [54].

The coupling capacitances are C_1 and C_2 are formed by excitation and pick-up electrodes, and R is the resistance of fluid between the two electrodes. The stray capacitances Cd_1 and Cd_2 are neglected as Cd_1 is small and the effect of Cd_2 is negligible in detecting conductivity. The simplified equivalent circuit is shown in Fig. 2.27(b).

The capacitive reactance and the inductive reactance were eliminated at the resonance frequency, and the reactance of the overall impedance of the AC path was zero according to series resonance principle [55]. The resonance frequency was found as,

$$f_r = \frac{1}{2\pi} \sqrt{\frac{C_1 + C_2}{LC_1C_2}} \tag{2.60}$$

The overall impedance of the AC path Z_r is equal to the resistance of the path containing fluid between two electrodes, $Z_r = R$. The resistance (R) was calculated by measuring AC current i_0, which is generated due to application of AC voltage V_{AC} on the excitation electrode. Therefore, the conductance C_r was determined as,

$$C_r = \frac{1}{R} = \frac{i}{V_i} \tag{2.61}$$

Based on this principle, the conductance of flowing fluid in a pipe was measured.

They used this technique to measure the void fraction of bubbly and slug flow regimes in millimeter-scale horizontal pipes of diameters 2.8, 3.9, 5.3, and 7 mm. The reference void fraction is calculated by,

$$\alpha_r = 1 - \frac{Q_l}{vA} \tag{2.62}$$

where Q_l is flow rate of liquid, A is cross-sectional area of pipe, and v is velocity of gas−liquid two-phase flow found by the images captured by the high-speed camera. The void fraction in any flow pattern is correlated to the dimensionless conductance $\left(G^* = \frac{G}{G_r}\right)$ by a linear relationship as observed in experiments in pipes, which is given by,

$$\alpha = aG^* + b \tag{2.63}$$

where a and b are the proportional coefficient and the intercept, respectively, and G^* is the conductance corresponding to the situation when the pipe is full of tap water (i.e., $\alpha = 0$). The values of a and b are different for different flow patterns. The following steps are observed:

1. Conductance signals of gas−liquid two-phase flow are acquired from C^4D.
2. Flow pattern map is identified through high-speed imaging and the corresponding void fraction model is adopted.
3. The void fraction is calculated based on the relation between α and G^* (Eq. (2.11)).

Their void fraction prediction lies within a maximum 7% of the mean, which shows the effectiveness of this technique.

2.2.2.2 Empirical Correlation for Void Fraction

In this section, various empirical correlations developed for calculating void fraction are discussed. Ali et al. [56] investigated the adiabatic two-phase flow (air—water) in narrow channels between flat plates that were 200 mm long and 80 mm wide with gap width of 1.465 and 0.778 mm. Six different orientations were studied: vertically upward, vertically downward, 45° inclined upward and downward, and horizontal flows between horizontal plates and between vertical plates. The experimental setup consisted of test section, air and water supply system, instruments, and data acquisition system. Gap spacer was placed between the plates to make a narrow channel, which had expanding inlet and outlets. There were two outlets through which the air—liquid mixture moved into the recirculation chamber.

Measurement of void fraction was performed by electrical conductivity method as shown in Fig. 2.28. Dual-system of probes and electrodes were required to measure the local, test section average and cross-sectional average void fractions. Electrodes were manufactured with solid copper bars with good surface finish, which were then plated with nickel and then overcoated with the gold thin layer for inertness. Two pairs of electrodes, each 12.5 × 76 × 3 mm, were placed 120 mm apart. Each pair of two electrodes were placed facing each other at a position for measuring cross-sectional void fraction. The probe was present in the middle of the upstream and downstream electrodes for measuring local void fraction.

The local void fraction was measured by a probe consisting of a pair of flattened chromel wires, which were flush mounted on the inside surfaces of the back and front plates facing each other as shown in Fig. 2.28(a). For the test section void fraction measurement, electrodes were shorted to form the ring electrode at both

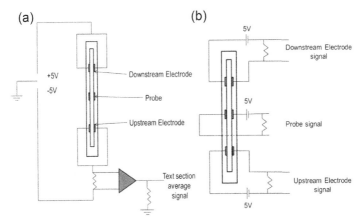

Figure 2.28 Electrical Circuit for measurement of (a) test section average void fraction and (b) cross section average void fraction [56].

inlet and outlet as shown in Fig. 2.28(b). Void fraction was calculated as the ratio of electrical conductivity of air—water mixture and electrical conductivity of single-phase water.

For all orientations except horizontal flow through vertical plates, flow patterns were found as bubbly, intermittent, rivulet, and annular. The boundaries between the intermittent and rivulet flow were found to be least dependent on gap width and orientations. However, three flow patterns that were observed in case of horizontal flow through vertical plates were bubbly, intermittent, and stratified-wavy. In this orientation at low superficial liquid velocity, the stratified flow was observed with negative level gradient. Even at high volumetric flux, the annular flow was not observed as it might be due to least interfacial area and shear.

The test section average void fraction data were properly correlated for all cases excluding horizontal flow between vertical plates as the function of the Lockhart—Martinelli parameter X, as previously depicted by Lowry and Kawaji [57]. The Lockhart—Martinelli parameter (Lockhart and Martinelli [65]) is the ratio of single-phase friction pressure drop of liquid and single-phase friction pressure drop of gas and is defined as follows,

$$X^2 = \frac{(\Delta P/\Delta Z)_{F,SPL}}{(\Delta P/\Delta Z)_{F,SPG}} \tag{2.64}$$

The correlation proposed by Chisholm and Laird [58] for two-phase flow in pipe was written as,

$$\alpha = 1 - \left(1 + \frac{C}{X} + \frac{1}{X^2}\right)^{-1} \tag{2.65}$$

It was predicted that for all orientations except horizontal flow between vertical plates, data were well related taking the value of C as 20. The effect of orientation was much less. However, for narrower gap channel, void fraction and values of C were found to be lower.

Another correlation widely used for evaluating the void fraction analysis in two-phase flow is the drift flux model (Zuber and Findlay [59]; Wallis [60]) and is given by,

$$U_G = C_0\langle j\rangle + V_{gj} \tag{2.66}$$

where $\langle j\rangle$ is the mixture mean velocity ($j_G + j_L$), U_G is the average gas velocity (j_G/α), C_0 is the distribution parameter, and V_{gj} is the mean drift velocity. C_0, distribution parameter, is accounted for the difference in flow velocity of liquid and gaseous phase, and void fraction also depends on this factor. Mean drift velocity V_{gj} was the difference between the velocity of gas and the mean velocity of mixture and was considered to be a function of bubble terminal rise velocity in a stagnant liquid.

Most of the data for all orientations except H—V were properly correlated by the drift flux model with $C_0 = 1.25$ and $V_{gj} = 0$. The small value of V_{gj} for the narrow channels was the cause of inability of the bubbles rising through the stagnant liquid due to the surface tension force. When $V_{gj} = 0$, the drift flux model was similar to the variable density single-fluid model of Bankoff [61] given by,

$$\alpha = K\beta \tag{2.67}$$

where the flow parameter K represents to the inverse of the distribution parameter C. Thus, a simple correlation was seen to represent void fraction data for different gap widths and for all orientations excluding for horizontal flow between vertical plates (H−V) and was given by,

$$\alpha = 0.8\beta \tag{2.68}$$

For the horizontal flow through vertical plates, both Lockhart−Martinelli type correlation and drift flux model were unable to correlate well the void fraction data. A strong mass velocity effect was observed for $j_L < 2$ m/s. The flow patterns observed were stratified and wavy-stratified at the lower liquid flow rates. For those low liquid velocity data were observed to be properly matching with a simplified stratified flow model, which was derived by Taitel and Dukler [62] through theoretical approach, but the interfacial shear was assumed to be neglected as briefly described next. Hence, momentum balances for the two phases may be combined to obtain as in stratified flow the pressure gradient dP/dx for phases was assumed to be equal,

$$\tau_{wG}(S_G/A_G) - \tau_{wL}(S_L/A_L) + \tau_i S_i \left(\frac{1}{A_L} + \frac{1}{A_G} \right) = 0 \tag{2.69}$$

where A_G is the cross-sectional area occupied by gas and A_L is the cross-sectional area occupied by liquid, S_G is the length of the channel perimeters in contact with the gas phase and S_L is the length of the channel perimeters in contact liquid phase, and S_i is the interfacial area per unit axial length. In a narrow channel, for stratified flow between vertical plates, S_i was assumed to be neglected and the following expression was written as,

$$S_G/A_G = S_L/A_L = 2/W \tag{2.70}$$

where W is the gap width. The momentum balance was then simplified to,

$$\tau_{wG} = \tau_{wL} \tag{2.71}$$

which was then written in terms of the average gas velocity, average liquid velocity, and friction factors as follows,

$$f_G\left(\rho_G U_G^2/2\right) = f_L\left(\rho_L U_L^2/2\right) \tag{2.72}$$

The friction factors could be replaced by using the Blasius type correlation, $f = C_1 Re^{-m}$ for both turbulent and laminar flows. The Lockhart—Martinelli parameter was then calculated using $U_G = j_G/\alpha$, $U_L = j_L/(1 - \alpha)$ to give,

$$X^2 = (1 - \alpha)^{2-m} \big/ \alpha \tag{2.73}$$

$$\alpha = 1 \big/ \left(1 + X^{2/(2-m)} \right) \tag{2.74}$$

The friction factor correlations were used for the present, and then the above equation for turbulent flow in both phases was simplified to,

$$\alpha = 1 \big/ \left(1 + X^{1.143} \right) \tag{2.75}$$

and for laminar flow in both phases,

$$\alpha = 1 \big/ \left(1 + X^2 \right) \tag{2.76}$$

Equations (2.81) and (2.82) were applied to the limiting case of stratified flow in horizontal flow between vertical plates. These were compared with the measured void fraction that occurred for the lowest flow rates in different channels overall j_G (0.15—16 m/s) for stratified flow.

Test section void fraction was well correlated with Lockhart—Martinelli and drift flux model for all orientation excluding the horizontal flow through the vertical plates. However, for horizontal flow through vertical plates, the stratified flux model is correlated by Taitel and Dukler [62] by neglecting the interfacial shear. Except for horizontal flow through vertical walls, the effect of different orientation and gap width was found to be very small on the flow patterns, void fractions, and frictional pressure drops. In horizontal flow between the vertical plates, the pressure drop and void fraction were highly affected by mass velocity.

Triplett et al. [63] experimentally evaluated the void fraction and two-phase frictional pressure drop through the microchannel. Air—water mixture was used to carry out the experiment. Different microchannel profiles used were circular with 1.1- and 1.45-mm inner diameter and semitriangular profile (one corner is smoothed) of hydraulic diameter of 1.09 and 1.49 mm. They found several flow patterns in circular test section viz. bubbly, slug, churn, slug-annular, and annular flow, whereas bubbly, slug, churn, slug-annular, and annular flow was found in the semitriangular test section. However, As surface tension force was dominant, stratified flow was found to be absent, however slug and churn flows were prominent with least velocity slip.

Void fraction was measured with the help of photographs to capture the form in the center of the test sections. In bubbly flow, the bubbles were assumed to be spherical or ellipsoidal. In slug flow, Taylor bubbles were considered to be cylindrical and spherical segments. In bubbly flow, the photograph was covered by a large number of bubbles so it was easier to evaluate volume-averaged void fraction. For slug flow pattern, multiple photographs were used to evaluate the void fraction. In annular flow pattern, bubbles were parted into cylinders and channel average void fraction was evaluated. Slug-annular flow void fraction was not considered due to its high uncertainty. In churn flow, an average of 0.5 of local void fraction was assumed.

They compared the homogeneous model, Chexal et al. [64] correlation, and Lockhart and Martinelli [65] correlation. Chexal et al. [64] correlation was based on the drift flux model, whereas the Lockhart–Martinelli correlation for void fraction was given by,

$$\frac{1-\alpha}{\alpha} = A\left(\frac{1-x}{x}\right)^p \left(\frac{\rho_G}{\rho_L}\right)^q \left(\frac{\mu_L}{\mu_G}\right)^r \tag{2.77}$$

where $A = 0.28$, $p = 0.64$, $q = 0.36$, and $r = 0.07$ for Lockhart–Martinelli and $A = 1$, $p = 0.74$, $q = 0.65$, and $r = 0.13$ for a correlation due to Baroczy [66].

They deduced that the void fraction increased with increasing superficial velocity of gas keeping the superficial velocity of liquid constant, and vice versa. Void fraction data were best predicted through homogeneous mixture model for bubbly and slug two-phase flow patterns. However, in case of churn and annular flow patterns, the homogeneous model overpredicted the experimental void fraction data due to the significant interphase slip that was likely due to the separation of the liquid and gas phases. Other correlations predicts satisfactorily for other flow patterns except for low superficial liquid velocity, U_{LS}, where it underpredicts the void fraction.

Mishima and Hibiki [66] studied the void fraction in Pyrex glass capillary tubes with inner diameters ranging from 1 to 4 mm. Air and demineralized water are used as working fluids. They observed bubbly flow, slug flow, churn flow, annular flow, and annular-mist flow. They correlated the measured void fraction using the drift-flux model where the gas velocity (v_G) is related to the mixture volumetric flux (j_G) in the following way,

$$v_G = \frac{j_G}{\varepsilon} = C_0 j + V_{Gj} \tag{2.78}$$

where C_0 and V_{Gj} are distribution parameter and drift velocity, respectively.

According to Ishii [67], the C_0 and V_{Gj} for round tube for different flow regimes can be written as.

For bubbly flow,

$$C_0 = 1.2 - 0.2\sqrt{\frac{\rho_G}{\rho_L}} \tag{2.79}$$

$$V_{Gj} = (1-\varepsilon)^{\frac{3}{2}}\sqrt{2}\left(\frac{\sigma g \Delta \rho}{\rho_L^2}\right)^{\frac{1}{4}} \tag{2.80}$$

For slug flow,

$$C_0 = 1.2 - 0.2\sqrt{\frac{\rho_G}{\rho_L}} \tag{2.81}$$

$$V_{Gj} = 0.35\left(\frac{dg\Delta\rho}{\rho_L}\right)^{\frac{1}{2}} \tag{2.82}$$

For churn flow,

$$C_0 = 1.2 - 0.2\sqrt{\frac{\rho_G}{\rho_L}} \tag{2.83}$$

$$V_{Gj} = \sqrt{2}\left(\frac{\sigma g\Delta\rho}{\rho_L^2}\right)^{\frac{1}{4}} \tag{2.84}$$

For annular flow,

$$C_0 = 1.0 \tag{2.85}$$

$$V_{Gj} = \frac{(1-\varepsilon)\left[j + \left(\frac{dg\Delta\rho(1-\varepsilon)}{0.015\rho_L}\right)^{\frac{1}{2}}\right]}{\varepsilon + 4\left(\frac{\rho_G}{\rho_L}\right)^{\frac{1}{2}}} \tag{2.86}$$

However, Mishima and Hibiki correlated their experimental data assuming zero drift velocity as,

$$v_G = C_0 j \tag{2.87}$$

where C_0 is found as 1.45, 1.31, 1.33, and 1.21 for tube diameter of 1.09, 2.10, 3.08, and 3.90 mm, respectively. They found that the distribution parameter decreases with inner diameter of the pipe in the bubbly and slug flow patterns and can be expressed as,

$$C_0 = 1.2 + 0.51e^{-0.691d} \tag{2.88}$$

where d is in mm. Their correlation is able to predict their experimental results and Kariyasaki et al. [68] data within the standard deviation of 2.5%. However, they excluded the experimental results for the case of $j > 30$ m/s corresponding to annular flow regime from the correlation.

Kawahara et al. [69] experimentally investigated the flow pattern, void fraction in a horizontally placed 100-μm inner diameter circular transparent microchannel made of fused silica. The length of the microchannel was 64.5 mm resulting in $L/D = 645$. The microchannel was connected to a T-junction of diameter 250 µm, which was connected to a mixing chamber of diameter 0.5 m. Deionized water and nitrogen gas were mixed in a mixing chamber which consisted of an internal passage of 0.5-mm diameter for liquid and gas flows through an external annular passage with a 0.2-mm gap.

They categorized four different flow patterns based on the probability of appearance and time-averaged void fraction: slug-ring flow, ring-slug flow, semiannular flow, and multiple flow.

The void fraction of liquid alone in the microchannel is taken as 0, whereas the void fraction for gas core flow with a smooth-thin liquid film is considered as 1 at low liquid flow rates. The time-averaged void fraction (ε) for low flow rate is defined as for each flow type as,

$$\varepsilon = \frac{\text{Number of gas core images}}{\text{Total number of images counted}} \tag{2.89}$$

For high flow rate, images of an extra flow pattern is recorded, which is gas core flow with a thick liquid film ($0 < \varepsilon < 1$). In this case, the time-averaged void fraction (ε) is calculated as,

$$\varepsilon = \frac{\text{Number of images of gas core flows with a thin liquid film} + \sum \varepsilon_{gc}}{\text{Total number of images counted}} \tag{2.90}$$

where ε_{gc} is gas core volume fraction.

They plotted the time-averaged void fraction results with respect to homogeneous void fraction (β) for different superficial liquid velocity (j_L). They found that the void fraction was not significantly dependent on j_L. The time-averaged void fraction in narrow channel did not correlate well with the homogeneous flow model ($\varepsilon = \beta$) and Armand [96] type of correlation ($\varepsilon = 0.8\beta$) as reported by Ali et al. [56]. The void fraction remained low compared with the homogeneous model for ($\beta < 0.8$), and it increases fast for $0.8 < \beta < 1$. This was attributed to the wall stress and stronger surface tension effect in narrow channel, which allows complete bridging of channel cross section by slow moving liquid; hence, the void fraction was low due to absence of gas phase. At low liquid and gas flow rates, the gas core flow was separated with the surrounding liquid film and a weak momentum couples them compared with large channels ($D_H \sim 1$ mm). Hence, the gas flows smoothly in the channel core, although the wall stress and stronger surface tension strongly affect the liquid. The time-averaged void fraction was correlated with the homogeneous void fraction as,

$$\alpha = \frac{0.03\beta^{0.5}}{1 - 0.97\beta^{0.5}} \tag{2.91}$$

Kawahara et al. [70] studied the effect of channel diameter and liquid properties on the void fraction in circular horizontal microchannels under adiabatic condition. As the surface tension effects is dominant in microchannels, two different fluids, water/nitrogen gas and ethanol−water/nitrogen gas mixtures, are used to evaluate that effect on void fraction. The effect of channel diameter on the void fraction was investigated by considering microchannels having 50-, 75-, 100-, and 251-μm inner diameters. Two experimental setups were constructed: (1) experimental apparatus 1 for 50- or 100-μm microchannels and (2) experimental apparatus 2 for 50-, 75-, 100-, or 251-μm microchannels. A pneumatic pump consisted of pressure vessels filled with liquid and connected to gas cylinder of nitrogen gas, and the mixture of water/nitrogen gas and ethanol−water/nitrogen gas was pumped to the microchannel. Depending on the resolution and frame rate requirement, two CCD cameras were used to capture the flow pattern in microchannel.

Authors determined the time-averaged void fraction by adding instantaneous void fraction for each image divided by the total number of images. The void fraction was assumed to be unity if a smooth thin liquid surrounds the gas core flow. The void fraction for the case with gas core surrounded by a thick liquid film was estimated by considering gas core as a cylinder with smaller radius than the microchannel radius. As a first step, the flow pattern was identified and the authors observed four flow patterns: single-phase liquid, tail of a glass slug, liquid-ring film, and serpentine-like gas core. They found that the time-averaged void fraction varies nonlinearly with the homogeneous void fraction $\left(\beta = \frac{j_G}{j_L+j_G}\right)$ for distilled water/nitrogen gas flow in microchannels with 50- and 100-μm diameter and can be correlated as,

$$\alpha = \frac{C_1 \beta^{0.5}}{1 - C_2 \beta^{0.5}} \tag{2.92}$$

where C_1 and C_2 are 0.03 and 0.97, respectively, for the 100-μm-diameter circular microchannel [69] and 0.02 and 0.98, respectively for the 50-μm-diameter circular microchannel for water/nitrogen gas flow. The void fraction remained low for $\beta < 0.8$ and rapidly increased for $0.8 < \beta < 1$; hence, the void fraction deviated from the homogeneous flow model or the Armand [96] type of correlation. The authors explained that there exists a large slip between liquid and vapor phases even at low gas flow rates. However, the void fraction varies linearly with homogeneous model or Armand [96] type of correlation in the case of 251-μm-diameter microchannels and is independent on ethanol concentrations. It was found that the void fraction was lowered by the liquid viscosity for 75- and 100-μm-diameter microchannels at high superficial gas velocities ($\beta > 0.8$). They concluded that the surface tension and liquid viscosity has less effect on the void fraction compared with the channel diameter.

Similar studies were performed by Kawahara et al. [71] to investigate the effect of surface tension and viscosity of fluid on bubble velocity, bubble length, void fraction and pressure drop under two-phase flow in an adiabatic experiment and the flow contraction effect on the inlet of the channel. The experimental test section consisted of a fused silica circular microchannel with an inner diameter of 250 μm and a total length of 99 mm. The effects of surface tension and viscosity was studied by considering distilled water, an aqueous solution of ethanol with different mass concentrations of 0%, 4.8%, 49%, and 100% as working liquids, and nitrogen as working gas. For ethanol, viscosity was found maximum at 48 wt% conc., and it is the same at 4.8 wt% and at 100 wt% conc. Some of the liquid properties are listed in Table 2.7. Table 2.8 depicts the volumetric fluxes of for liquid (j_L) and gas (j_G) phases, where j_G is calculated based on the gas density evaluated at the system pressure and liquid temperature at the midpoint of the channel.

The high-speed video camera was used to find bubble velocity in the test section. Flow image pixels were recorded at 8000 frames/s with a resolution of 160×78 with a shutter speed of 1/160,000 s. Recorded images were transferred to a computer where image processing was performed using software. The bubble velocity is calculated as,

$$u_G = \Delta Z f \qquad (2.93)$$

where f is frame rate and ΔZ is the moving distance of the bubble nose.

The flow contraction effect was examined by two mixers of inner diameters of 250 and 500 μm that were used at microchannels with a diameter of 250 μm. The void fraction was determined by,

$$\alpha = \frac{u_G}{j_G} \qquad (2.94)$$

where j_G is the volumetric flux of the gas phase at the midpoint of the test section.

Table 2.7 Properties of Liquids Used in Experiments

Working Liquids	Density (kg/m³)	Viscosity (m Pa/s)	Surface Tension (N/m)
Distilled water	996.5 ± 1.7	0.92 ± 0.1	0.072 ± 0.001
Ethanol 4.8 wt%	989.4 ± 1.6	1.19 ± 0.2	0.060 ± 0.001
Ethanol 49 wt%	910.9 ± 5.3	2.43 ± 0.5	0.028 ± 0.001
Ethanol 100 wt%	785.7 ± 7.5	1.16 ± 0.2	0.022 ± 0.001

Table 2.8 Volumetric Fluxes for Liquid and Gas

Working Liquid	j_L (m/s)	j_G (m/s)
Distilled water	0.22−1.43	0.04−1.24
Ethanol 4.8 wt%	0.11−1.08	0.07−0.96
Ethanol 49 wt%	0.21−0.91	0.04−1.77
Ethanol 100 wt%	0.22−1.52	0.02−1.33

They found that the void fraction data for distilled water and 4.8 wt% ethanol solution remain between the lines for homogeneous flow model ($\alpha = \beta$) and Armand [96] correlation, which was given by,

$$\alpha = \frac{1}{C_A}\beta \tag{2.95}$$

with $C_A = 1.2$. However, lower void fraction was found for 49 wt% ethanol solution and pure ethanol compared with that for distilled water and 4.8 wt% ethanol solutions. It was found that the void fraction with flow contraction was lower than that without flow contraction and follows equation $\alpha = \frac{0.03\beta^{0.5}}{1-0.97\beta^{0.5}}$ for 49 wt% ethanol solution. They attributed this on faster bubble velocity, which was due to elongation of bubble in the central region as a result to flow contraction.

Saisorn et al. [72] investigated the flow characteristics of two-phase flow in circular microchannel. The test section was made of fused silica. Experiment was conducted taking three inner diameters of 0.53, 0.22, and 0.15 mm with corresponding length as 320, 120, and 104 mm. Air, nitrogen, water, and deionized water were used as working fluids in the system. The superficial velocity of gas and superficial velocity of liquid were varied in a range of 0.37−42.36 and 0.005−3.04 m/s, respectively. Flow visualization was facilitated using stereo-zoom microscope and high-speed camera. Void fraction was determined using image processing analysis. The following observations were made by the authors:

1. For channel diameter 0.53 mm, four different flow patterns were depicted as slug flow, throat-annular flow, churn flow, and annular-rivulet flow. Slug flow, which occurred at a relatively low air velocity, was characterized by elongated bubble flowing in the axial direction. Throat-annular flow was considered as a unique flow pattern in microchannels due to the fact that this flow pattern was never observed in ordinary channel size under the gravity conditions. Churn flow was characterized by disruption near the bubble tail of the slug flow pattern. Annular−rivulet flow corresponded to the situation in which both annular flow and rivulet flow appeared interchangeably.

2. For channel diameter 0.22 mm, the observed flow patterns were throat-annular flow, annular flow, and annular-rivulet flow. It should be noted that annular flow was characterized by the stable flow of the water film on the tube wall with continuous air flow in the tube core, whereas annular-rivulet flow was simultaneous appearance of both annular flow and rivulet flow. The rivulet flow was a flow pattern characterized by river-like water stream flow pattern on the tube surface.

3. For channel diameter 0.15 mm, four different flow patterns were observed. These flow patterns were liquid-alone flow, throat-annular flow, serpentine-like gas core flow that was characterized by small ripple formation on the gas−liquid interface, and annular flow. It was noted that such four flow patterns were observed periodically alternate pattern even at a given flow condition.

Void fraction was calculated using image processing technique. Image processing analysis was considered assuming symmetrical volumes that included ellipsoidal, cylindrical, and spherical formed by gas–liquid interface. It was observed that the void fraction in 0.53-mm-diameter channel agreed well with homogeneous model, i.e.,

$$\alpha = \beta \tag{2.96}$$

For 0.22-mm-diameter channel, Armand [96] type correlation is agreed well, i.e.,

$$\alpha = 0.833\beta \tag{2.97}$$

For 0.15-mm-diameter channel, the void fraction can be predicted using Kawahara correlation agreed well, i.e.,

$$\alpha = \frac{C_1\beta^{0.5}}{1 - C_2\beta^{0.5}} \tag{2.98}$$

where β is volumetric quality, C_1 is 0.036, and C_2 is 0.945.

Woldesemayat and Ghajar [73] evaluated the performance of 68 correlations for predicting void fraction in horizontal and upward inclined pipes. Due to the complexity and lack of understanding of the physics of two-phase flow, several empirical correlations for predicting void fraction are developed. They categorized all available correlations into four groups, viz. (a) slip ratio, (b) $K\varepsilon_H$, (c) drift flux, and (d) general correlations. They considered the data of Eaton [74], Beggs [75], Spedding and Nguyen [76], Mukherjee [77], Minami and Brill [78], Franca and Lahey [79], Abdulmajeed [80], and Sujumnong [81] while testing the performance of the void fraction correlations. Table 2.9 shows the range of diameters, inclination angles, number of data points, and fluids considered in their study.

Authors pointed out that the Dix [82] correlation is able to consistently predict void fraction over the entire range and the prediction of highest number of void fraction data also falls within the 5% error index. Hence, this correlation is taken for further improvement and it becomes after introducing two correction factors into the drift velocity expression,

$$\varepsilon = \frac{U_{SG}}{U_{SG}\left(1 + \left(\frac{U_{SL}}{U_{SG}}\right)^{\left(\frac{\rho_G}{\rho_L}\right)^{0.1}}\right) + 2.9\left[\frac{gD\sigma(1+\cos\theta)(\rho_L-\rho_G)}{\rho_L^2}\right]^{0.25}(1.22 + 1.22\sin\theta)^{\frac{P_{atm}}{P_{system}}}} \tag{2.99}$$

Xiong and Chung [83] studied the gas–liquid flow patterns and void fractions in microchannels under adiabatic condition using nitrogen and deionized water. They evaluated the microchannel size effect on the flow regime maps and void fraction with the rectangular channels with hydraulic diameters of 0.209, 0.412, and 0.622 mm. The superficial gas and liquid velocities varied from 0.06–72.3 to 0.02–7.13 m/s, respectively.

Table 2.9 Databased Used in the Analyses

Literature	Orientation of Channel	Channel Internal Diameter (mm)	Working Fluids	Measurement Technique	Data Points Used
Eaton [74]	Horizontal	52.5, 102.26	Natural	Quick-closing Valve	237
Beggs [75]	Horizontal, vertical, uphill (5°, 10°, 15°, 20°, 35°, 55°)	25.4, 38.1	Gas (nitrogen)—water	Quick-closing valve	291
Spedding and Nguyen [76]	Horizontal, vertical, uphill (2.75°, 20.75°, 45°, 70°)	45.5	Air—water	Quick-closing valve	1383
Mukherjee [77]	Horizontal, vertical, uphill (5°, 20°, 30°, 50°, 70°, 80°)	38.1	Air—kerosene	Capacitance probes	558
Minami and Brill [78]	Horizontal	77.93	Air—water and air—kerosene	Quick-closing valve	54 and 57
Franca and Lahey [79]	Horizontal	19	Air—water	Quick-closing valve	80
Abdulmajeed [80]	Horizontal	50.8	Air—kerosene	Quick-closing valve	83
Sujumnong [81]	Vertical	12.7	Air—water	Quick-closing valve	101

They visualized several flow patterns; however, only four basic flow patterns were classified based on Weber number, which is the ratio of inertia force to surface tension force. In bubbly slug flow regime, the dominant force is surface tension force compared with inertial force. The importance of two forces is same in slug-ring flow regime. The weightage of inertial force is more in dispersed-churn flow, although both forces dominate. Inertial force dominated annular flow regime is characterized by smooth liquid−gas interface due to weak interaction at liquid gas interface. They flow regime map was plotted using liquid and gas Weber number instead of superficial velocities of water and liquid, which is used traditionally. They concluded that the flow regime maps are relatively similar for the larger channels with hydraulic diameters of 0.622 and 0.412 mm compared with the smallest channel (0.209 mm). In the smallest channel, the bubbly flow is replaced by the slug flow for lower liquid and gas Weber numbers defined based on superficial liquid, superficial gas flow. The dispersed-churn flow is also found to be absent due to strong surface tension force, which prevents breakdown of slugs and the interruption of the gas−liquid interface.

In their study, authors showed the relationship between the homogeneous void fraction and the time-averaged void fraction. The homogeneous void fraction (β) is calculated by considering same liquid and gas velocities in a dynamic equilibrium condition and is defined as the ratio between the superficial gas velocity (j_G) and the sum of superficial gas and liquid velocities ($j_G + j_L$).

$$\beta = \frac{j_G}{j_L + j_G} \tag{2.100}$$

However, the homogeneous void fraction can differ from the actual void fraction due to unequal liquid and gas velocities in different flow regimes. Hence, the time-averaged void fraction is calculated from instantaneous void fractions, which is a ratio between the volume of gas to that whole region on each image field obtained by high-speed CCD camera. The time-averaged void fraction is defined as,

$$\alpha = \frac{\sum_{n=1}^{N} \alpha_n}{N} = \frac{\sum_{i=1}^{N_l} \alpha_{l,i} + \sum_{j=1}^{N_g} \alpha_{g,j} + \sum_{k=1}^{N_m} \alpha_{m,k}}{N_l + N_g + N_m} \tag{2.101}$$

where $\alpha_{l,i}$, $\alpha_{g,j}$, and $\alpha_{m,k}$ are the estimated void fractions for the pure liquid type, gas core with a smooth interface type and any other type, respectively. N is total number of the recorded images, and N_l, N_g, and N_m are number of the images of liquid, gas core with a smooth interface, and other types, respectively.

They reported that the time-averaged void fraction is not in good agreement with the correlations [56] $\alpha = \beta$ and $\alpha = 0.8\beta$. The nonlinear relation between the homogeneous and time-averaged void fractions in case of smaller channels is attributed to the strong surface tension force compared with the inertial force, which prevents breakup of connection between slugs, thereby resulting in bubbly slug flow and a lower time-averaged void fraction. Based on their experimental results and data

from Kawahara et al. [69], an empirical correlation of the time-averaged void fraction
was developed for microchannels with hydraulic diameters less than 1 mm,

$$\alpha = \frac{C\beta^{0.5}}{1 - (1 - C)\beta^{0.5}} \tag{2.102}$$

$$C = \frac{0.266}{1 + 13.8e^{-6.88D_h}} \tag{2.103}$$

The proposed correlation shows a deviation of $\pm 35\%$ with the results of Xiong and
Chung [83] and Kawahara et al. [69] where only 2 of 66 data are outside the range.

Saisorn and Wongwises [84] experimentally studied void fraction, two-phase flow
pattern and two-phase frictional pressure drop in a circular horizontal microchannel using air−water system. The test section consisted of a fused silica tube with inner diameter of 530 μm and length of 320 mm.

They observed four flow patterns in the channel: slug flow, throat-annular flow,
churn flow, and annular-rivulet flow. The flow pattern map is plotted with superficial
liquid velocity (j_L) as ordinate and superficial gas velocity (j_G) as abscissa at average
ambient condition (1.013 bars, 30 °C).

In their study, the void fraction in the microchannel was obtained for three flow
patterns viz. slug flow, throat-annular flow, and annular-rivulet flow as the flow
was highly disruptive in nature in the fourth flow regime (i.e., churn flow). The volumetric void fraction was calculated based on the image analysis of the different
geometrical shapes assumed by the liquid−gas interface in each flow pattern. They
found that their results agree well with the homogeneous flow model ($\alpha = \beta$) and
with the results reported by Chung and Kawaji [85] for a mixture of nitrogen gas
and deionized water.

Shedd [86] measured the pressure drop and void fraction of R-410 at saturation
temperature of 50 °C for three different horizontally oriented single tubes of diameter
0.5, 1, and 3 mm. The void fraction in the channel was measured by a nonintrusive
technique (i.e., capacitance sensor technique using ring sensors). In the analysis, the
gravitational effect was neglected as experiment was strictly applicable for horizontal
flow orientations.

The author qualitatively divided the flow pattern into four regimes viz. annular,
intermittent, stratified, and stratified wavy in all channels. A new model was proposed
to take into account of the void fraction following the homogeneous model until the
flow pattern is liquid slug. The transition from slug to annular flow was based on
the consideration of the constant turbulent−turbulent Lockhart-Martinelli parameter:
$X_{tt} = 1$. Hence, the void fraction model could be correlated to the homogeneous model
till $X_{tt} = 1$ and then gradually reduced to zero while transitioning to a separated flow
pattern. The void fraction (ε_{mc}) is given by,

$$\varepsilon_{mc} = \xi\varepsilon_{hem} + (1 - \xi)\varepsilon_{RA} \tag{2.104}$$

where

$$
\xi = \begin{cases} 1 & \text{if } X_{tt} > 1 \\ (1 - X_{tt}) & \text{if } X_{tt} \leq 1 \end{cases}
\tag{2.105}
$$

$$
X_{tt} = \left(\frac{1-x}{x} \right)^{0.875} \left(\frac{\rho_g}{\rho_l} \right)^{0.5} \left(\frac{\mu_l}{\mu_g} \right)^{0.125}
\tag{2.106}
$$

$$
\varepsilon_{hem} = \frac{1}{1 + \left(\frac{1-x}{x} \right) \frac{\rho_g}{\rho_l}}
\tag{2.107}
$$

$$
\varepsilon_{RA} = \frac{x}{\rho_g} \left\{ (1 + 0.12(1-x)) \left(\frac{x}{\rho_g} + \frac{1-x}{\rho_l} \right) + \frac{1.18(1-x) \left[g\sigma \left(\rho_l - \rho_g \right) \right]^{0.25}}{G_{tot} \rho_l^{0.5}} \right\}^{-1}
\tag{2.108}
$$

However, the starting of transition between ε_{hem} and ε_{RA} could not be represented by these equations as a function of tube diameter. Hence, the capillary length model is included to take into account of surface tension effect while flow transits to annular flow regime,

$$
\xi = \begin{cases} 1 & \text{if } \zeta > 1 \\ 1 - \zeta & \text{if } \zeta \leq 1 \end{cases}
\tag{2.109}
$$

where $\zeta = X_{tt} \left(\frac{\lambda}{D} \right)$
$$\tag{2.110}$$

and the capillary length is defined as,

$$
\lambda = \sqrt{\frac{\sigma}{\rho_l g}}
\tag{2.111}
$$

Nino et al. [87] investigated the characteristics of void fraction for rectangular multiple channel aluminum microchannels. Hydraulic diameter of 6-port microchannels and 14-port microchannels were taken as 1.54 and 1.02 mm. Refrigerants used were R134a and R410A with mass flux ranging from 100 to 300 kg/s m^2. Slug and elongated flow bubble or plug flow, annular flow, only vapor, and only liquid were the flow configurations that were observed in microchannels. Slug flow, only liquid, and only vapor flows were observed at lower quality (x lower than 0.4 for $G = 100$ kg/s m^2 and x is lower than 0.2 for $G = 300$ kg/s m^2). But annular flow was observed when mass flux and quality were increased. Some studies indicated that void fraction was dependent on mass flux as well as the properties of two-phase flowing through the pipe.

The void fraction correlation based on slip ratio is given as,

$$\alpha = \cfrac{1}{1 + \frac{1-x}{x}\left(\frac{\rho_{vapor}}{\rho_{liquid}}\right)S} \tag{2.112}$$

Slip ratio (S) is defined as the ratio of vapor velocity and liquid velocity and it is expressed mathematically as,

$$S = \frac{V_{vapor}}{V_{liquid}} \tag{2.113}$$

where x is quality of mixture and ρ is density (kg/m^3).

In the ideal homogeneous model, in which the vapor and the liquid phase were assumed to travel at the same velocity (model suitable for intermittent flow regimes where bubbles of vapor travels at the same velocity as the slugs or plugs of liquid), the slip ratio was equal to 1.

However, in case of stratified or annular flow regime, the bulk velocities of each phase was found to be different and was assumed as separated flow. It was because the velocity of vapor core was found to be higher than the liquid ring. For example, Rigot [88] predicted the slip ratio value to be 2.

Slip ratio was found to be increasing with increase in hydraulic diameter for low quality. Slip ratio was found to be related to density ratio. At constant mass flux, velocity of vapor decreased with increase in vapor density. When slip ratio was carefully observed, it was seen that it had two inflexion points when varied with quality. Slip ratio started from a low quality flow and increased to maximum. And, again, when quality was increased, it decreased to minimum, and when quality approached to 1, it increased to infinitely large. Slip ratio was found to be the function of bulk velocities of vapor core and liquid ring. Void fraction results deduced that it was dependent on hydraulic diameter and mass flux.

Choi et al. [89] performed the experiment of liquid water and nitrogen gas two-phase flow in glass rectangular microchannel to study the flow pattern, pressure drop, and void fraction under adiabatic condition. The width and height of the microchannels were taken as 510 × 470, 608 × 410, 501 × 237 and 503 × 85 μm. The aspect ratios were 0.92, 0.67, 0.47, and 0.16, and corresponding hydraulic diameters were 490, 490, 322, and 143 μm. Void fraction was calculated using the following expression,

$$\alpha = \frac{j_G}{u_B} \tag{2.114}$$

where u_B is a bubble velocity. They found that the void fraction varied linearly with volumetric quality (β) for the aspect ratio of channel 0.47. Their experimental results matched well with Armand [96] type correlation (as given in the following equation) rather than nonlinear type correlations such as Chung and Kawaji [85] correlation.

$$\alpha = C_A \beta = C_A \frac{j_G}{j_G + j_L} \tag{2.115}$$

where $C_A = 0.833$. The coefficient C_A was found to depend on aspect ratio of the channel. C_A increased with the decrease in aspect ratio because reduction of edge portion (corner effect) in liquid film, resulting in homogeneous two-phase flow. Table 2.10 summarized C_A for different aspect ratios.

Jassim et al. [90] developed probabilistic two-phase flow map, pressure drop, and void fraction model for six-port microchannel to make the result more accurate and find the simplified method of determining void fraction and pressure drop. Model was prepared for R134a, R410A, and air–water in six-port microchannels at 10 °C saturation temperatures and qualities ranging from 0 to 1, and mass fluxes were in the range between 50 and 300 kg/m^2 s. The total pressure drop and void fraction were determined as the summation of time fractions of flow regimes, which was then multiplied by the corresponding model for that flow regime. They noted that there are three two-phase flow maps:

1. Baker/Mandhane type [91]: Baker has taken superficial vapor mass flux times along vertical axes as fluid property scaling factor and superficial liquid mass flux times along horizontal axes as different fluid property scaling factor. However, Mandhane later developed a similar map with air water data, but used superficial gas and liquid velocities on the horizontal and vertical axes, respectively.
2. Taitel–Dukler type [92]: They developed a mechanistic type flow map with the Lockhart-Martinelli parameter along horizontal axis and a modified Froude rate times a transition criteria on the vertical axis.
3. Steiner type [93]: Steiner style flow maps was consisting of quality along horizontal axis and mass flux along vertical axis.

Chung and Kawaji [85] used the average flow velocity in the microchannel along the vertical axis instead of mass flux, unlike the Steiner type.

Flow maps indicated a particular flow regime at a given flow condition with lines dividing transitions. However, according to Coleman, Garimella [93], and Hajal et al. [94], there exists more than one flow regime at boundaries. Hence, it was difficult to implement maps into models as maps could not be readily represented by

Table 2.10 C_A **for Different Aspect Ratio of Channels**

Aspect Ratio	C_A
0.16	1
0.47	0.89
0.67	0.86
0.92	0.82

continuity function for all quality ranges. Therefore, probabilistic representation of flow regime may be better for describing flow. Nino [95] represented the two-phase flow mapping in horizontal microchannels in a very different manner by recording the time fraction for each flow regime and in each channel at a given mass flux and quality. This was obtained by capturing numerous pictures for a given mass flux and quality at a periodic intervals.

In their work on six-port microchannels, curve fits were used for liquid, intermittent, annular, and vapor flow regime time fractions for probabilistic flow regime maps of R410A, R134a, and air−water as working fluids. These curve fit relations made the time fraction as the physical limits. The liquid time fraction curve fit was proposed as,

$$F_{liq} = (1 - x)^a \tag{2.116}$$

It contains the physical limits of the liquid time fraction with a time fraction of 1 at a quality of 0 and a time fraction of 0 at a quality of 1. The intermittent time fraction curve fit can be written as,

$$F_{int} = (1 - x)^{(bx^c)} - (1 - x)^d \tag{2.117}$$

In Eq. (2.117), the physical limits are, at a quality of both 0 and 1, a time fraction is 0. Similarly, the vapor time fraction curve and annular curve fits were given respectively by,

$$F_{vap} = x^g \tag{2.118}$$

$$F_{ann} = 1 - F_{liq} - F_{int} - F_{vap} \tag{2.119}$$

The physical limits of Eq. (2.118) correspond to a time fraction of 0 at a quality of 0 and a time fraction of 1 at a quality of 1. Equation (2.119) is obtained from the definition of total time fraction, which is equal to 1 for all flow regimes. The constants in the curve fits of R410A, R134a, and air−water are listed in Table 2.11(a−c), respectively.

Table 2.11a Time Fraction Curve Fit Values of Flow Regime for R410A at 10 °C

Constants	$G = 50$ kg/m^2 s	$G = 100$ kg/m^2 s	$G = 200$ kg/m^2 s	$G = 300$ kg/m^2 s
a	10.08	25.77	28.35	27.87
b	0.66	1.88	23.86	47.32
c	2.24	2.15	2.81	1.98
d	10.08	25.77	28.35	27.87
g	3.38	6.80	37.92	65.58

Table 2.11b **Time Fraction Curve Fit Values of Flow Regime for R134a at 10 °C**

Constants	G = 50 kg/m² s	G = 100 kg/m² s	G = 200 kg/m² s	G = 300 kg/m² s
a	25.23	36.99	50.92	55.68
b	0.66	3.27	31.82	60.17
c	0.01	0.27	0.62	1.00
d	25.23	36.99	50.92	55.68
g	2.97	12.81	37.75	51.21

Table 2.11c **Time Fraction Curve Fit Values of Flow Regime for Air–Water at 10 °C**

Constants	G = 50 kg/m² s	G = 100 kg/m² s	G = 200 kg/m² s	G = 300 kg/m² s
a	30.60	71.09	111.02	118.28
b	1.20	1.21	29.17	54.34
c	4.04	0.19	0.57	0.93
d	1.62	2.90	22.81	16.86
g	6.40	9.91	21.67	37.94

Void fraction was modeled by finding the summation of the time fraction of a flow regime multiplied by a void fraction model for the corresponding flow regime as shown in the following equation,

$$\alpha_{total} = F_{liq}\alpha_{liq} + F_{int}\alpha_{int} + F_{vap}\alpha_{vap} + F_{ann}\alpha_{ann} \qquad (2.120)$$

$$\alpha_{liq} = \begin{cases} 0 & for\ liq \\ 1 & for\ vap \end{cases} \qquad (2.121)$$

Armand [96] void fraction model is used for the intermittent flow void fraction model,

$$\alpha_{int} = \frac{(0.833 + 0.167x)x\left(\frac{1}{\rho_v}\right)}{(1-x)\left(\frac{1}{\rho_l}\right) + x\left(\frac{1}{\rho_v}\right)} \qquad (2.122)$$

The equation for the annular flow void fraction model, developed by Nino, was given as,

$$\alpha_{ann} = \left[1 + \left(X_{tt} + \frac{1}{We_v^{1.3}} \right) \left(\frac{\rho_l}{\rho_v} \right)^{0.9} \right]^{-0.06} \tag{2.123}$$

Void fraction data were found to be close from the probabilistic flow mapping—based void fraction model. However, more accurate void fraction data could be obtained.

Cioncolini and Thome [97] developed a new methodology to evaluate void fraction in annular two-phase flow in both micro-scale and macro-scale channels. There were 2673 data points collected from 29 various literature reviews for eight different gas—liquid and vapor—liquid combinations (water—steam, R410a, water—nitrogen, water—argon, water—air, alcohol—air water plus alcohol—air, and kerosene—air), for tube diameters ranging from 1.05 to 45.5 mm and for both circular and noncircular channels. The new prediction methodology was simplified to a greater extent in comparison to the most existing correlations, as it depended on quality of vapor, gas-to-liquid density ratio, and represented the available data points better than the existing prediction methods. Hence, from the study it was clear that there was no macro-to-micro-scale transition for annular flows, which was at least observed around the diameters of about 1.0 mm.

After many literature reviews, new prediction methods for void fraction were deduced. As noted by Beattie and Sugawara (1986) and Brennen (2005), the drift-flux approach relevant for modeling annular flows. In fact, drift-flux models were designed to handle distributed and unseparated two-phase flows. One phase was continuous and the other phase was dispersed and vigorous relative motion between the two phases was caused by gravity which acted an external force field. After investigating 68 different void fraction predicting methods as assessed by Woldesemayat and Ghajar [73], it was seen that many different parameters were influencing the void fraction, including quality of vapor, densities and viscosities of both the phases, surface tension, operating pressure P, mass flux, tube diameter, and acceleration due to gravity and could be written as

$$\varepsilon = \varepsilon \left(x, \rho_l, \rho_g, \mu_l, \mu_g, \sigma, P, d, G, g \right) \tag{2.124}$$

The effect of acceleration due to gravity was negligible in the shear-driven annular flow and both the viscosity ratio and the surface tension cover limiting range. Inclusion of the acceleration due to gravity g, viscosity ratio $\mu_g \mu_l^{-1}$, and surface tension σ into the new prediction method was not relevant. Hence, these parameters could be removed from the equation. Increased vapor quality depicted that more percentage of gaseous phase was present and thereby higher void fraction was found at higher quality.

Simple modeling approach for the void fraction was proposed as,

$$\varepsilon = \varepsilon\left(x, \rho_g \rho_l^{-1}\right) \tag{2.125}$$

When density ratio was low, the rate of growth was a strong function of quality of vapor. The growth was very quick at lower vapor qualities and very slow at higher vapor qualities. However, when the density ratio increased, the dependence of the growth rate on vapor quality became very less and the void fraction growth became more uniform when vapor quality was changed.

In the limit of $\rho_g \rho_l^{-1} \to 1^-$, the void fraction tending to the vapor quality $\varepsilon \to x^+$ and growth rate lost its dependency on vapor quality. Therefore, it was deduced that the void fraction should have a strictly positive first order derivative and a nonpositive second order derivative with respect to vapor quality (i.e., $\partial\varepsilon/\partial x > 0$ and $\partial^2\varepsilon/\partial x^2 \le 0$). There was inappropriate variation of void fraction with independent variables of this equation.

The Hill function was proposed during the modeling of biochemical kinetics (Hill [98]) and commonly used in mathematical physiology (Murray [99]). In the present context, the Hill function was a three-parameter function as follows,

$$\varepsilon = \frac{hx^n}{1 + kx^n} \quad h, \ k > 0, \ 0 < n < 1 \tag{2.126}$$

The upper bound on the parameter n ensured that the second order derivative $\partial^2\varepsilon/\partial x^2 \le 0$ was nonpositive, as required study context. Asymptotical consistency was required that as $x \to 1^-$ then $\varepsilon \to 1^-$, and this condition allowed reducing the number of parameters in the above equation from three to two as follows,

$$\varepsilon = \frac{hx^n}{1 + (1-h)x^n} \quad h > 0, \ 0 < n < 1 \tag{2.127}$$

where the parameters h and n are functions of the density ratio $\rho_g \rho_l^{-1}$ as follows:

$$h = a + (1-a)\left(\rho_g \rho_l^{-1}\right)^\alpha \tag{2.128}$$

$$n = b + (1-b)\left(\rho_g \rho_l^{-1}\right)^\beta \tag{2.129}$$

where $a = -2.129$, $b = 0.3487$, $\alpha = -0.2186$, and $\beta = 0.5150$.

This new prediction method was specifically designed for evaluating void fraction for annular flows and void fraction values were ranging between 0.7 and 1.0. However, the asymptotical consistency for $x \to 0^+$ was not actually appropriate in the present study. According to Shedd [86], with other flow patterns such as bubbly flow and elongated bubble or slug flows, when channel diameter was decreased the void fraction behavior approached toward homogeneous mixture that indicated a reduction in the size of channel, which prevented the phases from forming a slip.

2.3   Liquid Film Thickness Measurement

2.3.1   Introduction

In the recent past, a great deal of interest has been laid on two-phase flow going through microchannels in respect to cooling of electronic devices, fusion reactors that require high heat dissipation. During two-phase flow boiling, different flow regimes are formed along the channel depending on the flow rate, channel dimensions, and fluid properties. It has been seen the heat transfer characteristics are strongly affected by the different flow regimes. Thus liquid film thickness is one of the important parameters that determine heat transfer. Hence, measurement of liquid film thickness has attracted attention of many researchers, and a great deal of research has been carried out to measure liquid film thickness, which is discussed in the following sections.

2.3.2   Methods of Measurement

Tibirica et al. [100] stated various methods for estimating liquid film thickness such as (1) acoustic, (2) electrical, and related to some radiation concept, and (3) optical method, which used reflection and refraction concept to evaluate film thickness. According to these classifications, some literature review was done basing on each measurement techniques.

2.3.2.1   Acoustics Method

Lu et al. [101] represented the process for measuring condensate film thickness by ultrasonic technique. Experimentally, they used R-113 and FC-72 to measure the condensate film thickness on the horizontal lower surface of the rectangular duct at different points in the axial direction. From the data obtained, they formulated power law, which related film thickness with axial distance from leading edge by regression analysis.

There were two loops (i.e., condensation and coolant loop) in the experimental setup. In condensation loop, vapor was formed in boiler, which was then pumped into the duct through a converging unit. Some portion of vapor was condensed on the bottom surface of the duct and the rest of the vapor was discharged out of the duct and condensed in an auxiliary condenser. Then, condensates were collected from the auxiliary condenser and from the outlet of the duct into a liquid tank, from which it was then pumped to the boiler. The test section was 1000 mm long, 40 mm wide, and 25 mm high. But the condensing surface was 914 mm long, and a 0.086-mm-long stainless steel plate was used for draining the condensate. The bottom condensing surface was made of copper plate of thickness 6.4 mm, 914 mm long, and 40 mm wide. The side and top plates were made of polycarbonate plastic sheets, which made possible visualization of the condensate flow pattern. The condensing surface temperature was measured by eight thermocouples that were embedded in the condensing surface. In the coolant loop, cooling water was moved from temperature-controlled bath into the cooling channel of the test section through a set of rotameters. Then cooling water was again returned to the coolant bath. The cooling channel was below the condensing surface. The sides of the channel were made of brass and the bottom walls of polycarbonate plastic.

Using ultrasonic transducer technique, the condensate film thickness was noted at various locations such as 50, 152, 254, 457, and 813 mm from the leading edge. The operational principle of ultrasonic technique was that the transducer emitted an ultrasonic signal, which was reflected at surface discontinuity. The condensate film thickness was calculated as the product of half the measured time duration between the reflections of the signal from the condensing surface and the liquid—vapor interface and the sonic velocity in the condensate. The time delay between the reflected signals was obtained by an oscilloscope. The analyzer transmitted 20-MHz signal frequency to the transducer. An oscilloscope captured the reflected signals and the signals from the analyzer as shown in Fig. 2.29. Time delay between the condensing surface— condensate interface reflection and the condensate—vapor interface reflection was 0.5 μs.

Calibration of transducer was done. Calibration instrument has a pointer attached to micrometer that was placed on the top of copper duct. There was loss in the intensity of reflected signal during initial calibration. Therefore, aluminum plugs were used in copper plate for improving the signal strength. The depth of the liquid in the container was obtained by moving the pointer attached to the micrometer. Time delay was obtained from the oscilloscope, and sonic velocity was obtained for different heights of the liquid at room temperature. The sonic velocity in a fluid is a function of the density

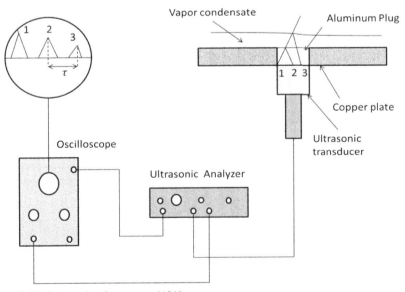

Figure 2.29 Schematic of apparatus [101].

of the fluid, which is dependent on temperature. Reid and Sherwood [102] represented the sonic velocity of liquid as

$$a = c\rho_l^3 \tag{2.130}$$

where ρ_l is the liquid density and c is a constant related to molecular weight of the fluid.

After measuring film thickness at five different points from the leading edge and assuming zero thickness at the leading edge, power law equation was obtained from regression analysis as such

$$\delta = c_1 x^n \tag{2.131}$$

where x is the distance from the leading edge.

The rate of increase of the film thickness reduced in condensate flow direction and film thickness decreased with the increase in velocity of vapor. Toward the trailing edge, the film thickness was hardly found to vary. However, for measuring liquid film thickness by ultrasonic transducer, there are a few assumptions that the film should be smooth or it should have small amplitude interfacial waves. Otherwise, use of a digital oscilloscope can help to evaluate film thickness in the presence of interfacial waves.

2.3.2.2   Electrical Methods

Lee et al. [103] used electrical conductance technique to evaluate local film thickness of the liquid in two-phase annular flow condition. They used electrodes in the inner walls of the pipe to measure the voltage difference between the neighboring electrodes. Accordingly, local film thickness was calculated through interpolation using the relationship between liquid film thickness and given voltage.

They used the electrical conductance method to evaluate the local film thickness of annular two-phase flow. In annular flow, the liquid and gas portion in the circular pipe is shown in Fig. 2.30.

Some L ($e_1, e_2, e_3,...,e_L$) numbers of electrodes were used for the investigation. Two current patterns were used; first, they excited electrodes in the sequence $(e_1, e_2), (e_3, e_4),...(e_L, e_1)$. Second, electrodes were excited in the sequence (e_1, e_3), $(e_2, e_4),...(e_L, e_2)$. Then, voltage difference was noted of the corresponding electrode pairs.

Normalized voltage difference ΔV^* was given in Eq. (2.132) with respect to the normalized liquid film thickness δ^* in Eq. (2.133) for different current patterns.

$$\Delta V^* = \frac{\Delta V - \Delta V_m}{\Delta V_m} \tag{2.132}$$

$$\delta^* = \frac{\delta}{r} \tag{2.133}$$

where ΔV is the voltage difference at corresponding liquid film thickness δ and ΔV_m represents the voltage difference at $\delta = r$, which is at zero void fraction.

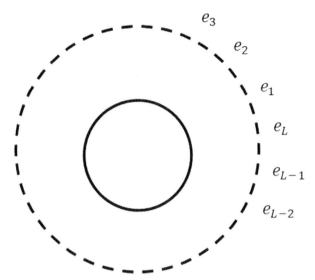

Figure 2.30 Concentric annular flow [103].

A look-up table that could relate ΔV^* to δ^* for two-phase annular flow was prepared after calibration. After that, local film thickness was determined from the lookup table by knowing the voltage difference. Conductivity is a temperature function; the voltage difference is inversely proportional to the conductivity. When temperature is known, the conductivity can be searched from handbooks and the voltage difference at reference temperature was used to form the lookup table that was calculated by finding the product of the ratio of the conductivity at the present temperature and the conductivity at reference temperature with voltage difference at the present temperature.

They proposed few steps to evaluate local film thickness for two-phase annular flow:

1. Lookup table was prepared consisting of normalized voltage differences between electrode pairs for different liquid film thickness.
2. For any annular flow, the voltage difference was measured.
3. Then, liquid film thickness was estimated from the lookup table by the interpolation method.

Conductance technique was used to evaluate the local liquid film thickness in two-phase annular flow. They used extensive numerical and static phantom experiment with 32 electrodes and found that the liquid film thickness within 5.5% error.

Thorncroft and Klausner [104] worked on the capacitance technique for measuring liquid film thickness or concentration of phase in two-phase flow. They used 12.7-mm square cross-sectional capacitance sensor for measuring film thickness. They used composite material analysis as shown in Fig. 2.31 and effective permittivity ratio to form capacitance relations for evaluating liquid film thickness in case of stratified or annular flow. And they compared their results from capacitance technique using the optical measurement technique with a CCD camera.

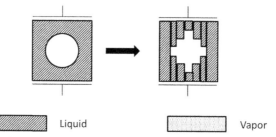

Figure 2.31 Conversion of vapor core into composite material [104].

They performed analysis on two parallel plates of area A, distance between the plates of d, and relative permittivity ε_r between. They considered ideal case $\sqrt{A} \gg d$ and electric field was kept constant and capacitance was given by

$$C = \frac{\varepsilon_0 \varepsilon_r A}{d} \tag{2.134}$$

where $\varepsilon_0 = 8.82 \times 10^{-12} \, C^2/N \, m^2 = 8.85 \, pf/m$ is the permittivity of free space. Composite material capacitance was calculated by constructing an equivalent circuit of series and parallel capacitances as shown in Fig. 2.32.

Equivalent capacitance was given by

$$C = C_1 + \left[\frac{1}{C_2} + \frac{1}{C_3} + \frac{1}{C_2} \right]^{-1} + C_1 \tag{2.135}$$

where C_1, C_2, C_3 were calculated using Eq. (2.134).

In stratified film measurement, a square channel was placed in two different orientations (i.e., upward orientation and side orientation), as shown in Fig. 2.33. C_u and C_s represent the capacitances in upward and sidewise orientations. Using Eq. (2.134), the ratio of the liquid permittivity to vapor permittivity was given by

$$\frac{\varepsilon_L}{\varepsilon_v} = \frac{C_L}{C_v} \tag{2.136}$$

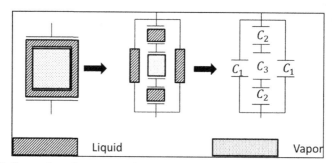

Figure 2.32 Conversion of vapor core into capacitance [104].

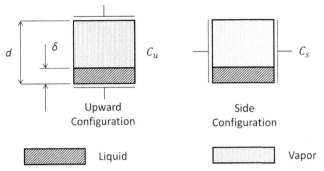

Figure 2.33 Upward and side configuration of capacitor [104].

where C_L is capacitance for liquid is filled sensor and C_v is capacitance for vapor-filled sensor.

Using composite analysis and Eqs (2.134) and (2.136), dimensionless capacitance in upward and sidewise orientation was given by

$$C_u^* = \frac{C_u - C_v}{C_L - C_v} = \frac{1}{\frac{C_L}{C_v} - 1}\left(\frac{\frac{C_L}{C_v}}{\frac{\delta}{d} + \frac{C_L}{C_v}\left(1 - \frac{\delta}{d}\right)} - 1\right) \tag{2.137}$$

$$\text{and} \quad C_s^* = \frac{C_s - C_v}{C_L - C_v} = \frac{\delta}{d} \tag{2.138}$$

where δ is the liquid film thickness and d is width of the channel.

In annular flow, the capacitance C_a^* was given by

$$C_a^* = \frac{C_a - C_v}{C_L - C_v} = \frac{2\left(\frac{\delta}{d}\right)}{\left(\frac{C_L}{C_v} - 1\right)}\left[\frac{C_L}{C_v} - \frac{1}{2\left(\frac{\delta}{d}\right) + \frac{C_L}{C_v}\left(1 - 2\left(\frac{\delta}{d}\right)\right)}\right] \tag{2.139}$$

These two capacitance models were the function of liquid film thickness. Experimental data and these capacitance model data matched well with each other. However, whether calculating liquid film thickness through capacitance technique in either stratified flow case or annular flow case, determining C_L and C_v is needed over a certain temperature range. The deviation in result for vertical annular flow was greater than the stratified flow result due to the difficulty in measuring thickness of the film with the camera.

2.3.2.3 Optical Methods

Steinbrenner et al. [105] developed the fluorescence imaging technique for measuring the water film thickness in hydrophilic channels with different air velocity and water

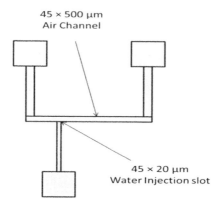

Figure 2.34 Micro-fabricated structure [105].

injection rate in order to study the enhancement of fuel delivery and convective cooling in polymer electrolyte membrane (PEM) fuel cells. Fluorescence imaging process and stratified flow modelling were used for evaluating film thickness and they were compared. The test section consisted of micro-fabricated silicon micro-structure which consisted of a 2.37 cm U-shaped air channel with rectangular cross-section (500 × 45 μm) and water injection was done through a 45 × 20 μm slot as shown in the Fig. 2.34.

The experimental setup is shown in Fig. 2.35. The micro-fabricated structure was kept facing down on the inverted microscope. Flow Images were captured with CCD camera (1392 × 1040 pixels). A 10× objective lens was used to capture full

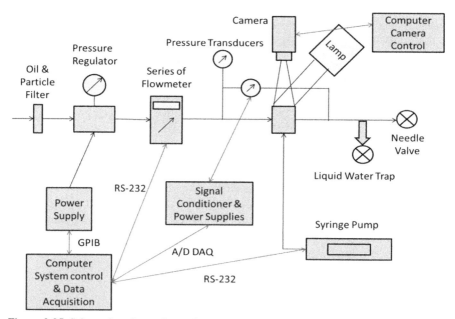

Figure 2.35 Schematics of experimental setup [105].

500-μm channel width in a single snap with each square pixel area as 0.465 by 0.465 μm. Downstream needle valve arrangement was done for varying air flow rates.

Fluorescent imaging was used to obtain detailed ideas about the water film extent within the channel. Fluorescence technique was used to get more distinguishable images on air–water interfaces than with white light imaging process. Water seeded with a 0.5 mmol/L concentration of fluorescein dye was injected into the channels to have fluorescent visualization. Fluorescent excitation was done with metal halide lamps.

Fluorescent images were captured at various flow rates of water for constant inlet air pressures and varying exit pressures. Flow images were captured when system was in steady state for each exit pressure. Between settings, the channel was evacuated by releasing the exit valve and water film was cleared with flowing air. This prevented the surface tension induced during hysteresis. Images were captured at a single location (i.e., halfway between the channel downstream bend and the water injection slot). The film thickness was measured nearer to the water injection area where a stable film was formed on the wall. Film thickness measurements were evaluated from the fluorescent flow images by the threshold intensity pixel counting method. The intensity profile was checked at a typical flow cross section after subtracting the average background intensity from each image. Two threshold intensity values were evaluated, which represented a low- and a high-intensity reading, which were the edges of the film. Images were processed for each threshold value such that for each pixel if intensity was greater than the threshold value, then 1 was assigned as a binary value; otherwise, the pixel was assigned 0. The film thickness at each channel cross section was easily evaluated by summing of a column of pixels. If there were more cross sections in each image, film thickness was averaged over all images.

A stratified flow model was predicted for estimating liquid film thickness in rectangular channel with air–water interface as shown in Fig. 2.36.

Film thickness was evaluated by equating air and water pressure gradients in the direction of flow with equal velocities and interfaces shear stress and was given as

$$\tau_{wG}\frac{P_G}{A_G} - \tau_{wL}\frac{P_L}{A_L} + \tau_i P_i \left(\frac{1}{A_L} + \frac{1}{A_G}\right) \tag{2.140}$$

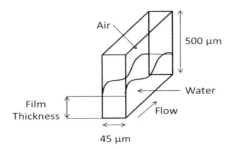

Figure 2.36 Liquid film in rectangular channel.

where A is cross-sectional area and P is the wetted perimeter. Both were dependent on film thickness. τ_{wG} and τ_{wL} represent wall shear stress in gas and liquid phase, respectively. τ_i is shear stress at the interface of two fluids.

$$\tau = \frac{1}{2}\rho u^2 f \tag{2.141}$$

The shear stresses was found dependent on channel geometry and two-phase (moving interface) interaction by modified friction factor,

$$fRe = C\psi \tag{2.142}$$

where C was based on the channel aspect ratio by an interface modification factor, ψ which was an analytical solution of stratified two-phase flow.

It was found that film thicknesses through fluorescent technique were consistent with stratified flow modeling. Liquid film thickness decreased nonlinearly with the increase in air velocity. Liquid film was very sensitive when film was thick. Therefore, little increases in air velocity resulted in a decrease of film thickness value. The water film thickness nonlinearly approached minimum film thickness value at high air flow rates. The water film filled a large channel portion, which formed stationary waves along channel length, which might lead to varying surface tension forces along the channel wall. Therefore, higher air flow rates should be used to suppress the stationary waves and the liquid film growth.

Nozhat [106] discussed the variation of liquid film thickness with small Reynolds number. Film thickness was measured using laser interferometry technique by flowing liquid inside the small-bore glass tube. The author proposed laser interferometry to measure liquid film thickness. The optical arrangement is shown in Fig. 2.37.

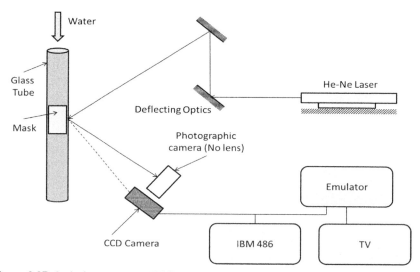

Figure 2.37 Optical arrangement [106].

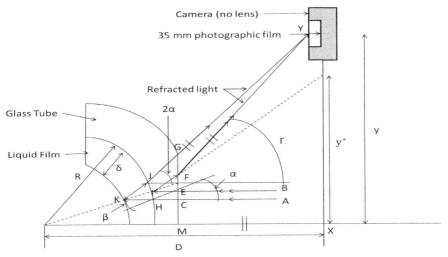

Figure 2.38 Portion of horizontal cross section of glass tube showing a falling liquid film [106].

Horizontal cross section of a vertical glass tube is shown in Fig. 2.38. Laser beam was incident on the glass tube. After refraction and reflection, two rays within the beam, ACHKJGY and BEIFY, were made to coincide at Y in a photographic camera. The beam made angle of incidence α and angle of refraction β when rays were passed from the glass tube to liquid film (i.e., H_2O). The glass tube bore was cylindrical in shape, but the exterior surface of the glass tube was made planar (between C and G) and the light was made to incident on planar region CG of exterior glass bore.

Assumption was made that $\delta/R \ll 1$. The inner bore of the circular section tube between J and H was assumed to be planar. The line OJ made an angle α with the ray OX, whereas JY made an angle 2α with OX. Derivation for liquid film thickness:

$$y = \overline{JM} + \overline{MX} \tan 2\alpha = R \sin \alpha + (D - R \cos \alpha) \tan 2\alpha \qquad (2.143)$$

$$y = R\alpha + (D - R)2\alpha \qquad (2.144)$$

$$\alpha = y/(2D - R) \qquad (2.145)$$

$$n_g \sin 2\alpha = n_a \sin \Gamma \qquad (2.146)$$

where n_g and n_a represent refractive index of glass and air, respectively.

$$y^*/\tan 2\alpha = y/\tan \Gamma \qquad (2.147)$$

$$y^* = y(n_a/n_g) \qquad (2.148)$$

Optical path difference was calculated with reference to Fig. 2.38.

$$OPD(\alpha) = \frac{2n_w\delta}{\cos\beta} - 2\delta\tan(\beta)\sin(\alpha)\,n_g \tag{2.149}$$

where n_w represents refractive index of water.

Using Snell's law, we obtained

$$\sin\alpha/\sin\beta = n_w/n_g \tag{2.150}$$

Substituting Eq. (2.150) in Eq. (2.149), we obtained

$$OPD(\alpha) = \frac{2n_w\delta}{\cos\beta} - 2n_w\delta\frac{\sin^2\beta}{\cos\beta} \tag{2.151}$$

$$OPD(\alpha) = \frac{2n_w\delta}{\cos\beta}\left(1 - \sin^2\beta\right) \tag{2.152}$$

$$OPD(\alpha) = 2n_w\delta\cos\beta \tag{2.153}$$

Now,

$$OPD(0) = 2n_w\delta \tag{2.154}$$

$$OPD(0) - OPD(\alpha) = 2n_w\delta(1 - \cos\beta) \tag{2.155}$$

where $\sin\beta/2 = (1 - \cos\beta/2)^{1/2}$. For small approximation $\beta^2/4 = (1 - \cos\beta/2)$. Therefore, using above approximation in Eq. (2.155), we obtained

$$OPD(0) - OPD(\alpha) = n_w\delta\beta^2 \tag{2.156}$$

From Eq. (2.150), β was substituted in Eq. (2.156) to obtain

$$OPD(0) - OPD(\alpha) = \delta\left(n_g^2/n_w\right)\alpha^2 \tag{2.157}$$

Bright or dark fringe appears on the screen when

$$OPD(\alpha_m) = m\lambda \tag{2.158}$$

where m is one or more number of successive fringes, and λ is the wavelength.

Equation (2.157) was used to obtain

$$OPD(0) - OPD(\alpha_m) = \Delta m\lambda \tag{2.159}$$

Then, liquid film was calculated as

$$\delta = \left(\Delta m \lambda / n_g^2\right)\left(n_w / \alpha^2\right) \tag{2.160}$$

Δ is the fringe range. The value (y^*) from Eq. (2.148) was used in Eq. (2.145) to evaluate the angle α after the measurement of y and D. Δm value was found from the interferograms.

However, this theory was based on certain thin liquid film approximation (i.e., $\delta/R \ll 1$). And this method was less effective with R provided that $R/D \ll 1$. Accuracy of measurement of liquid film thickness was $\pm 4\%$ for small flow rates (i.e., with $Re \le 20$).

Anastasiou et al. [107] used the μ-PIV method to measure the liquid film thickness. The μ-PIV method is commonly used to measure velocity fields, but it can also be effectively used to measure the liquid film thickness. The μ-PIV system consists of a light source, high-speed CCD camera, a timer box to synchronize the camera, and the light source. Since micro-scale dimensions are investigated, a microscope with appropriate objective lenses is also integrated with the system. Fluorescent particles are seeded into the test fluid for identification of the liquid phase. When viewed through a microscope, these fluorescent particles appear as light spots in the dark background. Hence, presence of these light spots corresponds to the liquid phase, whereas dark images refer to either gas phase or channel wall. Images are taken on different planes as shown in Fig. 2.39. The plane on which first light spots are obtained (plane A) is taken as the initial point. Then, the plane is moved gradually until no dark spots are obtained (plane C). The plane on which no light spots are visible resembles the bottom wall. The distance between plane C and bottom wall of the microchannel is the liquid film thickness.

Eain et al. [108] used a nonintrusive optical technique to measure the liquid film thickness. The schematic of their experimental set is shown in Fig. 2.40. The slug generated with the aid of the syringe pumps and the segmenter was supplied to the Teflon tubing of 1-m length and 1.59-mm internal diameter. To ensure that fully

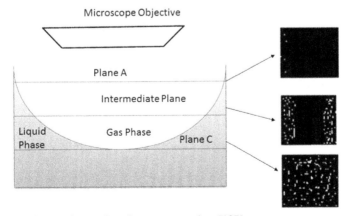

Figure 2.39 Different planes where images were taken [107].

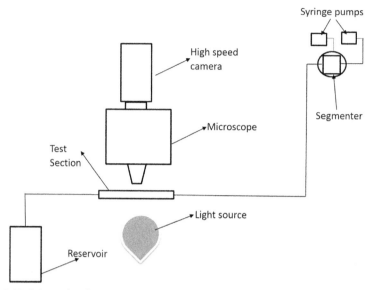

Figure 2.40 Schematic of the experimental set up [108].

developed flow has been achieved, the image of the flow was taken at 0.75 m downstream. A high-speed camera mounted on a microscope was used to capture the image of the flow. The images were recorded at a frequency of 5000 Hz and with an exposure of 122 μs. The captured images were then analyzed in MATLAB where a custom code was used to measure the liquid film thickness.

Han et al. [109] measured liquid film thickness under adiabatic conditions in parallel channels of heights 0.1, 0.3, and 0.5 mm. In their study, they used interferometer method with ethanol as the working fluid. Schematic of the interferometer is shown in Fig. 2.41. The light reflected from the wall surface, and the liquid–air interface creates different fringe patterns depending on the film thickness. A constructive interference (white fringe) is obtained when twice the liquid film thickness is an even multiple of

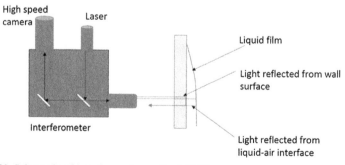

Figure 2.41 Schematic of interferometer method [109].

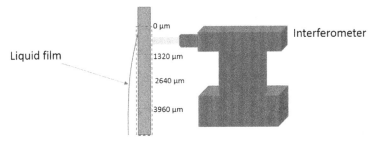

Figure 2.42 Different positions were observations were recorded [109].

the half the laser wavelength divided by the refractive index of the fluid. Likewise, a destructive interference (black fringe) is obtained for odd multiples. The liquid film thickness can be expressed as:

$$\delta = 2m\frac{\lambda_l}{4\eta_f} \quad (\text{Constructive interference}) \tag{2.161}$$

$$\delta = (2m+1)\frac{\lambda_l}{4\eta_f} \quad (\text{Destructive interference}) \tag{2.162}$$

where m is number of fringes, η_f is refractive index of the working fluid, and λ_l is wavelength of the laser.

The size of the fringe pattern image in their study was 0.88×0.66 mm. As the size of the image was limited, images were captured at several positions along the flow direction as shown in Fig. 2.42.

Fang et al. [110] calculated the liquid film thickness by an optical interference method. The schematic of their experiment is shown in Fig. 2.43. The microchannel is illuminated by a beam of monochromatic green light projected from a fluorescent lamp which first passes through a 520-nm filter and then is partially reflected by a 50% beam splitter before reaching the microchannel through objective of the

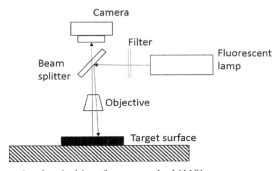

Figure 2.43 Schematic of optical interference method [110].

microscope. The light then reflected from the microchannel is passes through the beam splitter and is captured by the camera. Due to interference of the reflected light by the solid–liquid and liquid–vapor interface fringe patterns are generated, which represents the contour of the liquid film thickness. The fringes are then extracted from raw interference image depending upon the pixel gradient. Then a two-dimensional interpolation is carried out to obtain a grid of points from the fringe to generate a smooth film thickness map.

Han and Shikazono [111] used laser displacement method to measure the liquid film thickness in five circular tubes having diameters 0.3, 0.5, 0.7, 1.0, and 1.3 mm using air, ethanol, water, and FC-40 as working fluids. The principle of laser displacement method can be explained with the use of Fig. 2.44. The objective lens is moved by the turning fork, and the position of the target surface is determined by the displacement of the objective lens. When the target surface is at the focus of the objective lens, the intensity of the reflected light becomes maximum in the light-receiving element. The liquid film thickness is obtained by continually vibrating the objective with an amplitude of 0.3 mm. The measured value of liquid film thickness is converted to DC voltage signal in the range ±10 V, which is then sent to the computer through GPIB interface and recorded with LabVIEW.

The path of the laser through the channel wall and liquid film is depicted in Fig. 2.45 (Han and Shikazono [112]). y_1 (wall thickness) was measured initially without flowing the liquid and then y_2 (distance from the outer bubble to liquid–air interface) was measured. The difference between these gives the liquid film thickness (Eq. (2.163)).

$$\delta = (y_2 - y_1) \frac{\tan \theta_{air}}{\tan \theta_f} \tag{2.163}$$

where δ is liquid film thickness, θ_{air} is angle of incidence for air (14.91° for present experiment), and θ_f is angle of incidence for working fluid.

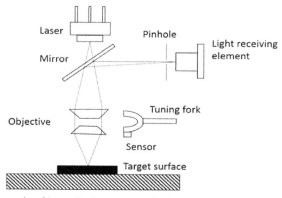

Figure 2.44 Schematic of laser displacement method [111].

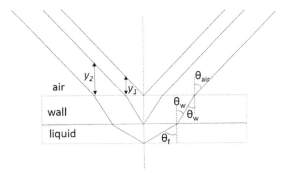

Figure 2.45 Path of laser through channel wall and liquid film [112].

θ_f can obtained from Snell's law as follows:

$$\theta_f = \sin^{-1}\left(\frac{\eta_w}{\eta_f}\sin\theta_w\right) \tag{2.164}$$

$$\theta_w = \sin^{-1}\left(\frac{\eta_{air}}{\eta_w}\sin\theta_{air}\right) \tag{2.165}$$

where η_{air}, η_w, and η_f are refractive index of air, channel wall, and working fluid, respectively, and θ_w is the angle of incidence for channel wall.

Fries et al. [113] measured the liquid film thickness by confocal laser scanning microscopy (LSM). The schematic of the experimental setup is shown in Fig. 2.46. In contrast to conventional microscope where the entire surface within the len's field of view is seen, confocal laser microscopy has limited field of view and scattered light from out-of-focus objects are restricted by the confocal pinhole from reaching the detector. By varying the focal distance of the microscope, layer-wise optical slicing of the target surface is done (Fig. 2.47), and a stack of slices are generated that can

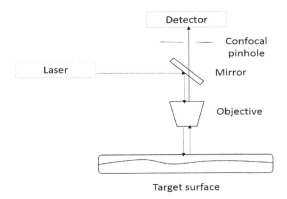

Figure 2.46 Schematic of confocal laser scanning microscopy [113].

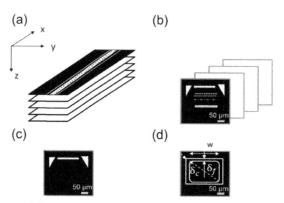

Figure 2.47 Scheme of data analysis for measurement of liquid film thickness [113].

be later summed to give a sharp, high-resolution image of the required surface. In their experiments, a total of 1024 pictures over a length of 420 μm were obtained. The regions having a permanent liquid film have higher intensity than a region composed of passing liquid and gas. Hence, by averaging all 1024 slices, a clear distinction of region having a permanent liquid film can be obtained (Fig. 2.47(c)). The film thickness at channel to δ_t and channel corner δ_c was determined with aid of a MATLAB program.

Gstoehl et al. [114] and Wang et al. [115] used a nonintrusive fluorescence image processing technique to measure the liquid film thickness. A fluorescent particle (rhodamine B) was injected into the test fluid. Laser is used to illuminate the liquid film, and its images are captured by a high-speed camera. A filter placed in front of the camera eliminates the reflections from the laser light. In the experimental analysis conducted by Gstoehl et al. [114] the field of view was 3300 × 3100 μm and the images had a resolution of 512 × 480 pixels. Hence, each pixel was about 6.4 μm (3300 μm/512). Images of test section with the liquid flowing in it were compared with the images of the dry test section. The liquid film thickness was obtained comparing the number of pixels of these two images and then multiplying with the pixel size.

References

[1]   V.P. Carey, Liquid Vapor Phase Change Phenomena, Taylor and Francis Group, New York, 2008.
[2]   Y.Y. Hsu, On the size range of active nucleation cavities on a heating surface, J. Heat Transfer 84 (3) (1962) 207−216.
[3]   R. Raj, J. Kim, J. Mcquillen, Subcooled pool boiling in variable gravity environments, J. Heat Transfer 131 (9) (2009) 09152.
[4]   A.E. Bergles, W.M. Rohsenow, The determination of forced-convection surface-boiling heat transfer, J. Heat Transfer 86 (3) (1964) 365−372.
[5]   T. Sato, H. Matsumura, On the conditions of incipient subcooled-boiling with forced convection, Bull. JSME 7 (26) (1964) 392−398.

[6]   E.J. Davis, G.H. Anderson, The incipience of nucleate boiling in forced convection flow, AIChE J. 12 (4) (1966) 774–780.

[7]   S. Kandlikar, V. Mizo, M. Cartwright, Bubble Nucleation and Growth Characteristics in Subcooled Flow Boiling of Water, American Society of Mechanical Engineers (ASME), 1997.

[8]   S.M. Ghiaasiaan, R.C. Chedester, Boiling incipience in microchannels, Int. J. Heat Mass Transfer 45 (23) (2002) 4599–4606.

[9]   W. Qu, I. Mudawar, Prediction and measurement of incipient boiling heat flux in micro-channel heat sinks, Int. J. Heat Mass Transfer 45 (19) (2002) 3933–3945.

[10] J. Li, P. Cheng, Bubble cavitation in a microchannel, Int. J. Heat Mass Transfer 47 (12) (2004) 2689–2698.

[11] D. Liu, P.S. Lee, S.V. Garimella, Prediction of the onset of nucleate boiling in microchannel flow, Int. J. Heat Mass Transfer 48 (25) (2005) 5134–5149.

[12] S.G. Kandlikar, Nucleation characteristics and stability considerations during flow boiling in microchannels, Exp. Thermal Fluid Sci. 30 (5) (2006) 441–447.

[13] R.K. Shah, A.L. London, Laminar flow forced convection in ducts: a source book for compact heat exchanger analytical data, Adv. Heat Transfer (1978) (Suppl. 1).

[14] S.S. Bertsch, E.A. Groll, S.V. Garimella, Effects of heat flux, mass flux, vapor quality, and saturation temperature on flow boiling heat transfer in microchannels, Int. J. Multiphase Flow 35 (2) (2009) 142–154.

[15] X.F. Peng, B.X. Wang, Evaporating space and fictitious boiling for internal evaporation of liquid, Sci. Found. China 2 (2) (1994) 55–59.

[16] L. Jiang, L. Wong, Y. Zohar, Phase change in microchannel heat sink under forced convection boiling, in: Micro Electro Mechanical Systems (MEMS). The Thirteenth Annual International Conference on, IEEE, 2000, pp. 397–402.

[17] X.F. Peng, H.Y. Hu, B.X. Wang, Boiling nucleation during liquid flow in microchannels, Int. J. Heat Mass Transfer 41 (1) (1998) 101–106.

[18] T. Harirchian, S.V. Garimella, Microchannel size effects on local flow boiling heat transfer to a dielectric fluid, Int. J. Heat Mass Transfer 51 (15) (2008) 3724–3735.

[19] N. Basu, G.R. Warrier, V.K. Dhir, Onset of nucleate boiling and active nucleation site density during subcooled flow boiling, J. Heat Transfer 124 (4) (2002) 717–728.

[20] J. Li, G.P. Peterson, Microscale heterogeneous boiling on smooth surfaces—from bubble nucleation to bubble dynamics, Int. J. Heat Mass Transfer 48 (21) (2005) 4316–4332.

[21] H. Müller-Steinhagen, N. Epstein, A.P. Watkinson, Effect of dissolved gases on subcooled flow boiling heat transfer, Chem. Eng. Process. Process Intensif. 23 (2) (1988) 115–124.

[22] M.E. Steinke, S.G. Kandlikar, Control and effect of dissolved air in water during flow boiling in microchannels, Int. J. Heat Mass Transfer 47 (8) (2004) 1925–1935.

[23] C.A. Ward, W.R. Johnson, R.D. Ventor, S. Ho, T.W. Forest, W.D. Fraser, Heterogeneous bubble nucleation and conditions for growth in a liquid–gas system of constant mass and volume, J. Appl. Phys. 54 (4) (1988) 1833–1843.

[24] A. Cioncolini, L. Santini, M.E. Ricotti, Effects of dissolved air on subcooled and saturated flow boiling of water in a small diameter tube at low pressure, Exp. Thermal Fluid Sci. 32 (1) (2007) 38–51.

[25] I. Hapke, H. Boye, J. Schmidt, Onset of nucleate boiling in minichannels, Int. J. Therm. Sci. 39 (4) (2000) 505–513.

[26] S.G. Kandlikar, M.E. Steinke, S. Tian, L.A. Campbell, High-speed photographic observation of flow boiling of water in parallel minichannels, in: 35th Proceedings of National Heat Transfer Conference, ASME, Anaheim, CA, 2001, pp. 675–684.

[27] C. Huh, J. Kim, M.H. Kim, Flow pattern transition instability during flow boiling in a single microchannel, Int. J. Heat Mass Transfer 50 (5) (2007) 1049−1060.

[28] L. Zhang, E.N. Wang, K.E. Goodson, T.W. Kenny, Phase change phenomena in silicon microchannels, Int. J. Heat Mass Transfer 48 (8) (2005) 1572−1582.

[29] B.J. Jones, J.P. McHale, S.V. Garimella, The influence of surface roughness on nucleate pool boiling heat transfer, J. Heat Transfer 131 (12) (2009) 121009.

[30] J.P. McHale, S.V. Garimella, T.S. Fisher, G.A. Powell, Pool boiling performance comparison of smooth and sintered copper surfaces with and without carbon nanotubes, Nanoscale Microscale Thermophys. Eng. 15 (3) (2011) 133−150.

[31] C.T. Lu, C. Pan, Convective boiling in a parallel microchannel heat sink with a diverging cross section and artificial nucleation sites, Exp. Thermal Fluid Sci. 35 (5) (2011) 810−815.

[32] A.K.M.M. Morshed, F. Yang, M.Y. Ali, J.A. Khan, C. Li, Enhanced flow boiling in microchannel with integration of nanowires, Appl. Therm. Eng. 32 (2012) 68−75.

[33] T. Alam, P.S. Lee, C.R. Yap, Effects of surface roughness on flow boiling in silicon microgap heat sinks, Int. J. Heat Mass Transfer 64 (2013) 28−41.

[34] P. Bai, T. Tang, B. Tang, Enhanced flow boiling in parallel microchannels with metallic porous Coating, Appl. Therm. Eng. 58 (2013) 291−297.

[35] D. Deng, Y. Tang, D. Liang, H. He, S. Yang, Flow boiling characteristics in porous heat sink with reentrant microchannels, Int. J. Heat Mass Transfer 70 (2014) 463−477.

[36] F. Yang, X. Dai, Y. Peles, P. Cheng, J. Khan, C. Li, Flow boiling phenomena in a single annular flow regimen in microchannels (I): characterization of flow boiling heat transfer, Int. J. Heat Mass Transfer 68 (2014) 703−715.

[37] C.S. Kumar, S. Suresh, L. Yang, Q. Yang, S. Aravind, Flow boiling heat transfer enhancement using carbon nanotube coatings, Appl. Therm. Eng. 65 (1) (2014) 166−175.

[38] M. Law, P.S. Lee, K. Balasubramanian, Experimental investigation of flow boiling heat transfer in novel oblique-finned microchannels, Int. J. Heat Mass Transfer 76 (2014) 419−431.

[39] D. Li, G.S. Wu, W. Wang, Y.D. Wang, D. Liu, D.C. Zhang, Y.F. Chen, G.P. Peterson, R. Yang, Enhancing flow boiling heat transfer in microchannels for thermal management with monolithically-integrated silicon nanowires, Nano Lett. 12 (7) (2012) 3385−3390.

[40] H.T. Phan, N. Caney, P. Marty, S. Colasson, Flow boiling of water on nanocoated surfaces in a microchannel, J. Heat Transfer 134 (2) (2012) 020901.

[41] B. Bourdon, P. Di Marco, R. Rioboo, M. Marengo, J. De Coninck, Enhancing the onset of pool boiling by wettability modification on nanometrically smooth surfaces, Int. Commun. Heat Mass Transfer 45 (2013) 11−15.

[42] H. Jo, M. Kaviany, S.H. Kim, M.H. Kim, Heterogeneous bubble nucleation on ideally-smooth horizontal heated surface, Int. J. Heat Mass Transfer 71 (2014) 149−157.

[43] S. Sarangi, J.A. Weibel, S.V. Garimella, Effect of particle size on surface-coating enhancement of pool boiling heat transfer, Int. J. Heat Mass Transfer 81 (2015) 103−113.

[44] S.G. Kandlikar, W.K. Kuan, D.A. Willistein, J. Borrelli, Stabilization of flow boiling in microchannels using pressure drop elements and fabricated nucleation sites, J. Heat Transfer 128 (4) (2006) 389−396.

[45] W.K. Kuan, S.G. Kandlikar, Experimental study on the effect of stabilization on flow boiling heat transfer in microchannels, Heat Transfer Eng. 28 (8−9) (2007) 746−752.

[46] S. Szczukiewicz, N. Borhani, J.R. Thome, Two-phase heat transfer and high-speed visualization of refrigerant flows in 100×100 μm^2 silicon multi-microchannels, Int. J. Refrig. 36 (2) (2013) 402−413.

[47] L. Xu, J. Xu, Nanofluid stabilizes and enhances convective boiling heat transfer in a single microchannel, Int. J. Heat Mass Transfer 55 (21) (2012) 5673−5686.

[48] M.P. Pujara, L. Kumar, A. Mogra, Two phase flow void fraction measurement using image processing technique, Int. J. Mech. Eng. Technol. 4 (3) (2013) 130−135.

[49] S.G. Singh, A. Jain, A. Sridharan, S.P. Duttagupta, A. Agrawal, Flow map and measurement of void fraction and heat transfer coefficient using an image analysis technique for flow boiling of water in a silicon microchannel, J. Micromech. Microeng. 19 (7) (2009) 075004.

[50] P. Gijsenbergh, R. Puers, Permittivity-based void fraction sensing for microfluidics, Sens. Actuators A Phys. 195 (2013) 64−70.

[51] S. Paranjape, S.N. Ritchey, S.V. Garimella, Electrical impedance-based void fraction measurement and flow regimen identification in microchannel flows under adiabatic conditions, Int. J. Multiphase Flow 42 (2012) 175−183.

[52] D. Fogg, R. Flynn, C. Hidrovo, L. Zhang, K. Goodson, Fluorescent imaging of void fraction in two-phase microchannels, in: 3rd International Symposium on Two-Phase Flow Modeling and Experimentation Pisa, Italy, 2004. Paper No SGK07.

[53] H. Ide, R. Kimura, M. Kawaji, Optical measurement of void fraction and bubble size distributions in a microchannel, Heat Transfer Eng. 28 (8−9) (2007) 713−719.

[54] Y. Zhou, Z. Huang, B. Wang, H. Ji, H. Li, A new method for void fraction measurement of gas−liquid two-phase flow in millimeter-scale pipe, Int. J. Multiphase Flow (2014).

[55] Z. Huang, J. Long, W. Xu, H. Ji, B. Wang, H. Li, Design of capacitively coupled contactless conductivity detection sensor, Flow Meas. Instrum. 27 (2012) 67−70.

[56] M.I. Ali, M. Sadatomi, M. Kawaji, Adiabatic two-phase flow in narrow channels between two flat plates, Can. J. Chem. Eng. 71 (5) (1993) 657−666.

[57] B. Lowry, M. Kawaji, Adiabatic vertical two-phase flow in narrow passages, AIChE Symp. Ser. 84 (263) (1988) 133−139.

[58] D. Chisholm, A.D.K. Laird, Two-phase flow in rough tubes, Trans. ASME 80 (2) (1958) 276−286.

[59] N. Zuber, J.A. Findlay, Average volumetric concentration in two-phase flow systems, J. Heat Transfer 87 (4) (1965) 453−468.

[60] G.B. Wallis, One-Dimensional Two-Phase Flow, McGraw-Hill Book Company, New York, 1969.

[61] S.G. Bankoff, A variable density single-fluid model for two-phase flow with particular reference to steam-water flow, J. Heat Transfer 82 (4) (1960) 265−272.

[62] Y. Taitel, A.E. Dukler, A theoretical approach to the Lockhart-Martinelli correlation for stratified flow, Int. J. Multiphase Flow 2 (5) (1976) 591−595.

[63] K.A. Triplett, S.M. Ghiaasiaan, S.I. Abdel-Khalik, A. LeMouel, B.N. McCord, Gas−liquid two-phase flow in microchannels: part II: void fraction and pressure drop, Int. J. Multiphase Flow 25 (3) (1999) 395−410.

[64] B. Chexal, M. Merilo, J. Maulbetsch, J. Horowitz, J. Harrison, J.C. Westacott, C. Peterson, W. Kastner, H. Schmidt, Void Fraction Technology for Design and Analysis, Electric Power Research Institute (EPRI) Distribution Center, Palo Alto, CA, 1997.

[65] R.W. Lockhart, R.C. Martinelli, Proposed correlation of data for isothermal two-phase, two-component flow in pipes, Chem. Eng. Prog. 45 (1) (1949) 39−48.

[66]   K. Mishima, T. Hibiki, Some characteristics of air-water two-phase flow in small diameter vertical tubes, Int. J. Multiphase Flow 22 (4) (1996) 703−712.

[67]   M. Ishii, One-dimensional Drift-flux Model and Constitutive Equations for Relative Motion between Phases in Various Two-phase Flow Regimes, No. ANL-77-47, Argonne National Lab, IL, USA, 1977.

[68]   A. Kariyasaki, T. Fukano, A. Ousaka, M. Kagawa, Isothermal air-water two-phase up and downward flows in a vertical capillary tube (1st report, flow pattern and void fraction), Trans. JSME (Ser. B) 58 (1992) 2684−2690 (in Japanese).

[69]   A. Kawahara, P.Y. Chung, M. Kawaji, Investigation of two-phase flow pattern, void fraction and pressure drop in a microchannel, Int. J. Multiphase Flow 28 (9) (2002) 1411−1435.

[70]   A. Kawahara, M. Sadatomi, K. Okayama, M. Kawaji, P.Y. Chung, Effects of channel diameter and liquid properties on void fraction in adiabatic two-phase flow through microchannels, Heat Transfer Eng. 26 (3) (2005) 13−19.

[71]   A. Kawahara, M. Sadatomi, K. Nei, H. Matsuo, Experimental study on bubble velocity, void fraction and pressure drop for gas−liquid two-phase flow in a circular micro-channel, Int. J. Heat Fluid Flow 30 (5) (2009) 831−841.

[72]   S. Saisorn, S. Wongwises, The effects of channel diameter on flow pattern, void fraction and pressure drop of two-phase air−water flow in circular micro-channels, Exp. Therm. Fluid Sci. 34 (4) (2010) 454−462.

[73]   M.A. Woldesemayat, A.J. Ghajar, Comparison of void fraction correlations for different flow patterns in horizontal and upward inclined pipes, Int. J. Multiphase Flow 33 (4) (2007) 347−370.

[74]   B.A. Eaton, The Prediction of Flow Patterns, Liquid Holdup and Pressure Losses Occurring during Continuous Two-Phase Flow in Horizontal Pipelines (Ph.D. Dissertation), The University of Texas, Austin, Department of Petroleum Engineering, 1966.

[75]   H.D. Beggs, An Experimental Study of Two Phase Flow in Inclined Pipes (Ph.D. Dissertation), The University of Tulsa, Tulsa, OK, Department of Petroleum Engineering, 1972.

[76]   P.L. Spedding, V.T. Nguyen, Data on Holdup, Pressure Loss and Flow Patterns for Two-Phase Air−Water Flow in an Inclined Pipe, University of Auckland, Auckland, New Zealand, 1976. Report Eng. 122.

[77]   H. Mukherjee, An Experimental Study of Inclined Two-Phase Flow (Ph.D. Dissertation), The University of Tulsa, Tulsa, OK, Department of Petroleum Engineering, 1979.

[78]   K. Minami, J.P. Brill, Liquid holdup in wet gas pipelines, SPE Prod. Eng. 5 (1987) 36−44.

[79]   F. França, R.T. Lahey, The use of drift-flux techniques for the analysis of horizontal two-phase flows, Int. J. Multiphase Flow 18 (6) (1992) 787−801.

[80]   G.H. Abdul-Majeed, Liquid holdup in horizontal two-phase gas-liquid flow, J. Petroleum Sci. Eng. 15 (2) (1996) 271−280.

[81]   M. Sujumnong, Heat Transfer, Pressure Drop and Void Fraction in Two-Phase, Two Component Flow in a Vertical Tube (Ph.D. Dissertation), The University of Manitoba, Winnipeg, Manitoba, Canada, Department of Mechanical and Industrial Engineering, 1997.

[82]   P. Coddington, R. Macian, A study of the performance of void fraction correlations used in the context of drift-flux two-phase flow models, Nucl. Eng. Design 215 (3) (2002) 199−216.

[83]   R. Xiong, J.N. Chung, An experimental study of the size effect on adiabatic gas-liquid two-phase flow patterns and void fraction in microchannels, Phys. Fluids (1994-Present) 19 (3) (2007) 033301.

[84] S. Saisorn, S. Wongwises, Flow pattern, void fraction and pressure drop of two-phase air—water flow in a horizontal circular micro-channel, Exp. Therm. Fluid Sci. 32 (3) (2008) 748—760.

[85] P.Y. Chung, M. Kawaji, The effect of channel diameter on adiabatic two-phase flow characteristics in microchannels, Int. J. Multiphase Flow 30 (7) (2004) 735—761.

[86] T.A. Shedd, Void Fraction and Pressure Drop Measurements for Refrigerant R410a Flows in Small Diameter Tubes, Preliminary AHRTI, 2010. Report No. 20110—01.

[87] V.G. Nino, P.S. Hrnjak, T.A. Newell, Analysis of void fraction in microchannels, in: International Refrigeration and Air Conditioning Conference, 2002. Paper 574.

[88] G. Rigot, Fluid capacity of an evaporator in direct expansion, Chaud-Froid Plomberie 328 (1973) 133—144.

[89] C.W. Choi, D.I. Yu, M.H. Kim, Adiabatic two-phase flow in rectangular microchannels with different aspect ratios: part I — flow pattern, pressure drop and void fraction, Int. J. Heat Mass Transfer 54 (1) (2011) 616—624.

[90] E.W. Jassim, T.A. Newell, Prediction of two-phase pressure drop and void fraction in microchannels using probabilistic flow regimen mapping, Int. J. Heat Mass Transfer 49 (15) (2006) 2446—2457.

[91] O. Baker, Simultaneous flow of oil and gas, Oil Gas J. 53 (1954) 185—195.

[92] Y. Taitel, A.E. Dukler, A model for predicting flow regimen transitions in horizontal and near horizontal gas-liquid flow, AIChE J. 22 (1) (1976) 47—55.

[93] J.W. Coleman, S. Garimella, Two-phase flow regimes in round, square and rect-angular tubes during condensation of refrigerant R134a, Int. J. Refrig. 26 (1) (2003) 117—128.

[94] J. El Hajal, J.R. Thome, A. Cavallini, Condensation in horizontal tubes, part 1: two-phase flow pattern map, Int. J. Heat Mass Transfer 46 (18) (2003) 3349—3363.

[95] V.G. Nino, Characterization of Two-Phase Flow in Microchannels (Ph.D. Thesis), University of Illinois, Urbana Champaign, IL, 2002.

[96] A.A. Armand, The resistance during the movement of a two-phase system in horizontal pipes, Izv. Vses. Teplotekh. Inst 1 (1946) 16—23.

[97] A. Cioncolini, J.R. Thome, Void fraction prediction in annular two-phase flow, Int. J. Multiphase Flow 43 (2012) 72—84.

[98] A.V. Hill, The possible effects of the aggregation of the molecules of haemoglobin on its dissociation curves, J. Physiol. (London) 40 (1910) 4—7.

[99] J.D. Murray, Mathematical Biology, Springer-Verlag, New York, 2002.

[100] C.B. Tibiriçá, F.J. do Nascimento, G. Ribatski, Film thickness measurement techniques applied to micro-scale two-phase flow systems, Exp. Therm. Fluid Sci. 34 (4) (2010) 463—473.

[101] Q. Lu, N.V. Suryanarayana, C. Christodoulu, Film thickness measurement with an ultrasonic transducer, Exp. Therm. Fluid Sci. 7 (4) (1993) 354—361.

[102] R.C. Reid, T.K. Sherwood, The Properties of Gases and Liquids, McGraw-Hill, New York, 1958.

[103] B.A. Lee, B.J. Yun, K.Y. Kim, S. Kim, Estimation of local liquid film thickness in two-phase annular flow, Nucl. Eng. Technol. 1 (2012) 71—78.

[104] G.E. Thorncroft, J.F. Klausner, A capacitance sensor for two-phase liquid film thickness measurements in a square duct, J. Fluids Eng. 119 (1) (1997) 164—169.

[105] J.E. Steinbrenner, C.H. Hidrovo, F.M. Wang, S. Vigneron, E.S. Lee, T.A. Kramer, C.H. Cheng, J.K. Eaton, K.E. Goodson, Measurement and modeling of liquid film thickness evolution in stratified two-phase microchannel flows, Appl. Therm. Eng. 27 (10) (2007) 1722—1727.

[106]   W.M. Nozhat, Measurement of liquid-film thickness by laser interferometry, Appl. Opt. 36 (30) (October 20, 1997) 7864–7869.

[107]   A.D. Anastasiou, C. Makatsoris, A. Gavriilidis, A.A. Mouza, Application of μ-PIV for investigating liquid film characteristics in an open inclined microchannel, Exp. Therm. Fluid Sci. 44 (2013) 90–99.

[108]   M.M.G. Eain, V. Egan, J. Punch, Film thickness measurements in liquid–liquid slug flow regimes, Int. J. Heat Fluid Flow 44 (2013) 515–523.

[109]   Y. Han, N. Shikazono, N. Kasagi, Measurement of liquid film thickness in a micro parallel channel with interferometer and laser focus displacement meter, Int. J. Multiphase Flow 37 (1) (2011) 36–45.

[110]   C. Fang, M. David, F.M. Wang, K.E. Goodson, Influence of film thickness and cross-sectional geometry on hydrophilic microchannel condensation, Int. J. Multiphase Flow 36 (8) (2010) 608–619.

[111]   Y. Han, N. Shikazono, Measurement of the liquid film thickness in micro tube slug flow, Int. J. Heat Fluid Flow 30 (5) (2009) 842–853.

[112]   Y. Han, N. Shikazono, Measurement of liquid film thickness in micro square channel, Int. J. Multiphase Flow 35 (10) (2009) 896–903.

[113]   D.M. Fries, F. Trachsel, P.R. von Rohr, Segmented gas–liquid flow characterization in rectangular microchannels, International Journal of Multiphase Flow 34 (12) (2008) 1108–1118.

[114]   D. Gstoehl, J.F. Roques, P. Crisinel, J.R. Thome, Measurement of falling film thickness around a horizontal tube using a laser measurement technique, Heat transfer engineering 25 (8) (2004) 28–34.

[115]   X. Wang, M. He, H. Fan, Y. Zhang, Measurement of falling film thickness around a horizontal tube using laser-induced fluorescence technique, J. Phys. Conf. Ser. IOP Publishing 147 (1) (2009) 012039.

Flow Patterns and Bubble Growth in Microchannels

Lixin Cheng
Department of Engineering, Aarhus University, Aarhus, Denmark

3.1 Introduction

Applications of microscale and nanoscale thermal and fluid transport phenomena involved in traditional industries and highly specialized fields such as micro-fabricated fluidic systems, microelectronics, automobile, cryogenics, aerospace technology, micro-chemical and bioreactors, micro-heat exchangers and electronic chips cooling using microchannels, and so on have been becoming especially important since the late twentieth century [1−9]. However, thermal and fluid transport phenomena are different from those of conventional scale or macroscale [10−24]. For instance, advance in micro-electronics technology continues to develop with surprisingly rapidity, and the thermal energy density of electronic devices to be dissipated is becoming much higher, up to 300 W/cm^2 or even higher [2−6,10−12,14−17]. It is essential to develop new high heat flux cooling technology to meet the challenging heat dissipation requirements. Flow boiling in microchannels using the latent heat to dissipate the high heat flux is the most efficient method to thermal management. However, flow boiling and gas−liquid two-phase flow characteristics in microchannels are different from those in conventional channels [2−6]. The channel confinement has a great effect on flow patterns, bubble growth, heat transfer, and pressure drop in flow boiling and gas−liquid two-phase flow processes [18−24]. Studies of flow boiling and two-phase flow phenomena in microchannels have exhibited contradictory results by various researchers. Therefore, there are many aspects to be clarified from both theoretical and applied aspects.

Flow patterns and bubble dynamics are very important to understand the flow boiling and two-phase flow characteristics and mechanisms in both macro and microchannels [2−6,10−13]. From a practical engineering point of view, one of the major design difficulties in dealing with gas−liquid flow is that the mass, momentum, and energy transfer rates and processes can be very sensitive to the geometric distribution or structures of the components within the flow, which are called flow patterns or flow regimes. For instance, for flow boiling in conventional channels, different heat transfer mechanisms are dominant according to the vapor quality range, and heat flux and mass velocity levels are also intrinsically relevant to the corresponding bubble flow pattern behavior. At low vapor qualities, nucleate boiling effects prevail, while at high vapor qualities and prior to the liquid dry-out, the heat transfer coefficient is mainly controlled by convective effects [25,26]. Figure 3.1 shows the flow patterns in diagrammatic form, the various flow patterns that may be encountered over the

Microchannel Phase Change Transport Phenomena. http://dx.doi.org/10.1016/B978-0-12-804318-9.00003-0

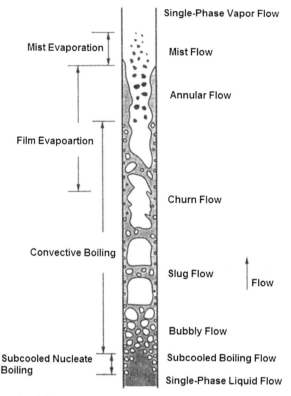

Figure 3.1 Schematic of flow patterns and the corresponding heat transfer mechanisms for upward flow boiling in a vertical tube [13].

length of a vertical tube heated by a uniform heat flux, together with the corresponding heat transfer regimes. Figure 3.2 shows a schematic representation of a horizontal tubular channel heated by a uniform heat flux and fed with subcooled liquid. Flow patterns formed during evaporation in a horizontal tube may be influenced by departures from thermodynamic and hydrodynamic equilibrium. Asymmetric phase distributions and stratification introduce additional complications. Important points to note from a heat transfer viewpoint are the possibility of intermittent drying and rewetting of the upper surfaces of the tube in slug and wavy flows and the progressive dry-out over long tube lengths of the upper circumference of the tube wall in annular flow. At higher inlet liquid velocities, the influence of gravity is less obvious, the phase distribution becomes more symmetric, and the flow patterns become closer to those as in vertical flow. Many studies on flow patterns and mechanisms have been conducted over the past decades as summarized by Cheng et al. [13] and they are the fundamental to investigating the flow patterns and flow pattern maps in microchannels, which must be better understand.

There are a number of studies regarding flow patterns in microchannels in the literature but most of these involve adiabatic conditions at the boundary between the fluid

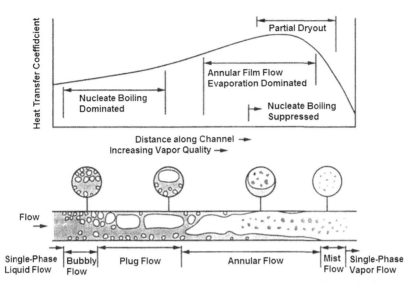

Figure 3.2 Schematic of flow patterns and the corresponding heat transfer mechanisms and qualitative variation of the heat transfer coefficients for flow boiling in a horizontal tube.

and the channel wall. Although there are some studies involving diabatic two-phase flow such as flow boiling, the study of flow boiling in microchannels is very complex and needs to be better understood. Furthermore, for flow boiling in multiple microchannels, the studies are much fewer because it involves complex bubble and flow pattern behaviors. The flow distribution and back flow had a significant effect on the observed flow patterns, called unstable flow patterns or flow regimes. Furthermore, the dry-out and flow instability during flow boiling make it difficult to document an accurate flow pattern map. Most of the available studies are based only on limited fluid type, channel geometry, and test parameter such as one or two saturation temperature, and thus the proposed flow maps based on limited fluids and conditions can only be applicable to limited fluids and flow conditions. There are disagreements regarding the observed flow patterns for similar test conditions by different researchers. It is necessary to address all these issues in the current research of flow patterns and bubble growth.

There are a number of other troublesome questions for flow pattern and bubble growth research. In single-phase flow, it is well established that an entrance length of 30–50 diameters is necessary to establish fully developed turbulent pipe flow. The corresponding entrance lengths for multiphase flow patterns are less well established, and it is possible that some of the reported experimental observations are for temporary or developing flow patterns. However, a number of researchers used very short channels in the flow pattern observation in microchannels. The inlet flow disturbance may be a big factor affecting the observed flow patterns. It becomes more serious for multiple microchannels due to instable flow effect. Moreover, the implicit assumption is often made that there exists a unique flow pattern for given fluids with given flow rates. It is by no means certain that this is the case.

Consequently, there may be several possible flow patterns whose occurrence may depend on the initial conditions, specifically on the manner in which the multiphase flow is generated. Some attributes of two-phase flow in microchannels are not fully understood, and there are inconsistencies among experimental observations, phenomenological interpretation, and theoretical models. Therefore, a well-designed test system is necessary but actually not always done in published work, and careful experimental observations of the two-phase flow boiling are required in order to better understand the flow pattern and bubble growth mechanisms in microchannels. It should also be noted that the adiabatic two-phase flow is different from the diabatic one (i.e., the one with heat addition such as flow boiling). Modeling of diabatic two-phase flow patterns presents a serious challenge, whereas the flow with heat transfer is even a more complex issue. Application of heat flux causes development of boiling and the flow development through various flow patterns, and the bubble growth behavior in confined channels becomes important to understand the two-phase flow and flow boiling behaviors.

The objectives of this chapter are to focus on the current research status on flow patterns and bubble growth in microchannels and to identify the research needs in the relevant topics. State-of-the-art fundamental research on gas−liquid two-phase flow patterns, flow pattern maps, and bubble growth in microchannels is presented. According to this review, recommendations on the future research directions have been given.

3.2 Criteria for Distinction of Macro and Microchannels

Due to the significant differences of transport phenomena in microchannels compared with conventional size channels or macro-scale channels, one very important issue should be clarified about the distinction between micro-scale channels and macro-scale channels. However, a universal agreement is not clearly established in the literature. Instead, there are various definitions on this issue, which are based on the engineering applications and bubble confinements.

Shah [27] defined a compact heat exchanger as an exchanger with a surface area density ratio >700 m^2/m^3. This limit translates into a hydraulic diameter of <6 mm. According to this definition, the distinction between macro- and micro-scale channels is 6 mm.

Mehendale et al. [28] defined various small and mini heat exchangers in terms of hydraulic diameter D_h, as:

- Micro heat exchanger: $D_h = 1−100 \ \mu m$.
- Meso heat exchanger: $D_h = 100 \ \mu m−1 \ mm$.
- Compact heat exchanger: $D_h = 1−6 \ mm$.
- Conventional heat exchanger: $D_h > 6 \ mm$.

According to this definition, the distinction between macro- and micro-scale channels is somewhere between 1 and 6 mm.

Based on engineering practice and application areas such as refrigeration industry in the small tonnage units, compact evaporators employed in automotive, aerospace, air separation and cryogenic industries, cooling elements in the field of microelectronics and micro-electro-mechanical-systems (MEMS), Kandlikar [5] defined the following ranges of hydraulic diameters D_h, which are attributed to different channels:

- Conventional channels: $D_h > 3$ mm.
- Minichannels: $D_h = 200$ μm-3 mm.
- Microchannels: $D_h = 10-200$ μm.

According to this definition, the distinction between small and conventional size channels is 3 mm.

There are several important dimensionless numbers that are used to represent the feature of fluid flow in micro-scale channels. According to these dimensionless numbers, the distinction between macro- and micro-scale channels may be classified as well. Triplett et al. [29] defined flow channels with hydraulic diameters D_h of the order, or smaller than, the Laplace constant L:

$$L = \sqrt{\frac{\sigma}{g(\rho_L - \rho_G)}} \tag{3.1}$$

as micro-scale channels, where σ is surface tension, g is gravitational acceleration, and ρ_L and ρ_G are liquid and gas/vapor densities, respectively.

Kew and Cornwell [30] earlier proposed the confinement number Co for the distinction of macro- and micro-scale channels, as

$$Co = \frac{1}{D_h} \sqrt{\frac{\sigma}{g(\rho_L - \rho_G)}} \tag{3.2}$$

which is actually based on the definition of the Laplace constant. When Co is less than 0.5, the channel is considered as microchannel.

Based on a linear stability analysis of stratified flow and the argument that neutral stability should consider a disturbance wavelength of the order of channel diameter, Brauner and Moalem-Maron [31] derived the Eotvös number $E\ddot{o}$ criterion for the dominance of surface tension for micro-scale channels:

$$E\ddot{o} = \frac{(2\pi)^2 \sigma}{(\rho_L - \rho_G)D_h^2 g} > 1 \tag{3.3}$$

The definition of a micro-scale channel is confusing because there are different criteria available as described earlier. Cheng and Mewes [14] made a comparison of these different criteria for micro-scale channels. Figure 3.3 shows their comparable results for water and CO_2, which shows the big difference among these criteria.

Harirchian and Garimella [32] proposed a micro- to macro-transitional criterion based on the convective confinement number, defined by them as the product between

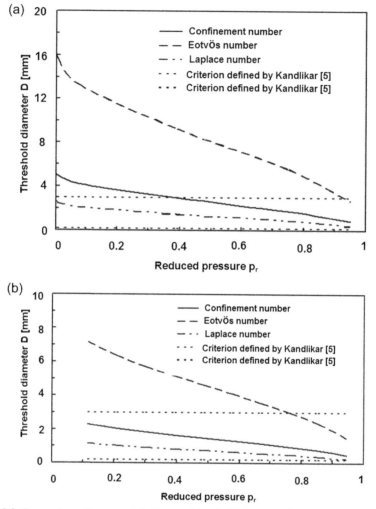

Figure 3.3 Comparison of various definitions of threshold diameters for micro-scale channels: (a) water, (b) CO_2, by Cheng and Mewes [14].

Bond and Reynolds assuming that the two-phase mixture flows as liquid. Their criterion is given as follows:

$$Bo^{0.5} \times \mathrm{Re} = \frac{1}{\mu_F}\left(\frac{g(\rho_L - \rho_G)}{\sigma}\right)^{0.5} GD^2 = 160 \qquad (3.4)$$

$$Bo = \left(\frac{g(\rho_L - \rho_G)}{\sigma}\right)^{0.5} \qquad (3.4a)$$

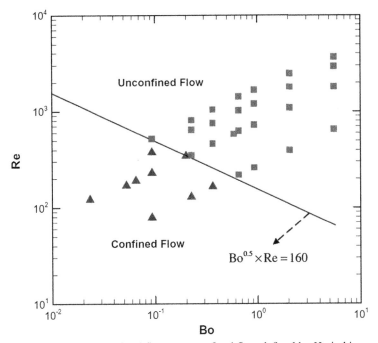

Figure 3.4 Transition from confined flow to unconfined flow defined by Harirchian and Garimella [32].

$$D = \sqrt{A_{cs}} \tag{3.4b}$$

where *Bo* is Bond number, *G* is mass velocity, *D* is length scale, and A_{cs} is cross-sectional area of microchannels.

Figure 3.4 shows their transition from confined flow to unconfined flow. This distinction is based on the flow characteristics, which is different from the aforementioned criteria. In fact, simply using a threshold diameter defined by these criteria does not capture the flow behavior in many cases. Thus, the criterion of Harirchian and Garimella seems to be a new concept but needs to be further validated with the experimental data.

Other new criteria have also been proposed by various researchers but in many cases, the available nondimensional number such as confinement number has been given different values based on individual observations and experimental data. For example, Ong and Thome [33] proposed a transitional criterion based on the confinement number *Co*. They observed a uniform liquid film along the tube perimeter during annular flows, as observed by them for confinement numbers greater than 1, suggests a micro-scale behavior while for confinement numbers lower than about 0.3, the liquid film is nonsymmetric. Isolated bubbles and bubbles coalescence flow pattern observed in small-diameter channels, characterized by confinement number greater than 1, are observed under micro-scale conditions, while for macro-scale conditions corresponding to a confinement number lower than 0.3, the plug-slug flow pattern is observed. This flow pattern presents elongated vapor bubbles, presenting strong buoyancy effects followed by liquid plugs. For confinement numbers between 1.0 and values within the range of 0.3−0.4, Ong and Thome [33] defined a mesoscale behavior. Ribatski [12] discussed the details of the criteria of including the aforementioned criteria in his review. It should be realized that the transition is a progress process. The most important thing is relate flow pattern behaviors to flow and heat transfer behaviors such as heat transfer, critical heat flux (CHF), and pressure drops as pointed out by Cheng et al. [13]. In fact, this has been validated by the flow pattern based CO_2 flow boiling model of Cheng et al. [34−36], which covers both macro and microchannels for CO_2 flow boiling heat transfer. Their generalized CO_2 heat transfer model based on flow patterns predicted both macro and microchannel flow patterns and heat transfer reasonably well. In particular, their CO_2 flow pattern map captured well the independent observed flow patterns in microchannels [42]. This gives a hint that the macro and microchannels may be determined according to the flow and heat transfer behavior. A mechanistic distinction criterion might be more of practice but needs to be developed in future. The transition between micro- and macrochannels has been neither very well defined nor seriously experimentally investigated. However, one may distinguish between flow boiling behaviors such as heat transfer and flow patterns etc. in macro and microchannels [13,43]. It will be useful from a practical point of view if they can be incorporated into heat transfer and pressure drop prediction methods such as those by Wojtan et al. [38,39] and Cheng et al. [34−36].

In this chapter, the distinction between macro and microchannels by the threshold diameter of 3 mm is adopted due to the lack of a well-established theory but is in line with that recommended by Kandlikar [5]. Using this threshold diameter enables more relevant studies to be included, and thus the different flow patterns and bubble growth characteristics in various channels with different sizes can be compared.

3.3 Fundamentals of Flow Patterns in Macro and Microchannels

Gas−liquid two-phase flow is a very complex physical process since they combine the characteristics of a deformable interface, channel shape, flow direction, and, in some cases, the compressibility of one of the phases. In addition to inertia, viscous and pressure forces present in single-phase and two-phase flows are also affected by interfacial tension forces, the wetting characteristics of the liquid on the tube wall (contact angle), and the exchange of mass, momentum and energy between the liquid and vapor phases. Depending on the operating conditions, such as pressure, temperature, mass velocity, adiabatic or diabatic flow, channel orientation (the effect of gravity, which in nonvertical channels tends to pull the liquid to the bottom of the channel), and fluid properties (widely different combinations of different classes of fluids such as air−water, steam−water, liquid and vapor phases of a fluid, and so on), various gas−liquid interfacial geometric configurations occur in two-phase flow systems. These are commonly referred to as flow patterns or flow regimes. Flow patterns provide important information to better understand the complexity of the two-phase flow and heat transfer mechanisms. Generally, the gas and liquid flow rates are constant for adiabatic flows, although in high-speed flow (as in critical flow), partial vaporization of the liquid may occur even though there is no heat addition. Diabatic two-phase flows with heat transfer occur during flow boiling, flow condensation, or gas—liquid two-phase flows with heat addition or removal. Furthermore, the channel size has a significant effect on the flow patterns and bubble behavior. The flow patterns and bubble growth in microchannels need to be understood.

The flow patterns in macrochannels have been well investigated and they are the basis for investigating flow patterns in microchannels. Flow patterns macrochannels are usually divided into groups consisting of bubbly, stratified, stratified-wavy, annular, and mist flow. Figure 3.5(a) shows the most commonly observed two-phase flow patterns in a vertical tube. Bubbly flow occurs when a relatively small quantity of gas or vapor is mixed with a moderate flow rate of liquid. Increasing the

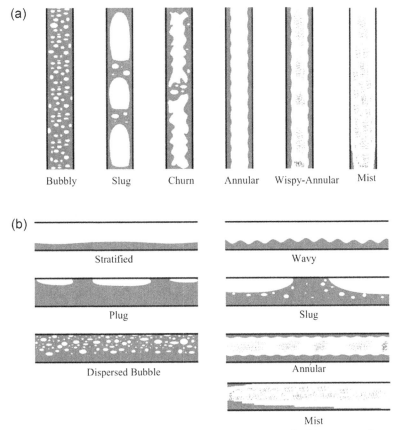

Figure 3.5 (a) Schematic of flow patterns in vertical upward gas liquid cocurrent flow; (b) schematic of flow patterns in horizontal gas liquid cocurrent flow.

gas flow rate may lead to plug flow, which some observers call Taylor bubble flow. With a further increase in gas flow rate, one may observe slug flow which consists of a regular train of large bubbles separated by liquid slugs. Each of these bubbles occupies nearly the entire channel cross section except for a thin liquid layer on the wall and their length is typically 1 to 2 times the channel diameter. An increase in both gas and liquid flow rates will lead to an unstable flow pattern, which is called churn flow. A relatively higher gas flow rate generates a wispy-annular flow pattern, which is not observed or recognized as such in many studies. Very high gas flow rates may cause some of the liquid flow to be entrained as droplets carried along with the continuous gas phase in annular flows. At even higher gas flow rates, all the liquid is sheared from the wall to form the mist flow regime. Figure 3.5(b) shows the most commonly observed flow patterns for cocurrent flow of gas and liquid in a horizontal tube. Two-phase flow patterns in a horizontal tube are similar to those in a vertical tube but distribution of the liquid is influenced by gravity. In general, most flow patterns in horizontal tubes show a nonsymmetric structure which is due to the effect of gravity on the different densities of the phases. This generates a tendency of stratification in the vertical direction, with the liquid having a tendency to occupy the lower part of the channel and the gas the upper part. Stratified flow is usually observed at relatively low flow rates of gas and liquid. As the gas and liquid flow rates are increased, the smooth interface of the liquid becomes rippled and wavy. This pattern is called a stratified wavy flow. If the liquid flow rate is further increased while the vapor flow is maintained low, an intermittent flow pattern will develop in which gas pockets or plugs are entrapped in the main liquid flow and then a plug flow will develop. If flow rates of gas and liquid are increase together, a so-called slug flow regime will develop. The main distinction between slug and plug flow is in the more pronounced nature of intermittent liquid mass separated by a larger gas bubble. With further increase in the gas flow alone, annular flow will develop. The gas flow in the core of an annular flow may entrain a portion of the liquid phase in the form of droplets and in some cases the liquid film may also entrain some small bubbles. At relatively large liquid flow rates, with little gas flow, one would observe the so-called dispersed bubble flow in which the liquid phase is in the dispersed form of the scattered bubbles. At very high gas flow rates, the mist flow is reached, which can begin at the top perimeter where the annular film is the thinnest and then progress downstream to the bottom perimeter.

Flow patterns depend on the physical properties of the fluids and their flow conditions such as mass flux, operation pressure, channel geometry and size, etc. It has been recognized that there is a complicated two-way coupling between the flow in each of the phases or components and the geometry of the flow (as well as the rates of change of that geometry). Furthermore, for diabatic condition such as flow boiling, understanding the bubble dynamics such as bubble generation and growth is extremely important. The complexity of flow patterns presents a major challenge in the study of gas−liquid flows, and there is much that remains to be done, especially for gas−liquid two-phase flow in microchannels.

An appropriate starting point is a phenomenological description of the flow patterns that are observed in common gas−liquid two-phase flows and flow boiling (in nearly all the studies for flow boiling, the vapor liquid two-phase flow are generated at diabatic conditions while the flow regimes were observed at adiabatic condition but this may be called quias-diabatic condition for flow boiling if less heat loss is considered). Various flow pattern identification methods have been used in the experimental studies, including direct visual observation and observation through high speed photography or camera, X-ray absorption, multi-beam gamma densitometry, signal processing of pressure fluctuations, void fraction fluctuations and light intensities, spectral distribution of wall pressure fluctuations, pressure gradient variations, neutron radiography, electrical conductance probes, etc. [13]. The observed results are often presented in the form of a flow pattern map, which is used to identify the flow patterns occurring at various flow conditions. Usually, only two flow parameters are used to define a coordinate system on which the boundaries between the different flow patterns are charted, such as the superficial gas and liquid velocities or mass velocity and vapor quality. Transition boundaries are then proposed to distinguish the location of the various flow regimes as in a classic map. It should be realized that most flow maps are only valid for a specific set of conditions and/or fluids, although efforts are made to propose generalized flow maps. Generally, flow pattern maps for other applications such as micro-scale channels have been proposed by modification of these leading flow maps or depicted according to the experimental data. Usually, the accuracy in determining transition lines on a flow map is in part dependent on the number of experiments carried out and on the adopted coordinate systems as well. There are many coordinate systems for flow pattern maps. The coordinates used in flow maps may be divided into three groups:

1. Phase velocities or fluxes: gas and liquid superficial velocities u_{GS} and u_{LS} in (m/s), or gas and liquid superficial mass fluxes G_{GS} and G_{LS} in (kg/m^2 s) and gas and liquid mass flow rates M_G and M_L in (kg/s). Use of these parameters, while undoubtedly being the most convenient, does not assure creation of a universal flow pattern map for different two-phase mixtures.
2. Quantities referring to the two-phase flow homogeneous model are the transformations of the parameters from group (1) such as total velocity u_T, total mass flux G_T, Froude number based on total velocity Fr_T, void fraction ε, and quality x, and they are only useful for the description of some flow pattern maps.
3. Parameters including the physical properties of phases such as liquid and gas Reynolds numbers Re_L and Re_G, Baker correction factors λ and ψ, gas and liquid kinetic energies E_G and E_L and others; this formulation gives the best possibility for attaining a universal flow pattern map.

As pointed out by Cheng et al. [13], the boundaries between the various flow patterns in a flow pattern map occur because a regime becomes unstable as the boundary is approached and growth of this instability causes transition to another flow pattern. However, these gas—liquid two-phase transitions can be rather unpredictable since they may depend on otherwise minor features of the flow, such as the roughness of the walls or the entrance conditions. Hence, the flow pattern boundaries are not distinctive lines but more poorly defined transition zones.

A number of flow pattern maps have been developed [13]. For instance, one of the leading empirical flow pattern maps for vertical upflows is that of Hewitt and Roberts [37]. On this map, the coordinates are the superficial momentum fluxes of the respective phases. Both air—water and steam—water data could be represented in terms of this plot, which thus covers a reasonably wide range of fluid physical properties. All the transitions are assumed to depend on the phase momentum fluxes. Wispy annular flow is a subcategory of annular flow, which occurs at high mass flux when the entrained drops are said to appear as wisps or elongated droplets.

There have been various attempts at a theoretical or semitheoretical description of flow pattern transitions. For such a description to be successful, it should be suitable for extrapolation to a wide range of conditions. Perhaps the most comprehensive treatment of flow pattern transitions in horizontal flow on a semitheoretical basis is that of Taitel and Dukler [38]. It has been proved successful in predicting a fairly wide range of system conditions. Figure 3.6 shows the Taitel and Dukler flow pattern map. The parameter groups, which are based on semitheoretical derivations for different flow pattern transitions in horizontal or slightly inclined channels (θ is the angle of inclination) are as follows:

$$X = \left[\frac{(dp/dz)_L}{(dp/dz)_G} \right]^{\frac{1}{2}} \tag{3.5}$$

$$Fr = \frac{G_G}{[\rho_G(\rho_L - \rho_G)Dg \cos \theta]^{\frac{1}{2}}} \tag{3.6}$$

$$T = \left[\frac{|(dp/dz)_L|}{g(\rho_L - \rho_G)\cos \theta} \right]^{\frac{1}{2}} \tag{3.7}$$

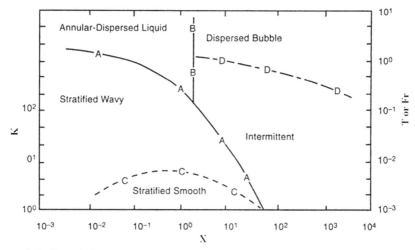

Figure 3.6 The Taitel and Dukler [38] flow pattern map for horizontal gas–liquid cocurrent flow: coordinates of curves A and B are Fr versus X, coordinates of curve C are K versus X and coordinates of curve D are T versus X.

$$K = Fr \left[\frac{G_L D}{\mu_L} \right]^{\frac{1}{2}} \qquad\qquad (3.8)$$

where X is the Martinelli parameter, $(dp/dz)_L$ is the frictional pressure gradient as if the liquid in the two-phase flow were flowing alone in the tube, $(dp/dz)_G$ is the frictional pressure gradient as if the gas in the two-phase flow were flowing alone in the tube, Fr is the Froude number, D is the tube diameter, g is the acceleration due to gravity, ρ_L is the liquid density, ρ_G is the gas density, and μ_L is the liquid viscosity. They suggested the K versus X coordinate with a theoretically derived boundary curve, C, for transition from stratified smooth to stratified wavy flow. The Fr versus X relationship was proposed for the transitions between stratified wavy, annular-dispersed (droplets), dispersed bubble and intermittent (plug or slug) flows. The theoretically determined transition curves A and B (at $X = 1.6$) between the said regimes were also given in those coordinates. Finally, T versus X was proposed for defining the transition between dispersed bubble and intermittent (plug or slug) flow regimes with the transition line D. The transition curves shown in Fig. 3.6 are for the case of zero inclination angle (horizontal). All the transition criteria used by Taitel and Dukler have some theoretical basis, although sometimes rather tenuous.

Most of the available flow pattern maps are for adiabatic flow conditions and they do not include dry-out regime, which occurs in flow boiling. In the case of diabatic two-phase flows such as flow boiling (evaporation), very few flow maps have been proposed. Important factors influencing these flows and their transitions are nucleate boiling, evaporation on what could otherwise be dry parts of the perimeter, and acceleration of the flows. For example, nucleate boiling in an annular film tends to increase the film thickness and change the void profile near the wall or vigorous nucleate boiling in an otherwise stratified flow can completely wet the upper perimeter, thus increasing liquid entrainment in the vapor core. It is desirable that diabatic flow pattern maps include the influences of heat flux and dry-out, etc. on the flow pattern transition boundaries. For instance, Cheng et al. [34] developed a flow pattern map for CO_2 evaporation inside horizontal tubes on the basis of the Wojtan et al. [39,40] map by modifying the I−A and A−D boundary transitions. More recently, Cheng et al. [35,36] further modified the A−D boundary transition, proposed a new D−M boundary transition, added a bubbly flow regime criterion (B stands for bubbly flow) and developed an updated flow boiling heat transfer model based on their flow map. Figure 3.7 shows the CO_2 flow pattern map of Cheng et al. [35,36] evaluated for the indicated test conditions of Yun et al. [41] and the corresponding heat transfer prediction based on their map, which captured their data very well.

It should be realized that there is an essential arbitrariness in the interpretation of flow pattern data and thus it is unlikely that perfect prediction methods will ever emerge [13]. Furthermore, there are other serious difficulties with most of the existing literature on flow pattern maps. One of the basic fluid mechanical problems is that these maps are often dimensional and therefore apply only to the specific pipe sizes and fluids employed by the investigator. A number of investigators have attempted to find generalized coordinates that would allow the map to cover different fluids

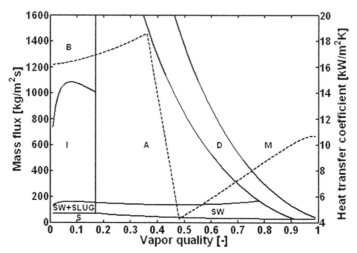

Figure 3.7 The CO_2 flow pattern map of Cheng et al. [35,36] evaluated for the test condition of Yun et al. [41]: $D_{eq} = 2$ mm, $G = 1500$ kg/m^2 s, $T_{sat} = 5$ °C, $q = 30$ kW/m^2 and the corresponding prediction of heat transfer coefficients (dashed line) (A stands for annular flow, B stands for bubbly flow I stands for intermittent flow, M stands for mist flow, S stands for stratified flow and SW stands for stratified-wavy flow. The stratified to stratified-wavy flow transition is designated as S-SW, the stratified-wavy to intermittent/annular flow transition is designated as SW-I/A, the intermittent to annular flow transition is designated as I-A).

and pipes of different sizes. However, such generalizations can only have limited value because several transitions are represented in most flow pattern maps and the corresponding instabilities are governed by different sets of fluid properties. For example, one transition might occur at a critical Weber number, whereas another boundary may be characterized by a particular Reynolds number. Hence, even for the simplest duct geometries, there exist no universal, dimensionless flow pattern maps that incorporate the full, parametric dependence of the boundaries on the fluid characteristics.

Flow patterns and flow pattern maps provide important information to better understand the complexity of the two-phase flow and flow boiling mechanisms [2−6,10−16,21−24]. However, flow patterns and bubble growth are still open research topics for flow boiling and two-phase flow in microchannels although there are a number of studies of two-phase flow patterns in microchannels. A number of these available studies involve adiabatic conditions at the boundary between the fluid and the channel wall. There are some studies involving flow boiling in microchannels, nearly all the flow pattern observations are based on adiabatic flow at the exit of the heated channels.

More recently, there are several studies concerning the flow pattern observation under diabtaic conditions such as Owhaib et al. [57], Celata et al. [67], and Tibiriçá and Ribatski [68]. A systematic knowledge has not yet fully established due to the limited studies. In flow boiling, the effect of mass flux, channel size, bubble generation and growth and heat flux has an effect on the flow patterns and bubble growth while for adiabatic flows such as air–water flow only flow rate of each component are considered. In most cases, flow patterns are observed and recorded through high-speed visualization and then presented in a flow regime map or compared with the flow pattern maps in conventional such as the Taitel and Dukler flow map and other maps available in the literature. In most cases, both available flow pattern maps in conventional and microchannels do not work well for individual observations. Furthermore, most of the published microchannel flow maps are only based on limited fluids or test parameters such as one or two saturation temperatures and presented in dimensional parameters which lack generalization. For flow boiling in microchannels, the surface tension dominates flow patterns as hydraulic diameter decreases, and thus the stratified flow basically does not exist in microchannels. Furthermore, confined bubbles in microchannels have an effect on flow pattern evolution and transitions. In general, the criteria for flow pattern transition with variations of vapor quality depend on mass flux during the saturation flow boiling in microchannels. That is, the pattern transition toward an annular flow occurs in the region of lower vapor quality as the mass flux becomes higher in microchannels compared with conventional channels. The different flow patterns have been observed by various researchers. For example, some researchers observed bubbly flow, while others did not. A universal flow pattern map is not yet available. Further, bubble dynamics is very important in understanding the two-phase flow patterns and heat transfer mechanisms during flow boiling in microchannels but such studies are very rare. In the following sections, studies on flow patterns, flow pattern map and bubble growth in microchannels are reviewed and discussed.

3.4 Flow Patterns and Flow Pattern Maps in Microchannels

In this section, highlights of experimental studies on flow patterns and flow pattern maps are summarized in micro-scale channels including channel diameter less than 3 mm. Selected microchannel studies are presented in Table 3.1 according to this definition. Both adiabatic and diabatic conditions have been investigated. Most studies used air–water or water–nitrogen as the working fluids. Some researchers used steam–water, nitrogen and its vapor, refrigerants and their vapors such as R134a, CO_2, and R123. Both single microchannels and multiple microchannels with various channel shapes and sizes are concerned.

3.4.1 Current Research Progress on Flow Patterns in Microchannels

Triplett et al. [29] conducted a systematic experimental investigation of air–water in microchannels with inner diameters of 1.1 and 1.45 mm for circular channels and with hydraulic diameters of 1.09 and 1.49 mm for semitriangular channels. The discernible flow patterns were bubbly, churn, slug, slug-annular, and annular flows as shown in Fig. 3.8. Furthermore, they compared their results to the criteria of Suo and Griffith [44]

Table 3.1 **Selected Experimental Studies on Flow Patterns in Microchannels in the Literature**

Authors/ References	Fluids	Test Channel Diameter(s) and Orientations	Main Research Contents
Triplett et al. [29]	Air−water	Circular channels, 1.1 and 1.45 mm, semi-triangular channels with hydraulic diameter $D_h = 1.09$ and 1.49 mm, horizontal.	Flow patterns were studied. The experimental data were compared with the existing data and flow maps.
Suo and Griffith [44]	Air−water, He, N_2/heptane	Circular tube, 1 and 1.6 mm, horizontal.	Two-phase flow patterns were observed. Surface tension dominates over gravity.
Revellin et al. [45] and Revellin and Thome [46,47]	R134a and R245fa evaporation, circular tubes	Circular tube, 0.509 and 0.8 mm, horizontal.	Flow patterns and transition were identified by optical technique and observed by high speed video. Diabatic flow maps were proposed according to the experimental data.
Cubaud and Ho [48]	Air−water	Square channels, 200 and 525 μm, horizontal.	Flow regimes were observed. A flow pattern map and the transition lines between flow regimes were drawn for the microchannels.
Coleman and Garimella [49]	Air−water	Circular and rectangular channels, 5.5 to 1.3 mm, horizontal.	The effects of tube diameter and surface tension on flow patterns were experimentally studied.
Chen and Garimella [50]	Dielectric fluid	Twenty-four microchannels, each with a square cross-section multi-square channels, each 0.389×0.389 mm, horizontal.	Flow patterns were observed with high-speed visualizations.

Table 3.1 **Selected Experimental Studies on Flow Patterns in Microchannels in the Literature—cont'd**

Authors/ References	Fluids	Test Channel Diameter(s) and Orientations	Main Research Contents
Zhao and Bi [51]	Air—water	Equilateral triangular channels, hydraulic diameter $D_h = 5.5$, 2.886, 1.443 and 0.866 mm, vertical.	Flow patterns were observed and the experimental data were compared with the flow pattern models.
Gasche [42]	CO_2 evaporation	Circular tube, 0.98 mm, horizontal.	Flow patterns were observed.
Pettersen [52]	CO_2 evaporation	Circular tube, 0.98 mm, horizontal.	Flow patterns were observed.
Lowry and Kawaji [53]	Air—water	Narrow passage between two flat plates with gaps: 0.5, 1 and 2 mm, vertical.	Flow patterns were studied and flow maps were constructed based on the experimental data.
Damianides and Westwater [54]	Air—water	Compact heat exchanger with an equivalent diameter $D_e = 1.74$ mm, several round tubes, 1, 2, 3, 4, and 5 mm, horizontal.	Flow patterns were determined by high-speed photography and fast-response pressure transducers. Flow maps were constructed.
Liu and Wang [55]	Air—water	Circular tubes with diameters of 1.47, 2.37, and 3.04 mm, vertical.	Flow patterns were observed and compared with the existing flow maps. A new map was proposed.
Chen et al. [56]	R134a evaporation	Circular tubes wit diameters of 1.1, 2.01, 2.88 and 4.26 mm vertical.	Flow patterns were observed and compared with the existing flow maps. A new map was proposed.
Yang and Shieh [57]	Air—water, R134a evaporation	Circular tube, 1—3 mm, horizontal.	Flow patterns were studied and the experimental data were compared with the available models.

Continued

Table 3.1 **Selected Experimental Studies on Flow Patterns in Microchannels in the Literature—cont'd**

Authors/ References	Fluids	Test Channel Diameter(s) and Orientations	Main Research Contents
Owhaib et al. [58]	R134a evaporation	Circular tube, 1.33 mm, vertical.	The flow patterns at high vapor qualities and the dry-out of the liquid film were visually studied.
Sobierska et al. [59]	Steam−water	Rectangular hydraulic diameter $D_h = 1.2$ mm, vertical.	Flow patterns were studied and compared with the existing criteria.
Yen et al. [60]	R123 evaporation	Circular channel, 0.21 mm and square channel, 0.214 mm, horizontal.	Visualizations of flow patterns with simultaneous measurement of heat transfer coefficients were performed.
Serizawa et al. [61]	Air−water, steam−water	Circular tubes, 0.02, 0.025, 0.05, and 0.1 mm, horizontal.	Flow patterns were observed and a flow map was constructed.
Ide et al. [62]	Air−water	Circular tubes, 1, 2.4, and 4.9 mm, rectangular, 1×1, 2×1, 5×1 and 9.9×1.1 mm, vertical upward and downward and horizontal.	The effects of the tube diameters and aspect ratios of the channels on flow patterns were studied.
Hetsroni et al. [63]	Air−water, water−steam	Multi-triangular channels, hydraulic diameter $D_h = 0.129$, 0.103, and 0.161 mm, horizontal.	Flow patterns were experimentally studied for both adiabatic and diabatic conditions.
Kawahara et al. [64]	Water−nitrogen	Circular channels, 0.1 mm, horizontal.	Flow patterns were observed and a flow map was constructed and compared with their existing flow map.

Table 3.1 **Selected Experimental Studies on Flow Patterns in Microchannels in the Literature—cont'd**

Authors/ References	Fluids	Test Channel Diameter(s) and Orientations	Main Research Contents
Chung and Kawaji [65]	Water−nitrogen	Circular channels, 0.53, 0.25, 0.1, and 0.05 mm, horizontal.	Flow patterns were observed and the effect of channel diameter on flow patterns was studied.
Fukano and Kariyasaki [66]	Air−water	Circular tubes, 1, 2.4 and 4.9 mm, vertical upward and downward, horizontal.	Flow patterns were experimentally studied and flow maps were constructed.
Celeta et al. [67]	FC-72 evaporation	Circular tube, 0.48 mm, horizontal.	Flow patterns were observed and compared with the existing flow maps.
Tibiriçá and Ribatski [68]	R134a and R245fa evaporation	Circular tube, 0.4 mm, horizontal.	Flow patterns were observed and compared with the existing flow maps.
Sir and Liu [69]	Air−water	Circular tubes, 100, 180, and 324 μm, horizontal.	Flow patterns were observed and compared with the existing flow maps. A new map was proposed.
Harirchia and Garimella [70]	FC-77	Parallel microchannels of width from 0.1 to 5.8 mm, all with depth of 0.4 mm, horizontal.	Flow visualization of flow patterns during flow boiling in microchannels.
Wang et al. [71]	Water evaporation.	Parallel trapezoidal microchannels with a hydraulic diameter of 186 μm, horizontal.	Flow visualization of flow patterns during flow boiling in microchannels.
Wang et al. [72]	Water evaporation	Parallel trapezoidal microchannels and a single microchannel with a hydraulic diameter of 186 μm, horizontal.	Flow visualization of flow patterns during flow boiling in microchannels.

Continued

Table 3.1 Selected Experimental Studies on Flow Patterns in Microchannels in the Literature—cont'd

Authors/ References	Fluids	Test Channel Diameter(s) and Orientations	Main Research Contents
Tuo and Hrnjak [73]	R134a evaporation	Parallel microchannels with the inner hydraulic diameter of 1.0 mm vertical.	Flow visualization of flow patterns during flow boiling in microchannels.

(a)

U_{LS} = 5.997 m/s
U_{GS} = 0.396 m/s

(b)

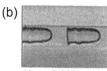

U_{LS} = 0.608 m/s
U_{GS} = 0.498 m/s

(c)

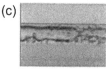

U_{LS} = 1.205 m/s
U_{GS} = 4.631 m/s

(d)

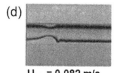

U_{LS} = 0.082 m/s
U_{GS} = 6.163 m/s

(e)

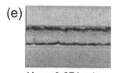

U_{LS} = 0.271 m/s
U_{GS} = 70.42 m/s

Figure 3.8 Photographs of flow patterns in a 1.1 mm diameter test section of Triplett et al. [29]. (a) Bubbly, (b) slug flow, (c) churn flow, (d) slug-annular flow, and (e) annular flow.

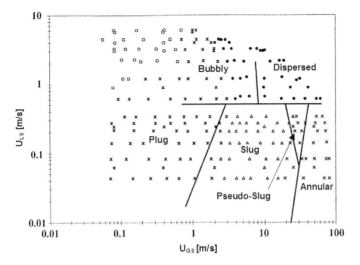

Figure 3.9 Comparison between the experimental flow patterns observed by Triplett et al. [29] for a 1.1 mm diameter circular test section to the experimental flow regime transition lines of Damianides and Westwater [54] based on a 1 mm diameter circular test section.

and the mechanistic model of Taitel et al. [38] for the flow pattern transition line leading to dispersed bubbly flow. The criteria of Suo and Griffith significantly disagreed with their data. The model of Taitel et al. [38] satisfactorily predicted the bubbly-slug transition line. However, transition from dispersed bubbly to churn flow was not captured. Figure 3.9 shows their observations compared with the experimental flow pattern transition lines of Damianides and Westwater [54] taken for a circular 1-mm-inner-diameter test section. The flow pattern names displayed on the figure represent the notation of Damianides and Westwater. The two data sets are in relative agreement with respect to slug and slug-annular flows (referred to as plug and slug, respectively, by Damianides and Westwater) and the flow conditions leading to annular flow.

Chen et al. [56] investigated the tube diameter effect on flow patterns and observed the flow patterns of R134a flow boiling in vertical circular tubes with four different diameters of 1.10, 2.01, 2.88, and 4.26 mm. Their flow patterns were observed in a glass tube connected to the heated tube at adiabatic condition. Figure 3.10 shows their observed flow patterns in the tubes of diameters of 1.10, 2.01, and 2.88 mm. They observed flow patterns include dispersed bubble, bubbly, confined bubble, slug, churn, annular, and mist flow. The flow patterns in the 2.88- and 4.26-mm tubes are similar to those typically described in normal size tubes. The smaller-diameter tubes of 1.10 and 2.01 mm exhibit strong effects of channel confinements on the flow patterns. The boundaries of slug to churn and churn to annular moved to higher vapor velocity while the dispersed bubble to bubbly

(a)

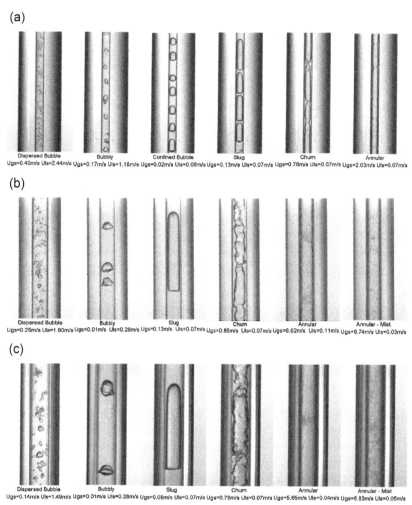

Dispersed Bubble	Bubbly	Confined Bubble	Slug	Churn	Annular
Ugs=0.40m/s Uls=2.44m/s	Ugs=0.17m/s Uls=1.18m/s	Ugs=0.02m/s Uls=0.06m/s	Ugs=0.13m/s Uls=0.07m/s	Ugs=0.76m/s Uls=0.07m/s	Ugs=2.03m/s Uls=0.07m/s

(b)

Dispersed Bubble	Bubbly	Slug	Churn	Annular	Annular - Mist
Ugs=0.25m/s Uls=1.90m/s	Ugs=0.01m/s Uls=0.28m/s	Ugs=0.13m/s Uls=0.07m/s	Ugs=0.85m/s Uls=0.07m/s	Ugs=6.62m/s Uls=0.11m/s	Ugs=6.74m/s Uls=0.03m/s

(c)

Dispersed Bubble	Bubbly	Slug	Churn	Annular	Annular - Mist
Ugs=0.14m/s Uls=1.49m/s	Ugs=0.01m/s Uls=0.28m/s	Ugs=0.08m/s Uls=0.07m/s	Ugs=0.79m/s Uls=0.07m/s	Ugs=5.65m/s Uls=0.04m/s	Ugs=6.83m/s Uls=0.05m/s

Figure 3.10 Photographs of flow patterns of R134a flow boiling in vertical circular tubes with three different diameters by Chen et al. [56]. (a) Flow patterns observed in the 1.10 mm internal diameter tube at 10 bar. (b) Flow patterns observed in the 2.01 mm internal diameter tube at 10 bar. (c) Flow patterns observed in the 2.88 mm internal diameter tube at 10 bar.

boundary moved to higher liquid velocity when the diameter changed from 4.26 to 1.1 mm. It seem that the tube diameter does not affect the dispersed bubble to churn and bubbly to slug. They compared their experimental observations to the Taitel and Dukler flow map [38] for conventional channels and the microchannel flow pattern map of Akbar et al. [91]. None of these flow pattern maps agrees to their observed flow patterns. It should be realized that both flow pattern maps are for horizontal channels. For the microchannel flow map, the channel diameter is much large than

0.25 mm for that map. This is an issue of how to define the confinement of microchannel. Although their smallest tube diameter is 1.1 mm, the flow behaviors are still like those in conventional tubes.

Liu and Wang [55] investigated flow patterns of upward air—water flow in circular channels with diameters of 1.47, 2.37, and 3.04 mm. Their observed flow patterns include bubbly, slug, Taylor, bubble-train slug, churn, and annular flows. They found that the effects of capillary diameter on the flow patterns were not remarkable. They compared their observed flow patterns to the flow pattern transition criteria of Taitel and Dukler [36] and Mishima and Ishii [92] for conventional tubes. None of these predicted their experimental data completely. The model of Taitel and Dukler predicted the transition from Taylor flow to churn flow and the transition from bubbly flow to churn flow well. However, there were big discrepancies between the model of Mishima and Ishii and their observed flow patterns. The model of Mishima and Ishii predicted the Taylor flow zone correctly (but not precisely) only at low liquid velocities. Furthermore, their flow patterns agreed with the flow patterns of Triplett et al. [29] while they did not agree well with the observed flow patterns of Zhao and Bi [51] in triangular channels. This is possibly due to the different capillary cross-sectional geometries.

Zhao and Bi [51] investigated upward air—water two-phase flow patterns in vertical equilateral triangular channels with hydraulic diameters of 2.886, 1.443, and 0.866 mm. Their observed flow patterns include dispersed bubbly flow, slug flow, churn flow, and annular flow in the channels having larger hydraulic diameters, 2 and 1.443 mm, which are similar to flow patterns in vertical conventional channels. For the 0.866-mm-diameter channel, dispersed bubbly flow pattern, characterized by randomly dispersed bubbles in continuous liquid phase, was not found, although the other typical flow patterns remained in the channel. Furthermore, they observed a so-called capillary bubbly flow pattern, characterized by a single train of bubbles, essentially ellipsoidal in shape and spanning almost the entire cross section of the channel, existed at low gas flow rates. Furthermore, they found that in the slug flow regime, slug-bubbles were substantially elongated. The transition boundary from slug flow to churn flow and from churn flow to annular flow in the flow regime map shifted to the right with the decrease in hydraulic diameter of the triangular channels. The flow pattern transition criteria of Taitel and Dukler [36] and Mishima and Ishii [92] did not predict their data satisfactorily.

More recently, Sur and Liu [69] investigated the effects of channel size and superficial phasic velocity on the flow patterns of air—water flow in circular microchannels with inner diameters of 100, 180, and 324 μm. They observed four basic flow patterns, namely, bubbly flow, slug flow, ring flow, and annular flow, as shown in Fig. 3.11. Their channel sizes are much smaller than the aforementioned studies by Chen et al. [56], Liu and Wang [55], and Zhao and Bi [51]. As the channel dimension decreases, the flow regime transition boundary lines shift mainly as the result of the force competition between the inertia and surface tension as shown in Fig. 3.12. They developed a new flow pattern map using the modified Weber numbers as the coordinates to unify

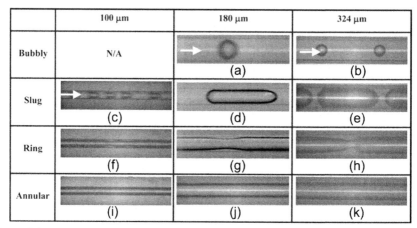

Figure 3.11 Photographs of two-phase flow patterns of air-water flow in horizontal circular microchannels of Sur and Liu [69].

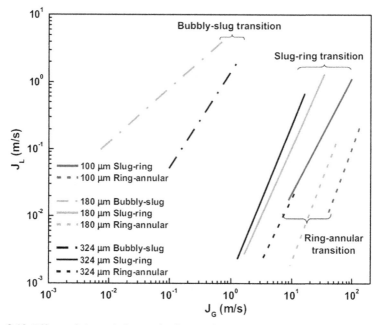

Figure 3.12 Effects of channel size on the flow regime transition lines by Sur and Liu [69].

the transition boundary lines between major flow regimes in microchannels of different sizes. Figure 3.13 shows their flow map based on their experimental data and Weber numbers. However, their flow pattern map has not yet been validated with independent data.

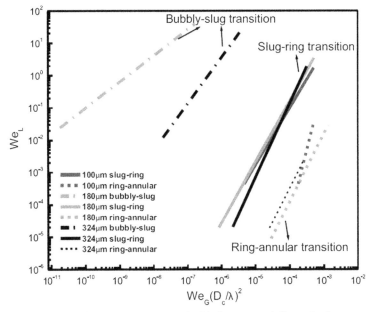

Figure 3.13 Two-phase flow map reconstructed with the proposed dimensionless parameters as the coordinates by Sur and Liu [69].

Different flow patterns have been observed in microchannels. For instance, Serizawa et al. [61] investigated flow patterns of air—water flow in circular tubes with inner diameters of 20, 25, and 100 µm and for steam—water flow in a 50-µm-inner-diameter circular tube. They observed dispersed bubbly flow, gas slug flow, liquid ring flow, liquid lump flow, annular flow, frothy or wispy annular flow, rivulet flow, and liquid droplets flow as shown in Fig. 3.14. They have found that two-phase flow patterns are sensitive to the surface conditions of the inner wall of the test tubes. A stable annular flow and gas slug formation with partially stable thin liquid film formed between the tube wall and gas slugs appeared at high velocities under carefully treated clean surface conditions. At lower velocities, dry and wet areas exist between gas slug and the tube wall. Figure 3.14 shows their observed flow patterns of air—water flow in a 100-µm-diameter tube. Compared with the flow patterns of Sur and Liu shown in Fig. 3.11, some different flow patterns were observed.

Cubaud and Ho [48] studied air—water flows in 200- and 525-µm^2 microchannels made of glass and silicon and observed bubbly flow, wedge flow, slug flow, annular flow, and dry flow. The newly defined wedge flow was proposed. As shown in Fig. 3.15, wedge flow consists of elongated bubbles, the size of which d is larger than the channel width h with partial dry-out of the film at the center of the walls downstream from the nose of the bubble. Wedge flow exhibits some differences from the Taylor bubbly flow. For a partially wetting system, as a function of the bubble velocity, the perimeter of the bubbles can dry out at the center face of the channel creating triple lines (liquid—gas—solid) while liquid still flows in the corners.

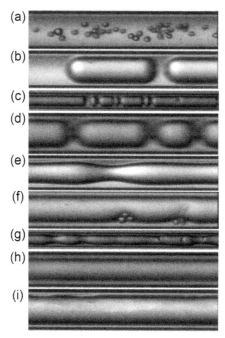

Figure 3.14 Air-water two-phase flow patterns in a 100 μm I.D. tube by Serizawa et al. [61]. (a) Bubbly flow, (b) slug flow, (c) transition, (d) skewed flow (Yakitori flow), (e) liquid ring flow, (f) frothy annular flow, (g) transition, (h) annular flow, and (i) rivulet flow.

In the channel sizes from 0.01 to 3 mm, flow patterns can be very different. For example, the effect of channel orientation (vertical or horizontal) tends to disappear with decreasing channel size. New flow patterns, such as wedge flow [48] and liquid lump flow [61], appear. Slug flows have much longer bubbles, reaching length to channel diameter ratios of 10−100 [45−47].

Similar to the cases in macro-scale channels, flow patterns at adiabatic conditions are different from those at diabatic conditions in micro-scale channels. Diabatic flows have been studied although only several studies are available. Flow patterns in a 0.509-mm microchannel for R245fa at 35 °C and 500 kg/m^2 s observed by Revellin and Thome [45−47] are shown in Fig. 3.16. Several transition regimes such as bubbly/slug flow, slug/semiannular flow, and semiannular flow have been defined according to their observations, the latter of which probably coincides with churn flow in macro-scale channels. Figure 3.17 shows their diabatic flow pattern maps observations and boundaries for R134a [46,47]. As already mentioned, only based on data in one or two channels and one fluid, such flow maps are possibly not applicable to other fluids and test conditions. Furthermore, dimensionless numbers are desirable parameters for generating flow pattern maps.

Celata et al. [67] investigated flow boiling patterns of FC-72 in a horizontal circular microchannel of 0.48 mm at mass fluxes ranging from 50 to 3000 kg/m^2 s. They observed bubbly flow, deformed bubbly flow, bubbly/slug flow, slug flow,

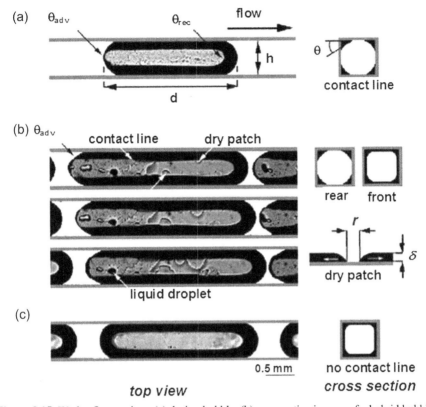

Figure 3.15 Wedge flow regime: (a) drying bubble, (b) consecutive images of a hybrid bubble, and (c) lubricated bubble, observed by Cubaud and Ho [48].

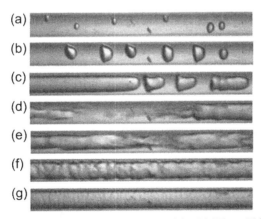

Figure 3.16 Flow patterns in a 0.509 mm microchannel for R245fa at 35 °C and 500 kg/m² s, observed by Revellin and Thome [46,47] at the exit of a micro-evaporator channel. (a) Bubbly flow at $x = 0.038$, (b) bubbly/slug flow at $x = 0.04$, (c) slug flow at $x = 0.043$, (d) slug/semi-annular flow at $x = 0.076$, (e) semi-annular flow at $x = 0.15$, (f) wavy annular flow at $x = 0.23$, and (g) smooth annular flow at $x = 0.23$.

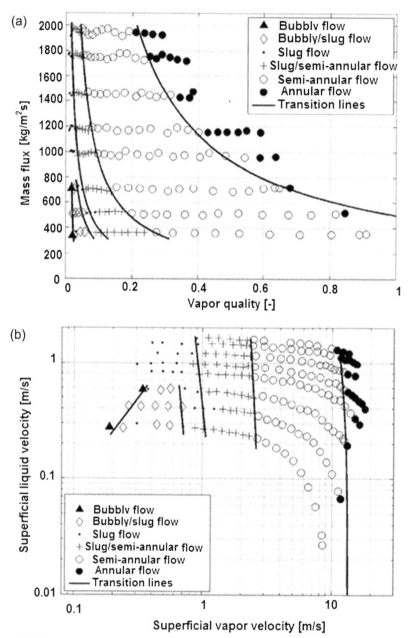

Figure 3.17 Flow pattern observations with experimental transition lines for R134a, $D = 0.509$ mm, $L = 70.7$ mm, $T_{sat} = 35\ °C$, $\Delta T_{sub} = 5\ °C$ using laser plotted in two different formats: (a) flow pattern observations with transition lines, (b) flow pattern map, by Revellin and Thome [46,47].

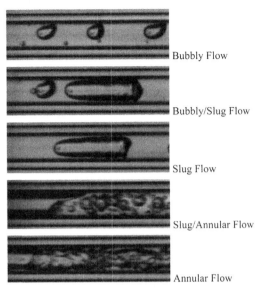

Figure 3.18 Flow patterns in a 0.48 mm microchannel for FC-72 flow boiling, observed by Celata [67] in a heated microchannel.

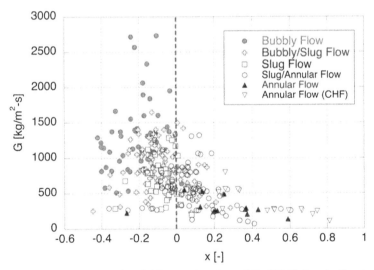

Figure 3.19 Flow patterns map in a 0.48 mm microchannel for FC-72 flow boiling by Celata [67] in a heated microchannel.

slug/annular flow, and annular flow. Figure 3.18 shows their observed flow patterns at diabatic conditions. Their experimental flow pattern map in Fig. 3.19 shows the presence of intermittent flow (slug flow) in the subcooled flow boiling region and the presence of instabilities associated mainly to intermediate flow patterns (bubbly/slug flow and slug/annular flow). Churn flow is not observed in the micochannel. They have found that the presence of a large number of tests with intermediate flow patterns is associated partly to the physical phenomena, and partly to the subjectivity of the classification methods. They compared of the flow patterns in the microchannel to those in larger tubes with inner diameters of 6.0, 4.0, and 2.0 mm and have found that the reduction of the tube diameter produces a significant effect on the transition bubbly to slug flow indicating confinement effects at mass flux lower than about 1250 kg/m^2 s. In particular, the presence of intermittent flow spreads toward the subcooled region. They compared their experimental flow patterns to the flow pattern maps of Mishima and Ishii [92], McQuillan and Whalley [93], and Ong and Thome [33]. Generally, these maps show unsatisfactory prediction capabilities of the bubbly to slug flow transition at low mass flux ($G < 1250$ kg/m^2 s). The models of Mishima and Ishii and Ong and Thome show a good prediction capability of the slug to annular flow transition, in the region of low mass flux ($G < 1250$ kg/m^2 s). The model of McQuillan and Whalley does not show a good agreement with the experimental results. The causes of failure can be attributed to transition criteria models that are not able to predict the observed behavior of the diabatic two-phase flow.

Tibiriçá and Ribatski [68] conducted fundamental investigation on flow patterns of flow boiling of R134a and R245fa in a horizontal circular microchannel with an inner diameter of 0.4 mm in order to better understand the flow pattern evolution in microchannels. Their observed flow patterns are bubble, slug, and annular flow, which are similar to those observed Celata et al. [61]. Stratified flow was not observed in microchannels. Furthermore, they observed static vapor slugs in horizontal tubes with three different inner diameters of 0.4, 1, and 2 mm as shown in Fig. 3.20. By increasing the diameter the interface starts to deform due to the gravity force. However, even with a diameter of 2.00 mm, the surface tension at the interface can hold the liquid pressure and avoid the formation of a stratified flow.

There are several studies on flow patterns during flow boiling patterns in multiple microchannels. Generally, flow patterns in multiple microchannels are similar to those in single microchannels, but flow instability becomes important in such channels. As such, reversible flow occurs. The complex flow instability issue is beyond the scope of this chapter. Simply considering the flow pattern in such cases, there

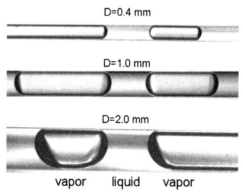

Figure 3.20 Static vapor slugs in horizontal channels. R245fa at 31 °C by Tibiriçá and Ribatski [68].

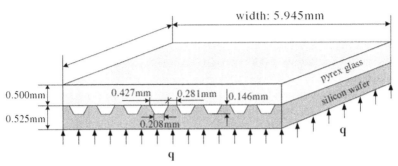

Figure 3.21 Arrangement of eight parallel microchannels having the same length and identical trapezoidal cross-sectional area and etched in a silicon substrate by Wang et al. [71].

are several studies relevant to flow patterns in multiple microchannels. Wang et al. [71] conducted flow visualization and measurement study on the effects of inlet/outlet configurations on flow boiling instabilities in eight parallel microchannels, having a length of 30 mm and a hydraulic diameter of 186 μm. The arrangement of their multiple test channels are shown in Fig. 3.21. Figure 3.22 shows their observed flow patterns in the parallel channels at steady flow condition. Elongated bubbles were observed due to the confinement of the microchannels. Other flow patterns include isolated bubble flow, coalescence bubble flow, and annular flow.

Harirchian and Garimella [70] conducted experiments with FC-77 to investigate the effects of channel size and mass flux on microchannel flow boiling regimes. The test sections are seven different silicon test pieces with parallel microchannels of widths ranging from 100 to 5850 μm, all with a depth of 400 μm. In general, their observed flow patterns include bubbly, slug, churn, wispy-annular, and annular flow—are identified. Flow patterns in the 100-μm- and 250-μm-wide microchannels are found to be similar, and differ from those in microchannels of width 400 μm and larger; the latter group showed similar flow patterns. As channel width increases, bubbly flow replaces slug flow and intermittent churn/wispy annular

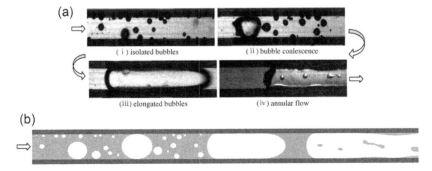

Figure 3.22 Photographs and sketch of flow patterns in steady bubbly/slug flow boiling regime in parallel microchannels ($D_h = 186$ μm) with the Type-B connection at $q = 364.68$ kW/m^2, $G = 124.03$ kg/m^2 s and $T_{in} = 35$ °C (i.e., $x_e = 0.359$) by Wang et al. [71]. (a) Photos of steady flow boiling pattern in parallel microchannels with the Type-C connection. (b) Sketch of steady flow boiling pattern in microchannels with the Type-C connection.

flow replaces intermittent churn/annular flow. For each microchannel size, as mass flux increases, the bubbles become smaller and more elongated in the bubbly region, and the liquid layer thickness in the wispy-annular and annular regimes decreases. Since the transition between specific flow patterns occurs at higher heat fluxes as the mass flux increases, different flow patterns were observed at a given heat flux for different mass fluxes. Figure 3.23 shows their observed flow patterns at a heat flux of 145 kW/m^2 for four different mass fluxes in the microchannels of width 400 μm. It can be seen that the intermittent flow formed for the mass fluxes of 225 and 630 kg/m^2 s, while the flow is still in the bubbly regime for the mass fluxes of 1050 and 1420 kg/m^2 s.

Hetsroni et al. [63] investigated flow boiling of water in the silicon triangular microchannels having hydraulic diameters of 103 and 129 μm. They observed periodic annular flow and the periodic dry steam flow (periodic wetting and rewetting phenomena). This explosive boiling was triggered by venting of elongated bubble due to very rapid expansion. They also observed dry-out in the channel. In the unstable flow boiling mode, Wang et al. [71,72] observed elongated bubbly/slug flow

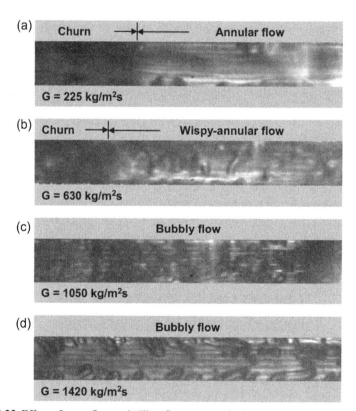

Figure 3.23 Effect of mass flux on boiling flow patterns in the 400 × 400 μm microchannels at heat flux of 145 kW/m^2 by Harirchia and Garimella [70]. (a) G = 225 kg/m^{2}s; (b) G = 630 kg/m^{2}s; (c) G = 1050 kg/m^{2}s and (d) G = 1420 kg/m^{2}s.

pattern. The flow pattern also behaved like an annular or semiannular flow, in which the thin liquid film evaporated between the vapor core and the heating wall. When long bubbles occupied the channel, they blocked the two-phase flow and caused higher pressure drop. Later, it was observed that upstream vapor plug broke through the liquid front, reaching the long bubble downstream. Local dry-out occurred due to the depletion of liquid film between vapor core and heating wall. Subsequent to this, annular/mist flow (alternating dry-out and rewetting phenomena) with imminent burnout was observed. Average wall temperature in bubbly/elongated bubbly/slug flow was lower than that in the semiannular flow and annular/mist flow.

It is obvious that the effects of surface tension are much more significant in microchannels than those present in macrochannels. Both the effects of surface tension and the effects of inertia have to be taken into account in presenting flow patterns in microchannels. The flow patterns, which are mainly surface-tension dominated, are characterized by large and elongated bubbles of gas, such as the bubbly, plug, and slug flows. Also, the flow patterns that are inertia-dominated include annular flows and dispersed flows. Tabatabai and Faghri [75] suggest that only patterns dominated by surface tension (bubble and slug) occur in tubes of 100 μm diameter or less. However, when the bubble size approaches the channel size, the result is the annular flow. The most difficult flow patterns are transition flows. On the one hand, there is the scarcity of data related to these transition flow patterns. On the other hand, the definitions of flow patterns are different from one research to another research [74,77].

In general, flow patterns in microchannels have not yet well understood although many studies have been conducted. Well-documented theoretically based flow pattern transition criteria and flow maps for micro-scale channels have not yet been established. Further efforts should be made to develop a generalized flow pattern map for microchannels, which may be applied to a wide range of conditions and fluids. The flow patterns observed during an experiment depends on several factors, such as flow rates, properties of the fluids used, channel inclination with respect to the horizontal, and channel geometry and size. As a general rule, the primary flow regimes observed in a pipe of horizontal inclination and large hydraulic diameter are bubbly flow, dispersed flow, stratified flow, slug flow, annular flow, and churn flow. These flow patterns have all been observed in microchannels, depending on the experiment, except for stratified flow, which, so far, has not been seen in any microchannel experiment. This exception can be explained by the suppression of the buoyancy occurring in a microchannel. Although the flow patterns are classified under the same name, some slight differences in the patterns can be observed between the macrochannels and the microchannels [74,77].

Most researchers constructed their flow pattern maps according to only their own experimental data, and hence these maps are only applicable to those specific conditions and fluids. Some researchers modified the generalized flow maps for macro-scale such as the Taitel and Dukler map [38] or others according to their own data. Some researchers simply compared their flow pattern data to the existing generalized flow pattern maps for macro-scale channels without further modifying these flow maps for micro-scale channels.

3.4.2   Proposed Flow Pattern Maps in Microchannels

As mentioned, it is necessary to use dimensionless numbers to construct a generalized flow pattern map. Several flow pattern maps have been developed in such a method as the one shown in Fig. 3.13.

Tabatabai and Faghri [75] proposed a flow pattern map to emphasize the importance of surface tension in two-phase flow in horizontal miniature and micro tubes. In fact, their map is a modified version of the Taitel and Dukler [38] map that incorporates the surface tension effect on the flow pattern transitions. So far, there is no independent experimental validation of observations for the same fluids under the same test conditions taken by different researchers.

In addition, the physical properties of fluids have a great effect on the flow regimes. For example, flow patterns of CO_2 at high reduced pressure are quite different from those of other refrigerants such as R22, R134a, and R410A. Thus, flow pattern maps developed from these fluids do not extrapolate well to CO_2 [33−36,39,40]. In addition, micro-scale channels often have noncircular shapes, such as triangular, square, rectangular, etc., and may be single- or multi-channel test sections, which greatly affect the flow patterns.

Ullmann and Brauner [76] studied the effect of the channel diameter on the mechanisms leading to flow pattern transitions. They proposed mechanistic models for adiabatic conditions and compared these with experimental maps from the literature. Their models indicate the controlling dimensionless groups and the critical values associated with various flow pattern transitions. With reducing the pipe diameter, the stratified flow region shrinks greatly and is limited to only a small region at very low liquid flow rates and relatively high gas flow rates. In the range where stratified flow may still exist, analysis of the predicted flow structure indicates that the distinction between stratified flow (curved interface) and annular flow is ambiguous. From a practical point of view of the transport phenomena involved, the flow structure can be considered as annular flow. Figure 3.24 shows the comparison of the predicted flow regime boundaries of the Ullmann and Brauner [76] flow pattern map to the horizontal 1-mm tube experimental data of Triplett et al. [47].

It should be remembered that diabatic flow boiling patterns are different from the adiabatic flow patterns. For example, high heat flux applications, such as cooling of microprocessors, bring in the onset of CHF through its corresponding critical vapor quality as an important flow pattern map transition. However, such a map has not yet been well developed [74,77].

Harirchian and Garimella [32] proposed a new criterion for physical confinement in microchannel flow boiling, termed the convective confinement number Eq. (3.4), which incorporates the effects of mass flux, as well as channel cross-sectional area and fluid properties. Furthermore, they developed a new comprehensive flow regime map for a wide range of experimental parameters and channel dimensions, along with quantitative transition criteria based on nondimensional boiling parameters. Figure 3.25 shows their comprehensive flow map developed according to the experimental results and flow visualizations performed with FC-77. The abscissa in this plot is the convective confinement number. The ordinate is a nondimensional form of the

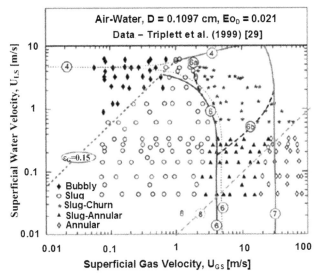

Figure 3.24 Comparison of the predicted flow regime boundaries of the Ullmann and Brauner [76] flow map in a horizontal 1 mm tube to the experimental data of Triplett et al. [29], where ε_G is the cross-sectional void fraction and $E\ddot{o}_D$ is Eotvös number.

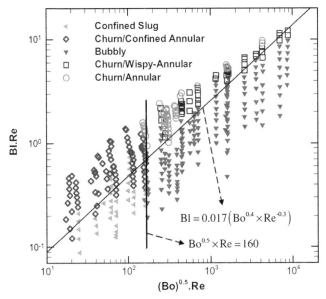

Figure 3.25 Comprehensive flow regime map for FC-77 by Harirchian and Garimella [32].

heat flux, $Bl \times Re$, which is proportional to $q_w \times D$. Their comprehensive flow map contains four distinct regions of confined slug flow, churn/confined annular flow, bubbly flow, and churn/annular/wispy-annular flow. The vertical transition line is given by Eq. (3.4), which represents the transition to confined flow. The other transition line is a curve fit to the points of transition from bubbly or slug flow to alternating churn/annular or churn/wispy-annular flow, given by

$$Bl = 0.017 \left(Bo^{0.4} \times Re^{-0.3} \right) \tag{3.9}$$

They compared experimental data from a range of other studies in the literature [50,63,71,94−97] to the comprehensive flow pattern map. Their flow map is able to represent the flow patterns for water and fluorocarbon liquids. Further validation of this map is needed with more relevant experimental data in multiple microchannels. It is not clear if this flow map works for single microchannels. This, it is of interest to validate this comprehensive map with experimental data in single microchannels.

Overall, generalized flow pattern has not yet developed. Nearly all the available flow maps have been proposed based on limited data. It is important to represent the data with proper dimensionless numbers. Furthermore, development of generalized flow map should be targeted in future [77].

3.5 Current Research Progress on Bubble Growth in Microchannels

Bubble dynamics is very important to understand the flow pattern evolution and transition mechanism during flow boiling in microchannels. In a homogeneous medium at macroscale, the bubble dynamics is governed by the well-known Rayleigh equation or its extended form. Plesset and Zwick [98] and Forster and Zuber [99] investigated the bubble growth and dynamics in boiling process and presented solutions for the time evolution of bubble radius. For the very early stage, bubble growth is limited by the inertial force. The bubble grows linearly with time with the proportional constant directly related to the square root of over-pressure divided by liquid density. However, bubble growth in microchannels is less investigated so far. A number of researchers observed bubbly flow pattern for microchannels while several researchers did not observe bubbly flow in their studies. For instance, Jiang et al. [100] performed visualization and measurements of flow boiling in silicon microchannels with triangular cross-sections for two different hydraulic diameters of 40 and 80 μm. They did not observe bubbly flow. Zhang et al. [101] reported mostly annular flow with a very thin layer of liquid on the channel walls and did not observe bubbly and slug flows. The reason why bubbly flow was not observed is that the bubbles have a very short lifespan as they either coalesce or grow to the channel size very quickly [22]. In order to understand flow pattern evolution and mechanisms in microchannels, it is essential to understand bubble growth in microchannels. In this section, the current research status on bubble growth in microchannels is reviewed. Table 3.2 lists several selected studies on bubble growth during flow boiling.

Table 3.2 Selected Studies on Bubble Growth and Bubble Dynamics in Microchannels in the Literature

Authors/ References	Fluids	Test Channel Diameter(s) and Orientations	Main Research Contents
Tibiriçá and Ribatski [68]	R134a and R245fa evaporation	Circular tube, 0.4 mm, horizontal.	Bubble departure diameter and frequency, and bubble growth were observed.
Li and Cheng [78]	Water evaporation	Microchannels in general.	The effects of microchannel size, mass flow rate, and heat flux on boiling incipience or bubble cavitation in a microchannel.
Balasubramanian, and Kandlikar [79]	Water evaporation	6 Parallel rectangular microchannels with a hydraulic diameter of 333 μm, horizontal.	The occurrence of thin film nucleate boiling, bubble nucleation and its subsequent growth into a slug in a channel.
Lee et al. [80]	Water evaporation	A single trapezoid microchannel with a hydraulic diameter of 41.3 μm, horizontal.	Bubble nucleation, growth, departure size, and frequency are observed using a high speed digital camera and analyzed.
Lee et al. [81]	Water evaporation	Two parallel trapezoidal micro channels with a hydraulic diameter of 47.7 μm for both channels, horizontal.	Bubble nucleation, growth, departure size, and frequency are observed using a high speed digital camera and analyzed.
Li and Peterson [82]	Water evaporation	Trapezoidal microchannel with a hydraulic diameter of 56 μm, horizontal.	The boiling nucleation temperature and two-phase flow patterns were observed and examined at different mass flow rates.

Continued

Table 3.2 **Selected Studies on Bubble Growth and Bubble Dynamics in Microchannels in the Literature—cont'd**

Authors/ References	Fluids	Test Channel Diameter(s) and Orientations	Main Research Contents
Cooke and Kandlikar [83]	Water evaporation	Five silicon chips with rectangular microchannels, horizontal.	The bubble nucleation and growth on the microchannel surfaces were investigated with high speed camera. Onset of nucleation, bubble growth and detachment were observed and analyzed by microscopic high-speed visualization under various flow rates and wall superheat conditions.
Lee et al. [84]	Water evaporation	Square or rectangular channels with width of 50 and 100 μm and height 46, 48, and 100 μm.	Onset of nucleation, bubble growth and detachment were observed and analyzed by microscopic high-speed visualization under various flow rates and wall superheat conditions.
Wang et al. [85] and Wang and Sefiane [86]	FC-72 and ethanol evaporation	Rectangular channels with hydraulic diameters of 571, 762, and 1454 μm, horizontal.	Bubble growth and flow patterns were observed using high-speed camera.
Bogojevic et al. [87]	Water evaporation	Microchannel heat sink contained 40 parallel, rectangular channels having hydraulic diameter of 194 μm.	Bubble growth rate and departure diameter under subcooled and saturated flow boiling are studied using a microscope and a high speed camera.

Table 3.2 **Selected Studies on Bubble Growth and Bubble Dynamics in Microchannels in the Literature—cont'd**

Authors/ References	Fluids	Test Channel Diameter(s) and Orientations	Main Research Contents
Yin et al. [88]	Water evaporation	Rectangular microchannel with 0.5 mm in width and 1.0 mm in height, horizontal.	Bubble growth under various mass flux, heat flux and inlet subcooling conditions was visualized using a high-speed CCD camera.
Wang and Sefiane [89]	FC-72	Rectangular channels with hydraulic diameters of 571, 762, and 1454 μm, horizontal.	Bubble growth and flow patterns were observed using high-speed camera.
Barber et al. [90]	n-Pentane evaporation of a single rectangular microchannel of hydraulic diameter 771 μm, horizontal	A single rectangular microchannel of a hydraulic diameter 771 μm, horizontal.	Bubble growth was investigated.

Lee et al. [80,81] documented the growth rate of bubbles in their microchannels. They found that the bubbles grow linearly with time and that the classical model is able to capture this behavior. The bubble departure from the wall is governed by the two opposing forces due to surface tension and drag force by the bulk flow.

Hetsroni et al. [63] studied bubble growth and temporal variation of bubble size in flow boiling of water in triangular silicon microchannels with hydraulic diameters of 103, 129, and 161 μm. A sequence of images showing bubble growth from their work is presented in Fig. 3.26. Figure 3.26(a) and (b) illustrate the bubble shape at different instants of time during growth and motion. It is the top view, observed through the transparent cover. The field of view is 2.4 mm in the streamwise direction and 2.2 mm in the spanwise direction, the flow moves from left to right. In these images our microchannels are shown marked by gray color or by gray color with the light regions. One can see that the vapor is generated in one microchannel only. Figure 3.26(a) shows the incipience of the bubble. The bubble is approximately spherical and occupies a small part of the triangular microchannel. During 0.001 s it grows and

(a)   (b)

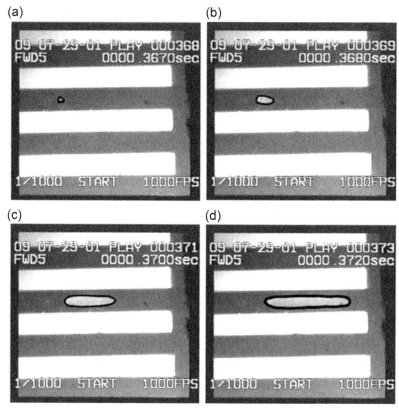

(c)   (d)

Figure 3.26 Bubble growth (superficial liquid velocity = 0.046 m/s, $q = 80$ kW/m^2) by Hetsroni et al. [63].

occupies about 0.3 of the cross section (Figure 3.26(b)). At that point, the bubble grows preferential in the axial direction (Figure 3.26(c)). In Figure 3.26(c), the bubble occupies about 0.7 of the cross section. Then the bubble moves into the exit manifold (Figure 3.26(d)).

Balasubramanian and Kandlikar [79] studied the growth of bubbles, formation of slugs, and periodic wetting and rewetting of the walls. They calculated the velocity of the liquid−vapor interface associated with bubble growth from the image sequences. The velocity was very high, in the order of 3.5 m/s. As the bubble nucleated and remained confined to the wall, it experienced a sudden reduction in its growth rate. After the bubble became confined to the channel wall, it started to grow toward the sides at a very high velocity. An increase in velocity at this stage was due to an increase in the evaporation rate of the bubble as it got closer to the channel walls.

Celata et al. [61] observed elongated bubbles formed by the coalescence of bubbles during their travel from the nucleation site toward the exit of the heated channel as shown in Fig. 3.27. During the motion of trains of bubbles, the larger bubbles

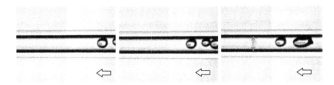

Figure 3.27 Bubbles coalescence in the 0.48 mm tube. The horizontal arrows indicate the direction of the flow, while the vertical double arrow indicates the distance of 0.48 mm by Celata et al. [67].

coming upward have the tendency to coalesce with the downward bubbles. This phenomenon occurs with very few bubbles flowing in the channel. The presence of the wake behind a bubble allows the coming bubbles to reduce the resistance of the motion and the coalescence has higher probability to occur. With this mechanism, the bubbles become larger during their motion from the nucleation site towards the channel exit, and as the bubble diameter reaches the channel diameter, the bubble starts to grow in the axial direction. With smaller-diameter channels, the number of coalesced bubbles needed to form a Taylor bubble is smaller than that in a larger tube. Once that the elongated bubbles are formed and flow in the channel, their length is increased further, by the evaporation of the liquid film between the bubble and the heated wall. Bubbles become very long and sometimes intermittent flow is identified as slug/annular flow. This mechanism has been observed many times during the video analysis of high speed movies and is considered the main cause for the formation of elongated bubbles. There is another mechanism that increases the bubble dimensions: it is the vaporization of the superheated liquid layer attached to the heated wall. The observations of the bubbles growing on their nucleation sites reveal that after detachment some bubbles slide attached to the heated wall increasing their volume.

Yin et al. [88] investigated a complete bubble growth process in the horizontally oriented microchannel, including the bubble nucleation, confinement, and elongation, is shown in Fig. 3.28. In the microchannel, working fluid flowed from right

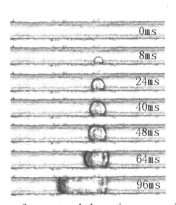

Figure 3.28 Typical bubble confinement and elongation process by Yin et al. [88].

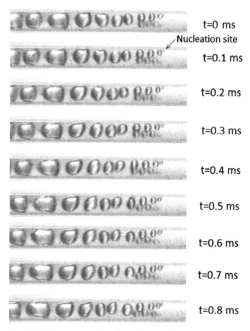

t=0 ms

Nucleation site

t=0.1 ms

t=0.2 ms

t=0.3 ms

t=0.4 ms

t=0.5 ms

t=0.6 ms

t=0.7 ms

t=0.8 ms

Figure 3.29 Growth of a bubble in a nucleation site. R134a, $D = 0.40$ mm, $T_{sat} = 31$ °C, $q_{est,fpt} = 75$ kW/m^2 by Tibiriçá and Ribatski [68].

to left and the bubble nucleated in the channel corner. It is noted that after nucleation the bubble remains its spherical shape until $t = 40$ ms. Bubble growth is constrained by the channel cross-section once the top of bubble approaches the channel wall, then its shape elongates along the channel flow direction (from $t = 48$ to 96 ms).

Tibiriçá and Ribatski [68] experimental investigation into the fundamental characteristics of flow boiling in micro-scale channels based on diabatic high-speed flow visualizations. They concluded from their study that (1) bubbles can detach from the wall with diameters much smaller than the tube diameter; (2) the bubble growth process has a square root time-dependence; and (3) bubble active nucleation sites are observed for all flow patterns. Figure 3.29 shows a sequence of images captured by the high-speed camera from a nucleation site with intervals of 0.1 ms. According to Fig. 3.29, the bubble nucleating on the site on the top right side of the tube (indicated by an arrow) has a life-time (waiting plus growing periods) of approximately 0.5 ms and a departure diameter of 80 µm.

Wang et al. [71] observed high-speed images of bubble nucleation process were obtained for different mass velocities, tube diameters, and fluids. Figure 3.30(a) shows that flow boiling exhibited nucleate boiling characteristics with small isolated bubbles at a negative local vapor quality ($x_e = -0.034$). Figure 3.30(b) shows that when the local vapor quality was nearly equal to zero ($x_e = 0.005$), the bubbles

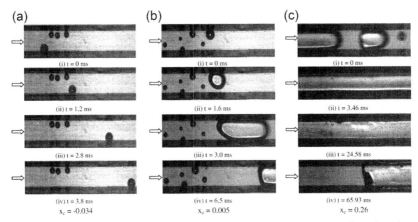

Figure 3.30 Photos of steady flow boiling patterns near outlet section of parallel microchannels ($D_h = 186$ μm) with the Type-C connection at $q = 364.68$ kW/m^2 and $T_{in} = 35$ °C: (a) $G = 682.11$ kg/m^2 s (i.e., $x_e = -0.034$), (b) $G = 471.32$ kg/m^2 s (i.e., $x_e = 0.005$), (c) $G = 156.04$ kg/m^2 s (i.e., $x_e = 0.26$), by Wang et al. [71].

filled the entire cross-section and then were flushed out of the channels. This resulted in the maximum value of heat transfer coefficient, as discussed below. Figure 3.30(c) shows that at an exit vapor quality of $x_e = 0.26$, local dry-out was observed instantaneously from $t = 24.6$ to 65.9 ms, and the lifetime of the dry-out increased with the increasing vapor quality. It can be speculated that the liquid film in the Taylor bubbles ruptures easily, causing dry-out in microchannels.

In summary, bubble growth from a heated wall is much more complicated with the presence of thermal boundary layer near the wall. Prior to the bubble growth, the bubble may stay on a cavity mouth waiting for the establishment of thermal boundary layer such that the liquid temperature at the bubble tip is higher than the vapor temperature inside the bubble. The bubble will then grow rapidly once such a bubble growth criterion is satisfied. The evaporation rate at the bubble interface will be nonuniform: large near the heating wall and small near the bubble tip. Moreover, in an ordinary sized channel, the rapid initial growth of bubble may create a microlayer between the bubble and the heating wall. The evaporation of microlayer has been demonstrated to have a significant effect on bubble growth from a heated wall. A bubble will depart from the heated wall after it grows to a certain size. For flow boiling in an ordinary sized channel, the major forces acting on a bubble includes buoyancy force, inertial force, surface tension, and drag due to bulk flow in the channel. Surface tension and inertial force tend to retain the bubble to the wall, while the drag and buoyancy forces tend to detach the bubble from the wall. In microchannels, it seems that the similar processes occur in the initial stage as that in macrochannel. However, when the bubble becomes larger than the channel diameter, it will be confined. The bubble coalesces is also important to understand microchannel but less information is available. Furthermore, systematic theoretical knowledge should be advanced.

3.6   Concluding Remarks

There remain many challenges in fundamental understanding of flow patterns and bubble growth in gas liquid two-phase flow in microchannels. Considerable experimental and theoretical research is still needed before reliable design tools become available. For instance, reliable prediction methods for flow boiling heat transfer in microchannels based on flow patterns and heat transfer mechanisms are needed as described in Chapter 4. Furthermore, the heat transfer mechanisms strongly depend on the flow patterns affected by the channel size in microchannels. Although simplified flow boiling heat transfer correlations can be obtained from the experimental data ignoring the details of the flow patterns in microchannels as described in Chapter 4, knowing the flow patterns allows more suitable assumptions to be applied for each particular flow pattern, leading to more accurate prediction methods. Therefore, methods for predicting the occurrence of the major two-phase flow patterns in microchannel are useful.

Current methods for predicting the flow patterns in microchannels are far from perfect. The difficulty and challenges arise out of the extremely varied morphological configurations, especially for flow boiling involving bubble growth which can be significantly affected by the channel confinement while the available experimental data show very different results from one study to another.

In general, flow patterns in microchannels have not yet completely understood although many studies have been conducted. Very different flow patterns have been observed by different researchers. Furthermore, well documented theoretically based flow pattern transition criteria and flow maps for micro-scale channels have not yet been established. Most researchers constructed their flow pattern maps according to only their own experimental data and hence these maps are only applicable to those specific conditions and fluids. Some researchers modified the generalized flow maps for macro-scale such as the Taitel and Dukler map or others according to their own data. Some researchers simply compared their flow pattern data to the existing generalized flow pattern maps for macro-scale channels without further modifying these flow maps for micro-scale channels. So far, no generalized flow pattern maps are available for microchannels.

Diabatic flow boiling patterns are different from the adiabatic flow patterns. For example, high heat flux applications, such as cooling of microprocessors, bring in the onset of CHF through its corresponding critical vapor quality as an important flow pattern map transition. However, such a map has not yet been well developed. Furthermore, unstable flow boiling may significantly affect the flow patterns and their transitions which are intrinsically related to bubble formation, growth and coalescing. Liquid film plays a major role in understanding both the bubble and flow pattern evolution in microchannel as bubble may also grow from the liquid film and affect the interface between the bubble and the liquid film. However, such a topic has been less investigated.

Bubble dynamics is the fundamental to understanding the flow pattern evolution and transition mechanism during flow boiling in microchannels. Similar bubble growth processes occur in the initial stage as that in macrochannel. However, the confinement effect on the bubble coalescing has not fully understood. In particular, the bubble dynamics at stable and unstable flow boiling conditions may also significantly affect the flow pattern involution but are seldom considered. Furthermore, systematic theoretical knowledge of bubble dynamics including bubble growth and coalescing in microchannels should be developed in future.

References

[1] L. Cheng, Microscale and nanoscale thermal and fluid transport phenomena: rapidly developing research fields, Int. J. Microscale Nanoscale Therm. Fluid Transp. Phenom. 1 (2010) 3–6.

[2] G.P. Celata, Microscale heat transfer in single- and two-phase flows: scaling, stability and transition, Int. J. Microscale Nanoscale Therm. Fluid Transp. Phenom. 1 (2010) 7–36.

[3] S.K. Saha, G. Zummo, G.P. Celata, Review on flow boiling in microchannels, Int. J. Microscale Nanoscale Therm. Fluid Transp. Phenom. 1 (2010) 111–178.

[4] J.R. Thome, The new frontier in heat transfer: microscale and nanoscale technologies, Heat Transfer Eng. 27 (9) (2006) 1–3.

[5] S.G. Kandlikar, Fundamental issues related to flow boiling in minichannels and microchannels, Exp. Therm. Fluid Sci. 26 (2002) 389–407.

[6] J.R. Thome, State-of-the art overview of boiling and two-phase flows in microchannels, Heat Transfer Eng. 27 (9) (2006) 4–19.

[7] L. Cheng, E.P. Bandarra Filho, J.R. Thome, Nanofluid two-phase flow and thermal physics: a new research frontier of nanotechnology and its challenges, J. Nanosci. Nanotechol. 8 (2008) 3315–3332.

[8] L. Cheng, L. Liu, Boiling and two phase flow phenomena of refrigerant-based nanofluids: fundamentals, applications and challenges, Int. J. Refrig. 36 (2013) 421–446.

[9] L. Cheng, Nanofluid heat transfer technologies, Recent Patents Eng. 3 (1) (2009) 1–7.

[10] L. Cheng, Fundamental issues of critical heat flux phenomena during flow boiling in microscale-channels and nucleate pool boiling in confined spaces, Heat Transfer Eng. 34 (2013) 1011–1043.

[11] J.R. Thome, Boiling in microchannels: a review of experiment and theory, Int. J. Heat Fluid Flow 25 (2004) 128–139.

[12] G. Ribatski, A critical overview on the recent literature concerning flow boiling and two-phase flows inside micro-scale channels, Exp. Heat Transfer 26 (2013) 198–246.

[13] L. Cheng, G. Ribatski, J.R. Thome, Gas-liquid two-phase flow patterns and flow pattern maps: fundamentals and applications, ASME Appl. Mech. Rev. 61 (2008) 050802.

[14] L. Cheng, D. Mewes, Review of two-phase flow and flow boiling of mixtures in small and mini channels, Int. J. Multiphase Flow 32 (2006) 183–207.

[15] G.P. Celata, S.K. Suha, G. Zummeo, Heat transfer characteristics of flow boiling in a single horizontal microchannel, Int. J. Therm. Sci. 49 (2010) 1086–1094.

[16] L. Cheng, Critical heat flux in microscale channels and confined Spaces: a review on experimental studies and prediction methods, Russ. J. Gen. Chem. 82 (12) (2012) 2116–2131.

[17] L. Cheng, J.R. Thome, Cooling of microprocessors using flow boiling of CO_2 in a micro-evaporator: preliminary analysis and performance comparison, Appl. Therm. Eng. 29 (2009) 2426−2432.

[18] C.B. Tibiriçá, G. Ribatski, An experimental study on flow boiling heat transfer of R134a in a 2.3 mm Tube, Int. J. Microscale Nanoscale Therm. Fluid Transp. Phenom. 1 (2010) 37−58.

[19] L. Cheng, L. Liu, Analysis and evaluation of gas-liquid two-phase frictional pressure drop prediction methods for microscale channels, Int. J. Microscale Nanoscale Therm. Fluid Transp. Phenom. 2 (2011) 259−280.

[20] L. Cheng, H. Zou, Evaluation of flow boiling heat transfer correlations with experimental data of R134a, R22, R410A and R245fa in microscale channels, Int. J. Microscale Nanoscale Therm. Fluid Transp. Phenom. 1 (2010) 363−380.

[21] L. Cheng, D. Mewes, A. Luke, Boiling phenomena with surfactants and polymeric additives: a state-of-the-art review, Int. J. Heat Mass Transfer 50 (2007) 2744−2771.

[22] G. Ribatski, L. Wojtan, J.R. Thome, An analysis of experimental data and prediction methods for two-phase frictional pressure drop and flow boiling heat transfer in microscale channels, Exp. Therm. Fluid Sci. 31 (2006) 1−19.

[23] J. Moreno Quibén, L. Cheng, R.J. da Silva Lima, J.R. Thome, Flow boiling in horizontal flattened tubes: part I—two-phase frictional pressure drop results and model, Int. J. Heat Mass Transfer 52 (2009) 3634−3644.

[24] J. Moreno Quibén, L. Cheng, R.J. da Silva Lima, J.R. Thome, Flow boiling in horizontal flattened tubes: part II—flow boiling heat transfer results and model, Int. J. Heat Mass Transfer 52 (2009) 3645−3653.

[25] J.G. Collier, J.R. Thome, Convective Boiling and Condensation, Oxford University Press Inc., New York, 1994.

[26] V.P. Carey, Liquid Vapor Phase Change Phenomena, Hemisphere, Washington, DC, 1992.

[27] R.K. Shah, Classification of heat exchangers, in: S. Kakac, A.E. Bergles, F. Mayinger (Eds.), Heat Exchangers: Thermal Hydraulic Fundamentals and Design, Hemisphere Publishing Corp., Washington, DC, 1986, pp. 9−46.

[28] S.S. Mehendale, A.M. Jacobi, R.K. Shah, Fluid flow and heat transfer at micro- and meso-scales with application to heat exchanger design, Appl. Mech. Rev. 53 (2000) 175−193.

[29] K.A. Triplett, S.M. Ghiaasiaan, S.I. Abdel-Khalik, D.L. Sadowski, Gas-liquid two-phase flow in microchannels. Part I: two-phase flow patterns, Int. J. Multiphase Flow 25 (1999) 377−394.

[30] P.A. Kew, K. Cornwell, Correlations for the prediction of boiling heat transfer in small-diameter channels, Appl. Therm. Eng. 17 (1997) 705−715.

[31] N. Brauner, D. Moalem-Maron, Identification of the range of small diameter conduits regarding two-phase flow pattern transitions, Int. Commun. Heat Mass Transfer 19 (1992) 29−39.

[32] T. Harirchian, S.V. Garimella, A comprehensive flow regime map for microchannel flow boiling with quantitative transition criteria, Int. J. Heat Mass Transfer 53 (2010) 2694−2702.

[33] C.L. Ong, J.R. Thome, Macro-to-microchannel transition in two-phase flow: part 1 − two-phase flow patterns and film thickness measurements, Exp. Therm. Fluid Sci. 35 (2011) 37−47.

[34] L. Cheng, G. Ribatski, L. Wojtan, J.R. Thome, New flow boiling heat transfer model and flow pattern map for carbon dioxide evaporating inside horizontal tubes, Int. J. Heat Mass Transfer 49 (2006) 4082–4094.

[35] L. Cheng, G. Ribatski, J. Quibén Moreno, J.R. Thome, New prediction methods for CO_2 evaporation inside tubes: part I – a two-phase flow pattern map and a flow pattern based phenomenological model for two-phase flow frictional pressure drops, Int. J. Heat Mass Transfer 51 (2008) 111–124.

[36] L. Cheng, G. Ribatski, J.R. Thome, New prediction methods for CO_2 evaporation inside tubes: part II – an updated general flow boiling heat transfer model based on flow patterns, Int. J. Heat Mass Transfer 51 (2008) 125–135.

[37] G.F. Hewitt, D.N. Roberts, Studies of Two-Phase Flow Patterns by Simultaneous X-ray and Flash Photography, Report AERE-M 2159, Atomic Energy Research Establishment, Harwell, 1969.

[38] Y. Taitel, A.E. Dukler, A model for predicting flow regime transitions in horizontal and near horizontal gas-liquid flow, AIChE J. 22 (1976) 47–55.

[39] L. Wojtan, T. Ursenbacher, J.R. Thome, Investigation of flow boiling in horizontal tubes: part I – a new diabatic two-phase flow pattern map, Int. J. Heat Mass Transfer 48 (2005) 2955–2969.

[40] L. Wojtan, T. Ursenbacher, J.R. Thome, Investigation of flow boiling in horizontal tubes: part II – development of a new heat transfer model for stratified-wavy, dryout and mist flow regimes, Int. J. Heat Mass Transfer 48 (2005) 2970–2985.

[41] R. Yun, Y. Kim, M.S. Kim, Flow boiling heat transfer of carbon dioxide in horizontal mini tubes, Int. J. Heat Fluid Flow 26 (2005) 801–809.

[42] J.L. Gasche, Carbon dioxide evaporation in a single micro-channel, J. Brazil Soc. Mech. Sci. Eng. 28 (1) (2006) 69–83.

[43] S.-M. Kim, I. Mudawar, Review of databases and predictive methods for heat transfer in condensing and boiling mini/micro-channel flows, Int. J. Heat Mass Transfer 77 (2014) 627–652.

[44] M. Suo, P. Griffith, Two-phase flow in capillary tubes, J. Basic Eng. 86 (1964) 576–582.

[45] R. Revellin, V. Dupont, T. Ursenbacher, J.R. Thome, I. Zun, Characterization of diabatic two-phase flows in microchannels: flow parameter results for R-134a in a 0.5 mm channel, Int. J. Multiphase Flow 32 (2006) 755–774.

[46] R. Revellin, J.R. Thome, Experimental investigation of R-134a and R-245fa two-phase flow in microchannels for different flow conditions, Int. J. Heat Fluid Flow 28 (2007) 63–71.

[47] R. Revellin, J.R. Thome, New type of diabatic flow pattern map for boiling heat transfer in microchannels, J. Micromech. Microeng. 17 (2007) 788–796.

[48] T. Cubaud, C.-M. Ho, Transport of bubbles in square microchannels, Phys. Fluids 16 (2004) 4575–4585.

[49] J.W. Coleman, S. Garimella, Characterization of two-phase flow patterns in small diameter round and rectangular tubes, Int. J. Heat Mass Transfer 42 (1999) 2869–2881.

[50] T. Chen, S.V. Garimella, Measurements and high-speed visualizations of flow boiling of a dielectric fluids in a silicon microchannel heat sink, Int. J. Multiphase Flow 32 (2006) 957–971.

[51] T.S. Zhao, Q.C. Bi, Co-current air-water two-phase flow patterns in vertical triangular microchannels, Int. J. Multiphase Flow 27 (2001) 765–782.

[52]   J. Pettersen, Flow vaporization of CO_2 in microchannel tubes, Exp. Therm. Fluid Sci.
       28 (2004) 111–121.
[53]   B. Lowry, M. Kawaji, Adiabatic vertical two-phase flow in narrow flow channels,
       AIChE Sym. Ser. Heat Transfer – Houston (1988) 133–139.
[54]   C.A. Damianides, J.W. Westwater, Two-phase flow patterns in a compact heat
       exchanger and in small tubes, in: Proc. 2nd U.K. National Conf. on Heat Transfer, vol.
       2, 1988, pp. 1257–1268.
[55]   D. Liu, S. Wang, Flow pattern and pressure drop of upward two-phase flow in vertical
       capillaries, Ind. Eng. Chem. Res. 47 (2008) 243–255.
[56]   L. Chen, Y.S. Tian, T.G. Karayiannis, The effect of tube diameter on vertical two-phase
       flow regimes in small tubes, Int. J. Heat Mass Transfer 49 (2006) 4220–4230.
[57]   C.Y. Yang, C.C. Shieh, Flow pattern of air-water and two-phase R-134a in small
       circular tubes, Int. J. Multiphase Flow 27 (2011) 1163–1177.
[58]   W. Owhaib, B. Palm, C. Martin-Callizo, Flow boiling visualization in a vertical circular
       minichannel at high vapor quality, Exp. Therm. Fluid Sci. 30 (2006) 755–763.
[59]   E. Sobierska, R. Kulenovic, R. Mertz, M. Groll, Experimental results of flow boiling of
       water in a vertical microchannels, Exp. Therm. Fluid Sci. 31 (2007) 111–119.
[60]   T.H. Yen, M. Shoji, F. Takemura, Y. Suzuki, N. Kasage, Visualization of convective
       boiling heat transfer in single microchannels with different shaped cross-sections, Int.
       J. Heat Mass Transfer 49 (2006) 3884–3894.
[61]   A. Serizawa, Z. Feng, Z. Kawara, Two-phase flow in microchannels, Exp. Therm. Fluid
       Sci. 26 (2002) 703–714.
[62]   H. Ide, A. Kariyasaki, T. Fukano, Fundamental data on the gas-liquid two-phase flow in
       microchannels, Int. J. Therm. Sci. 46 (2007) 519–530.
[63]   G. Hetsroni, A. Mosyak, Z. Segal, E. Pogrebnyak, Two-phase flow patterns in parallel
       micro-channels, Int. J. Multiphase Flow 29 (2003) 341–360.
[64]   A. Kawahara, P.M.-Y. Chung, M. Kawaji, Investigation of two-phase flow pattern, void
       fraction and pressure drop in microchannel, Int. J. Multiphase Flow 28 (2002)
       1411–1435.
[65]   P.M.-Y. Chung, M. Kawaji, The effect of channel diameter on adiabatic two-phase flow
       characteristics in microchannels, Int. J. Multiphase Flow 30 (2004) 735–761.
[66]   T. Fukano, A. Kariyasaki, Characteristics of gas-liquid two-phase flow in a capillary
       tube, Nucl. Eng. Design 141 (1993) 59–68.
[67]   G.P. Celata, M. Cumo, D. Dossevi, R.T.M. Jilisen, S.K. Saha, G. Zummo, Flow pattern
       analysis of flow boiling inside a 0.48 mm microtube, Int. J. Therm. Sci. 58 (2012) 1–8.
[68]   C.B. Tibiriçá, G. Ribatski, Flow patterns and bubble departure fundamental charac-
       teristics during flow boiling in microscale channels, Exp. Therm. Fluid Sci. 59 (2014)
       152–165.
[69]   A. Sur, D. Liu, Adiabatic air-water two-phase flow in circular microchannels, Int.
       J. Therm. Sci. 53 (2012) 18–34.
[70]   T. Harirchian, S.V. Garimella, Effects of channel dimension, heat flux, and mass flux
       on flow boiling regimes in microchannels, Int. J. Multiphase Flow 35 (2009)
       349–362.
[71]   G. Wang, P. Cheng, A.E. Bergles, Effects of inlet/outlet configurations on flow boiling
       instability in parallel microchannels, Int. J. Heat Mass Transfer 51 (2008) 2267–2281.
[72]   G.D. Wang, P. Cheng, H.Y. Wu, Unstable and stable flow boiling in parallel micro-
       channels and in a single microchannel, Int. J. Heat Mass Transfer 50 (2007)
       4297–4310.

[73] H. Tuo, P. Hrnjak, Visualization and measurement of periodic reverse flow and boiling fluctuations in a microchannel evaporator of an air-conditioning system, Int. J. Heat Mass Transfer 71 (2014) 639−652.

[74] W.-C. Liu, C.-Y. Yang, Two-phase flow visualization and heat transfer performance of convective boiling in micro heat exchangers, Exp. Therm. Fluid Sci. 57 (2014) 358−364.

[75] A. Tabatabai, A. Faghri, A new two-phase flow map and transition boundary accounting for surface tension effects in horizontal miniature and micro tubes, J. Heat Transfer 123 (2001) 958−968.

[76] A. Ullmann, N. Brauner, The prediction of flow pattern maps in microchannels, Multiphase Sci. Tech. 19 (1) (2007) 49−73.

[77] J.R. Thome, A. Bar-Cohen, R. Revellin, I. Zun, Unified mechanistic multiscale mapping of two-phase flow patterns in microchannels, Exp. Therm. Fluid Sci. 44 (2013) 1−22.

[78] J. Li, P. Cheng, Bubble cavitation in a microchannel, Int. J. Heat Mass Transfer 47 (2004) 2689−2698.

[79] P. Balasubramanian, S.G. Kandlikar, Experimental study of flow patterns, pressure drop and flow instabilities in parallel rectangular minichannels, Heat Transfer Eng. 26 (3) (2005) 20−27.

[80] P.C. Lee, F.G. Tseng, C. Pan, Bubble dynamics in microchannels. Part I: single microchannel, Int. J. Heat Mass Transfer 47 (2004) 5575−5589.

[81] P.C. Lee, F.G. Tseng, C. Pan, Bubble dynamics in microchannels. Part II: two parallel microchannel, Int. J. Heat Mass Transfer 47 (2004) 5575−5589.

[82] J. Li, G.P. Peterson, Boiling nucleation and two phase flow patterns in forced liquid flow in microchannels, Int. J. Heat Mass Transfer 48 (2005) 4797−4810.

[83] D. Cooke, S.G. Kandlikar, Pool boiling heat transfer and bubble dynamics over plain and enhanced microchannels, J. Heat Transfer 133 (2011) 052902−052911.

[84] J.Y. Lee, M.-H. Kim, M. Kaviany, S.Y. Son, Bubble nucleation in microchannel flow boiling using single artificial cavity, Int. J. Heat Mass Transfer 54 (2011) 5139−5148.

[85] Y. Wang, K. Sefiane, S. Harmand, Flow boiling in high-aspect ratio mini- and microchannels with FC-72 and ethanol: experimental results and heat transfer correlation assessments, Exp. Therm. Fluid Sci. 36 (2012) 93−106.

[86] Y. Wang, K. Sefiane, Effects of heat flux, vapour quality, channel hydraulic diameter on flow boiling heat transfer in variable aspect ratio microchannels using transparent heating, Int. J. Heat Mass Transfer 55 (2012) 2235−2243.

[87] D. Bogojevic, K. Sefiane, G. Duursma, A.J. Walton, Bubble dynamics and flow boiling instabilities in microchannels, Int. J. Heat Mass Transfer 58 (2013) 663−675.

[88] L. Yin, L. Jia, P. Guan, D. Liu, Experimental investigation on bubble confinement and elongation in microchannel flow boiling, Exp. Therm. Fluid Sci. 54 (2014) 290−296.

[89] Y. Wang, K. Sefiane, Single bubble geometry evolution in micro-scale space, Int. J. Therm. Sci. 67 (2013) 31−40.

[90] J. Barber, D. Brutin, K. Sefiane, J.L. Gardarein, L. Tadrist, Unsteady-state fluctuations analysis during bubble growth in a "rectangular" microchannel, Int. J. Heat Mass Transfer 54 (2011) 4784−4795.

[91] M.K. Akbar, D.A. Plummer, S.M. Ghiaasiaan, On gas−liquid two-phase flow regimes in microchannels, Int. J. Multiphase Flow 29 (2003) 855−865.

[92] K. Mishima, M. Ishill, Flow regime transition criteria for upward two-phase flow in vertical tubes, Int. J. Heat Mass Transfer 27 (1984) 723−737.

[93]   K.W. McQuillan, P.B. Whalley, Flow patterns in vertical two-phase flow, Int. J. Multiphase Flow 11 (1985) 161−175.

[94]   M.E. Steinke, S.G. Kandlikar, Flow boiling and pressure drop in parallel flow micro-channels, in: First International Conference on Microchannel and Minichannels, ICMM2003-1070, Rochester, NY, 2003.

[95]   S.V. Garimella, V. Singhal, D. Liu, On-chip thermal management with microchannel heat sinks and integrated micropumps, Proc. IEEE 94 (8) (2006) 1534−1548.

[96]   G. Hetsroni, A. Mosyak, Z. Segal, G. Ziskind, A uniform temperature heat sink for cooling of electronic devices, Int. J. Heat Mass Transfer 45 (2002) 3275−3286.

[97]   H.Y. Zhang, D. Pinjala, T.N. Wong, Experimental characterization of flow boiling heat dissipation in a microchannel heat sink with different orientations, in: Proceeding of 7th electronics Packaging technology Conference, EPTC, vol. 2, 2005, pp. 670−676.

[98]   M.S. Plesset, S.A. Zwick, The growth of vapor bubbles in superheated liquids, J. Appl. Phys. 25 (4) (1954) 493−500.

[99]   H.K. Forster, N. Zuber, Dynamics of vapour bubbles and boiling heat transfer, AIChE J. 1 (1955) 531−535.

[100] L. Jiang, M. Wong, Y. Zohar, Phase change in microchannel heat sinks with integrated temperature sensors, J. Microelectromech. Syst. 8 (1999) 358−365.

[101] L. Zhang, E.N. Wang, K.E. Goodson, T.W. Kenny, Phase change phenomena in silicon microchannels, Int. J. Heat Mass Transfer 48 (2005) 1572−1582.

Flow Boiling Heat Transfer with Models in Microchannels

Lixin Cheng
Department of Engineering, Aarhus University, Aarhus, Denmark

4.1 Introduction

Application of micro-scale and nano-scale thermal and fluid transport phenomena involved in traditional industries and highly specialized fields such as micro-fabricated fluidic systems, microelectronics, automobile, cryogenics, aerospace technology, micro-chemical and bioreactors, micro-heat exchangers and electronic chips cooling using microchannels, etc. have been becoming especially important since the late twentieth century [1−9]. With the rapid miniaturization of devices to micro-scale and nano-scale, new technologies taking these advances are faced with very serious heat dissipation problems per unit volume. For instance, the heat dissipation of current and future microprocessors has reached very high heat flux and new cooling systems are required to satisfy the increasingly high heat flux requirement. Even with the incorporation of heat transfer enhancements, such as thermal interface materials or heat spreaders, the power dissipation in integrated circuits and other electronics equipment has reached a threshold where conventional air-cooling technologies can no longer be relied on to effectively maintain reliable operating conditions. The micro-electronics technology continues to develop with surprisingly rapidity and the thermal energy density of electronic devices to be dissipated is becoming much higher up to 300 W/cm^2 or even higher [2−6,10−12,14−17]. One possible solution is to use forced vaporization in microchannels (e.g., multiple microchannels made in silicon or copper cooling elements attached to CPUs, or directly in the silicon chip itself) by making use of the high heat transfer performance of flow boiling in microchannels [18−24]. In this aspect, flow boiling in multichannel evaporators is one of the most promising heat transfer mechanisms for electronics cooling by utilizing the latent heat of evaporation of a fluid to extract the heat in an energy efficient manner. As a result of the enhanced thermal performance compared to other processes, better axial temperature uniformity, reduced coolant flow rates, and thus smaller pumping powers are obtained in flow boiling and two-phase flow cooling technology. Therefore, flow boiling two-phase flow cooling provides an excellent opportunity to meet the challenge of removing continuously increasing high heat fluxes dissipated in modern CPUs. However, understanding of the fundamental of flow boiling and two-phase phenomena and their mechanisms in microchannels is still a big challenge for the future development of flow boiling two-phase cooling systems. So far, there are contradictory experimental results of flow boiling heat transfer in microchannels. An understanding of heat transfer mechanisms associated with microchannels flow boiling has not yet been completely

Microchannel Phase Change Transport Phenomena. http://dx.doi.org/10.1016/B978-0-12-804318-9.00004-2

achieved due to many complicated affecting factors in microchannel flow boiling where channel size, fluid type, surface conditions, capillary forces, experimental accuracy, and others are more crucial, but no systematic knowledge has been achieved.

Over the past years, flow boiling in microchannels has become one of the "hottest" research topics in heat transfer as a highly efficient cooling technology, as it has numerous advantages of high heat transfer performance, chip temperature uniformity, hot spots cooling capability, and more. However, flow boiling heat transfer characteristics and mechanisms in microchannels are different from those in conventional channels [25−34]. The channel confinement has a significant on the flow boiling heat transfer characteristics and mechanisms. The available studies of flow boiling and two-phase flow phenomena in microchannels have exhibited contradictory results [10−20]. Although a large amount of experimental work, theory, and prediction methods for flow boiling in microchannels have been conducted over the past few years, due to the large discrepancies between experiment results from different researchers, systematic knowledge in microchannels has not yet completely achieved. Several correlations and models have been proposed for microchannel flow boiling, but most of these were only based on limited test fluids, channel shapes and diameters under limited test conditions such as one or two saturation temperatures or limited mass flux range. Thus, the extrapolation of these correlations and models other fluids and channels do not work properly in most cases. It must be mentioned here that differences among the experimental data from independent laboratories can be related to several aspects, including different surface roughness, channel dimension uncertainties, improper data reduction methods, flow boiling instabilities, improper designed test facility, test sections, and experiments. In some cases, the published results are unreasonable such as too high heat transfer coefficients, complete wrong heat transfer behaviors and trends, and others. For instance, anomaly heat transfer trends are presented, but they cannot be explained according to the proposed heat transfer mechanisms in some reports, although it is said that such mechanisms account for the heat transfer behaviors. Actually, such studies only present misleading information for other researchers. It must be mentioned here as well-performed and -documented research work should be published. Well-designed and careful experiments should be performed. For example, energy balance should be carefully checked before the experiments. Further, validation of the whole test facility and measurement system should also be conducted. In many cases, the extremely important information is missing in some published reports. Very tiny errors in the measurements and improper data reduction may result in different experimental results. Some correlations were simply developed by simply regressing the individual experimental data, which lack heat transfer mechanisms and generality. Mechanistic prediction should be targeted by incorporating the heat transfer mechanisms, which are actually related to bubble and flow pattern behaviors. Thus, it is essential to address these issues here, although many reviews on this important topic are available in the literature but seldom did they mention such important issues.

Despite increasingly reported research in microchannel flow boiling, there are some new emerging issues regarding experimentation that provide local, transient insight into microchannel flow boiling heat transfer, insight into the heat transfer mechanisms, and improvement of its heat transfer models. In general, the main issues of microchannel flow boiling research are as follows:

1. For flow boiling heat transfer, big discrepancies among experimental results from independent studies at similar conditions have been recognized. Furthermore, different heat transfer trends have been identified. Some extremely high or low heat transfer coefficients have been reported but they may not be correct.
2. The two main flow boiling heat transfer mechanisms in conventional channels are generally used to explain the experimental results in microchannels but some anomaly trends cannot be reasonably explained.
3. The existing correlations and models poorly predict the independent microchannel flow boiling heat transfer data. Reliable universal prediction methods are not available.
4. Mechanistic prediction methods and models are not well developed, and they should be related to the bubble and flow pattern phenomena during flow boiling in microchannels.
5. The channel size, shape, and surface condition such as roughness effect on the flow boiling heat transfer behaviors and prediction methods have not been well investigated.

Despite the large number of reports available in microchannel flow boiling, many aspects still need to be better explained in order to provide a complete understanding of local two-phase flow boiling characteristics. Such knowledge is essential to develop more reliable prediction methods that can be used for designing new high-performance microchannel heat spreaders for microelectronic and power electronic applications.

The objectives of this chapter are to focus on the fundamental issues and state-of-the-art of flow boiling studies in microchannel flow boiling. This chapter presents analysis of experimental results in flow boiling, the heat transfer mechanisms, and the relevant prediction methods in microchannels. Further research needs have been identified according to these review and analysis of the current research in this important field.

4.2 Flow Boiling Heat Transfer in Microchannels

4.2.1 Fundamental Issues in Microchannel Flow Boiling

As mentioned in Chapter 3, the distinction between macro- and microchannels by the threshold diameter of 3 mm is adopted here due to the lack of a well-established theory, but is in line with that recommended by Kandlikar [5]. Using this threshold diameter enables more relevant studies to be included and thus the different flow boiling characteristics in various channels with different sizes and shapes can be compared and analyzed. Furthermore, the channel size effect on the heat transfer mechanisms may be analyzed.

Flow boiling heat transfer in conventional channels is governed by two basic mechanisms of nucleate boiling dominant process (relating to the formation of vapor bubbles at the tube wall surface) and convection dominant process (relating to conduction and convection through a thin liquid film with evaporation at the liquid−vapor interface) [25,26,35,36]. The heat transfer mechanisms are intrinsically related to the bubble and flow pattern behaviors as indicated in Figs 3.1 and 3.2 in Chapter 3. As such, several flow pattern based flow boiling models have been developed for conventional channels [23,24,27−34]. The gravity becomes important for flow boiling in horizontal conventional channels and affects the flow patterns and heat transfer behavior. The flow boiling heat transfer is strongly dependent on the heat flux in nucleation dominant boiling, while the heat transfer is less dependent on the heat flux and strongly dependent on the mass flux and vapor quality in convection dominant boiling. For simplicity, one may assume that these boiling mechanisms function independently of one another. In fact, the flow boiling mechanisms can coexist as the thermodynamic vapor quality increases, where the convective boiling gradually suppress the nucleate boiling. Therefore, nucleate and convective boiling contributions can be superimposed by very complex mechanisms.

However, for flow boiling in microchannels, both mass flux and heat flux can affect the boiling process significantly, depending on the channel sizes and shapes, fluid type, and operation conditions [43−84]. The inlet subcooling may also play a role in the microchannel flow boiling heat transfer mechanisms but less investigation in this aspect is available in the literature. Although a large number of studies suggest the two flow boiling heat transfer mechanisms in microchannels, which are actually similar to those in conventional channels, different microchannel flow boiling heat transfer trends have been observed for similar test channels and conditions by different researchers, which sometimes cannot be explained by a single mechanism. Therefore, the dominant heat transfer mechanisms still need to be well clarified and they should be related to the relevant bubble and flow regime behaviors in microchannels. Figure 4.1 show schematically the two dominant flow boiling heat transfer mechanisms in microchannels: nucleate boiling dominant heat transfer and convective boiling dominant heat transfer [43]. However, the actual heat transfer mechanisms in microchannels are much more complex than the two mechanisms. The channel size and shape effects on flow boiling heat transfer and the mechanisms become more important as they have a significant effect on the corresponding bubble evaluation flow patterns as presented in Chapter 3. For instance, the bubbly and the elongated slug regimes are said to exhibit the characteristics of the nucleate boiling while the annular regime exhibits the convective boiling trend. This is similar to what is observed in conventional sized (macro) channels. In conventional channel flow boiling, different heat transfer mechanisms are dominant according to the vapor quality range, heat flux, and mass flux levels. At low vapor qualities, nucleate boiling effects prevail, while at high vapor qualities and prior to the liquid dry-out, the heat transfer coefficient is mainly controlled by convective effects. These heat transfer mechanisms are commonly considered when developing flow boiling heat transfer correlations in conventional channels such as the Chen correlation [37] and others [38−42].

However, nearly all conventional flow boiling correlations have been found to be inadequate to predict the flow boiling heat transfer data in the microchannels [12,13,18,20,22]. On a general basis, although by chance they sometimes work for a particular data set, the failure of these methods to accurately predict the heat transfer coefficient in microchannels means that more complex flow boiling heat transfer mechanisms dominate the flow boiling processes in microchannels. This is due to

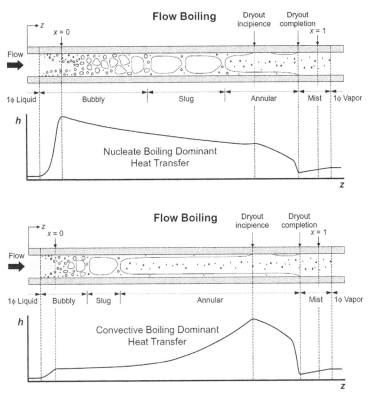

Figure 4.1 Schematic of flow patterns and the corresponding heat transfer mechanisms for upward flow boiling in a vertical tube [43].

significant differences in the phase change phenomena in the transition region between macro- and microchannels and also in the bubble growth and flow pattern evolution affected by the microchannels. Many extrapolations of macrochannel prediction methods to microchannel flow boiling conditions were performed without a sound physical basis and clearly understanding of the fundamental issues such as bubble dynamics and flow patterns representing the corresponding flow boiling heat transfer mechanisms in microchannels as explained in Chapter 3. Apparently these fundamental issues have not been well solved although experimental data are continuously published. Careful research work, deep analysis of the experimental data, and development of reasonable mechanisms governing the microchannel flow boiling phenomena are urgently needed.

Despite numerous investigations in the field of microchannel flow boiling for many years, the characteristics of flow boiling still need to be better clarified. Different trends of heat transfer coefficient with respect to quality, mass velocity, and heat flux have been reported in the literature. Figure 4.2 shows the heat transfer coefficient heat transfer coefficient h versus vapor quality x behaviors identified in the literature by Agostini and Thome [44] and illustrated by Ribatski et al. [22] in their review. The most common

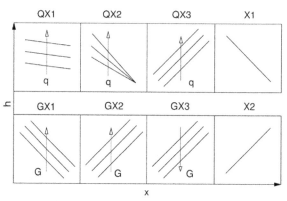

Figure 4.2 Heat transfer coefficient h versus vapor quality x behaviors identified in the literature by Agostini and Thome [44] and illustrated by Ribatski in their review [22].

of these is that heat transfer coefficient decreases with an increase in quality and hydraulic diameter, and heat transfer coefficient increases with mass velocity for a given quality. Other different heat transfer trends are also found as illustrated in Fig. 4.2. Just to show two comparisons of the experimental heat transfer coefficients obtained at almost similar test conditions by different researchers here as in Figs 4.3 and 4.4 [22]. Figure 4.3 shows the comparison for R22 flow boiling in microchannels. It can be

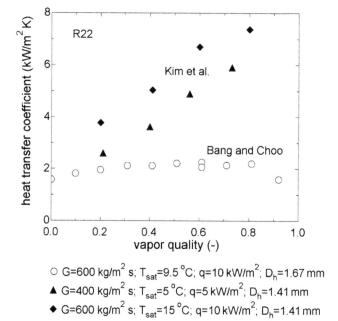

Figure 4.3 Comparison of the experimental results of Bang and Choo [64] (blank symbols) and Kim et al. [65] (filled symbols) and illustrated by Ribatski in their review [22].

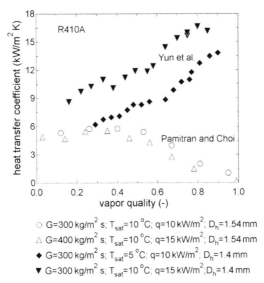

Figure 4.4 Comparison of the experimental results of Pamitran and Choi [52] (blank symbols) and Yun et al. [66] (filled symbols) and illustrated by Ribatski in their review [22].

seen that heat transfer coefficient increases from 3 to 8 kW/m^{2}K for vapor qualities from 0.2 to 0.8 according to Kim et al. [65], while for Bang and Choo [64] heat transfer coefficient presents an almost constant value of 2 kW/m^{2}K. Figure 4.4 shows the comparison between the data of Yun et al. [66] and Pamitran and Choi [52] for R410A flow boiling in microchannels, revealing remarkable discrepancies. According to Yun et al. [66], heat transfer coefficient increases with x until a vapor quality of 0.8 while for Pamitran and Choi [52] heat transfer coefficient is almost constant until vapor qualities of 0.4 and then decreases monotonically with vapor quality. Furthermore, at $x = 0.4$, $T_{sat} = 10\,°C$ and $q = 15$ kW/m^2, Yun et al. [66] obtained heat transfer coefficients nearly 3 times those obtained by Pamitran and Choi [52] and up to 10 times at larger vapor quality. The higher mass flux tested by Pamitran and Choi [52] does not seem to be related to such differences since the effects of mass flux on heat transfer coefficient were almost negligible according to these authors and neither a possible transition from micro- to macro-scale behavior, since both studies were performed for almost the same hydraulic diameter and at similar experimental conditions. Both Kim et al. [65] and Yun et al. [66] performed experiments in rectangular multichannels and obtained similar increasing trends in heat transfer coefficient versus vapor quality while Bang and Choo [64] and Pamitran and Choi [52] used a single circular channel and also got nearly flat and then declining variations in heat transfer coefficient versus vapor quality. In a square channel, it can be speculated that due to surface tension effects the liquid flow is concentrated on the corners of the channel, which may result in a thinner film on the regions between corners. This behavior may yield a higher heat transfer coefficient in square channels. However, this does not seem to explain the massive differences displayed in their studies. The different heat transfer trends make it difficult to explain them according to the flow boiling heat transfer mechanisms. Table 4.1 list selected studies on flow boiling heat transfer in microchannels. It can be seen the different results and mechanisms have been obtained in these studies as summarized in Table 4.1.

It is necessary to further analyze the available studies in microchannel flow boiling to identify the existing problems and further research needs. In this section, several representative experimental studies of flow boiling heat transfer in microchannels are reviewed at first. Then, the heat transfer mechanism are analyzed and summarized.

4.2.2   Current Research Progress on Flow Boiling Heat Transfer and Mechanisms in Microchannels

Experiments of flow boiling in microchannels have been extensively conducted over the past years. Both single and multiple microchannels with various shapes such as rectangular, trapezoidal, circular, and triangular have been used. Various fluids such as water, nitrogen, CO_2, refrigerants, dielectric fluids, and others have been employed as the working fluid in the experiments under different test conditions. Table 4.1 summarizes the studies on microchannel flow boiling heat transfer and mechanisms in selected publications.

Lazarek and Black [47] measured the local heat transfer coefficients for saturated flow boiling of R-113 in a vertical circular tube with an internal diameter of 3.1 mm, a heated length of 12.6 cm over the static pressure range of 1.3−4.1 bar, and mass flux from 140 to 740 kg/m^2s. They have found that the wall heat transfer process is controlled by nucleate boiling dominant mechanism for the independence of the heat transfer coefficient with the local vapor quality. They proposed a correlation for the local heat transfer coefficients using the Nusselt number as a function of the liquid Reynolds number and boiling number.

Bao et al. [48] conducted experiments on flow boiling heat transfer for Freon R-11 and HCFC123 flowing through a horizontal circular tube with an inner diameter of 1.95 mm. Their test parameter ranges are: mass fluxes from 50 to 1800 kg/m^2s, heat fluxes from 5 to 200 kW/m^2, and saturation pressures from 200 to 500 kPa. They have found that the flow boiling heat transfer coefficients are independent of mass flux and vapor quality but are strongly affected by the heat flux and saturation pressure. Figure 4.5 shows the variation of their measured heat transfer coefficient versus vapor quality for different mass fluxes (Fig. 4.5(a)) and heat flux (Fig. 4.5(b)). They have found the heat transfer behaving as the nucleate boiling regime up to vapor quality of 70%, as shown in Fig. 4.5. Their results are similar to those found by Lazarek and Black [47] using R-113 in a 3.1 mm diameter tube, by Tran et al. [68] using R-12 in a 2.46-mm-diameter horizontal tube, and by Owhaib et al. [69] using R-134a in vertical tubes with inner diameters (IDs) of 0.826, 1.224, and 1.7 mm.

Huo et al. [70], using R-134a in 2.01- and 4.26-mm tubes, have found a more complex behavior especially for vapor quality, x, larger than 20%, as shown in Fig. 4.6, where heat transfer coefficient is plotted versus vapor quality for different heat fluxes. Similar results have been found by Yan and Lin [71] using R-134a in a 2.0-mm 28 parallel tubes channel, and by Lin et al. [73] using R-141b in a 1.1-mm circular

Table 4.1 Selected Studies on Flow Boiling Heat Transfer and Mechanisms in Microchannels in the Literature

References	Fluid, Parameter Ranges G (kg/m^2 s), q (kW/m^2), T(°C)/P (bar)	Channel Size D_h (mm), Substrate, Geometry	Remarks
Lazarek and Black [47]	R-113, $G = 125-750$, $q = 14-380$, $P = 1.3-4.1$	$D_h = 3.1$, stainless steel, circular, vertical	• Nucleate boiling was dominant. • Heat transfer coefficient was strongly dependent on heat flux and independent of mass flux and vapor quality
Bao et al. [48]	R11/HCFC123, $G = 50-1800$, $q = 5-200$, $P = 20-50$	$D_h = 1.95$, copper, circular. Horizontal	• Nucleate boiling was dominant • Heat transfer coefficient was a strong function of heat flux and system pressure, less effects of mass flux and vapor quality
Warrier et al. [49]	FC-84, $G = 557-1600$, $q = 0-59.9$, $T = 26, 40, 60$	$D_h = 0.75$, G-10 fiber glass, 5 parallel rectangular, horizontal	• Heat transfer coefficient was correlated as a function of boiling number alone
Qu and Mudawar [58]	Water, $G = 135-402$, $T = 30, 60$	$D_h = 0.349$, parallel rectangular, horizontal 0.13×0.71 mm	• Convective boiling was dominant • Heat transfer coefficient was dependent of mass flux and vapor quality but independent of heat flux
Lee and Mudawar [51]	R-134a, $G = 127-654$, $q = 159-938$, $T = -18.1-24.7$	$D_h = 0.35$, copper, parallel rectangular 231 μm × 713 μm, hrozontal	• A correlation with vapor quality x in 3 regions: x < 0.05, $0.05 < x < 0.55$ and x > 0.55 was developed
Saitoh et al. [59]	R-134a, $G = 150-450$, $q = 5-39$, $T = 5, 10, 15$	$D_h = 0.51$, 1.12, 3.1, circular, horizontal	• Forced convective heat transfer decreases with decreasing D_h

Continued

Table 4.1 Selected Studies on Flow Boiling Heat Transfer and Mechanisms in Microchannels in the Literature—cont'd

References	Fluid, Parameter Ranges G (kg/m² s), q (kW/m²), T (°C)/P (bar)	Channel Size D_h (mm), Substrate, Geometry	Remarks
Pamitran et al. [52]	R-410A, $G = 300-600$, $q = 10-30$, $T = 10$	$D_h = 1.5, 3.0$, stainless steel, circular, horizontal,	• Nucleate boiling dominates at low x region
Bertsch et al. [53,60]	R-134a, $G = 20.3-81$, $q = 0-200$, $P = 4-7.5$	$D_h = 1.09$, copper, 17 parallel rectangular, horizontal	• Heat transfer coefficient varies with x but not with P_{sat}
In and Jeong [61]	R-134a, R123, $G = 314-470$, $q = 10-20$, $P = 1.58-2.08$, $9-11$	$D_h = 0.19$, stainless steel, circular, horizontal	• Nucleate boiling dominates for R-134a until its suppression at high vapor quality
Ong and Thome (2009)	R134a, R236a, R245fa, $G = 200-1600$, $q = 2.3-250$, $T = 31$	$D_h = 1.03$, stainless steel, circular, horizontal,	• Heat transfer coefficient has a strong function of heat flux and mass flux • Convection dominates at high vapor quality in the annular regime
Wang et al. [62]	R134a, $G = 310-860$, $q = 21-50$, $P = 0.65-0.75$	$D_h = 1.3$, stainless steel, circular, horizontal	• It demonstrates the dependences of heat transfer coefficient on mass flux, heat flux, saturation pressure and vapor quality
Tibirica and Ribatski [57]	R134a, R245fa, $G = 50-700$, $q = 5-55$, $T = 22, 31, 41$	$D_h = 2.3$, stainless steel, circular, horizontal	• Heat transfer coefficient is a function of heat flux, mass flux and vapor quality and generally increases with increasing saturation temperature

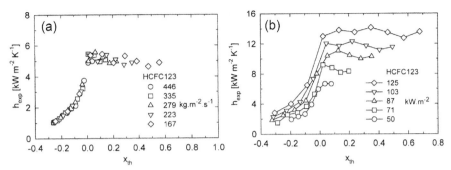

Figure 4.5 Flow boiling data for R-11 and R-123 in 1.95 mm tube plotted versus vapor quality, for different mass flux (left) and heat flux (right), Bao et al. [48].

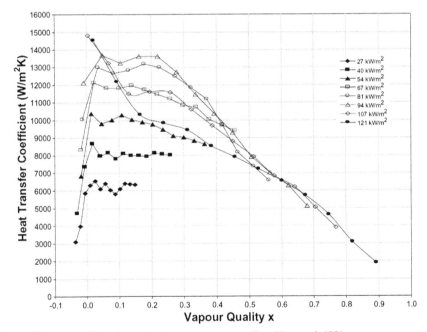

Figure 4.6 Flow boiling data plotted versus vapor quality, Huo et al. [70].

tube. Hu and Kim [72] have argued that major flow pattern is similar with annular flow in their study, which would not support the nucleate boiling dominant mechanism associated with an independency of heat transfer coefficient on mass flux or vapor quality in their experimental data. It has to be mentioned here that a 4.26-mm-ID tube falls into the macrochannel range, but their heat transfer data cannot be explained by the well-documented flow boiling mechanisms.

Warrier et al. [49] took FC-84 as test fluid flowing through five parallel channels with each channel having the following dimensions: hydraulic diameter of 0.75 mm and length to diameter ratio of 409.8. Their test parameter ranges are mass fluxes from 557 to 1600 kg/m^{2}s, heat fluxes from 0 to 59.9 kW/m^2 and the inlet liquid temperatures of 26, 40, and 60 °C. Based on the flow boiling heat transfer data, they proposed a correlation for saturated flow boiling heat transfer only using the boiling number. Their saturated flow boiling correlation is valid in their test ranges: $0.00027 \leq Bo \leq 0.00089$ and $0.03 \leq x \leq 0.55$.

Zhang et al. [50] conducted an extensive comparison of the existing correlations for flow boiling heat transfer in minichannels. The have found that the available flow boiling heat transfer correlations do not work for microchannels. They have found that a common feature of flow boiling heat transfer in many minichannels is liquid-laminar and gas-turbulent flow, and thus it may not be applicable in principle to predict the heat transfer coefficients by using the general existing correlations developed for liquid-turbulent and gas-turbulent flow conditions, such as the Chen [37] flow boiling correlation. They proposed a new correlation for microchannel flow boiling based on various existing experimental data in the literature.

Lee and Mudawar [51] conduced experiments on flow boiling heat transfer of R-134a in horizontal rectangular parallel microchannels with 231-μm-wide and 713-μm-deep groves in a copper block. They have found that flow boiling heat transfer is associated with different mechanisms for low, medium and high vapor qualities. Bubbly flow and nucleate boiling occur only at low qualities less than 0.05 and low heat fluxes, which suggests the heat transfer mechanism is nucleate boiling dominant. At medium qualities from 0.05 to 0.55 or high qualities larger than 0.55 and high heat fluxes, the flow boiling heat transfer is dominated by annular film evaporation. They proposed a new flow boiling heat transfer correlation for these three ranges of vapor qualities, which is recommended for both R-134a and water flow boiling in microchannels.

Pamitran et al. [52] investigated the convective boiling heat transfer of binary mixture refrigerant R-410A in horizontal circular tubes with inner diameters of 1.5- and 3.0-mm lengths of 1500 and 3000 mm, respectively. Local heat transfer coefficients were obtained for heat fluxes from 10 to 30 kW/m^2, mass fluxes from 300 to 600 kg/m^{2}s, a saturation temperature of 10 °C and vapor quality up to 1.0. They proposed a new flow boiling heat transfer correlation based on the superposition principle. It should be mentioned here that their test mass flux range is rather narrow and their correlation is only based on one fluid. Thus, it is necessary to valid their correlation with independent data.

Bertsch et al. [53] proposed a new heat transfer correlation for flow boiling in microchannels using a database of 3899 data points from 14 studies in the literature. Their database covers 12 different wetting and nonwetting fluids, the hydraulic diameters from 0.16 to 2.92 mm, mass fluxes from 20 to 3000 kg/m^{2}s, heat fluxes from 4 to 1150 kW/m^2, saturation temperatures from 94 to 97 °C, and vapor qualities from 0 to 1. They compared their correlation to the database. Although some of the data sets show opposing trends with respect to some parameters, the mean absolute error is still acceptable with their proposed correlation. However, a good prediction method should not be able to capture the heat transfer trends but also be able to explain the physical phenomena and mechanisms governing these variations. Therefore, effort should be made to develop mechanistic or phenomenological prediction methods, which should based on the well understanding flow boiling mechanisms and the corresponding flow regime behaviors.

Ong and Thome [54] investigated flow boiling heat transfer with three refrigerants (R134a, R236a, and R245fa) in a 1.030-mm horizontal circular channel. Their test parameter ranges are mass fluxes from 200 to 1600 kg/m²s, heat fluxes from 2.3 to 250 kW/m², and saturation temperature of 31 °C. They have found that the local saturated flow boiling heat transfer coefficients display a heat flux and a mass flux dependency, the changes in heat transfer trends correspond well with flow regime transitions and the fluid properties significantly influence the heat transfer characteristics of the fluid. However, it should be mentioned here that the physical properties of a fluid are related to saturation temperature or pressure. Based only on one test saturation temperature, it is difficult to say how the physical properties would affect the heat transfer behavior and mechanisms. It is thus necessary to well define a range of saturation temperatures and other test parameters when designing test runs in an experiment campaign.

Sun and Mishima [55] evaluated 13 prediction methods for flow boiling heat transfer in microchannels against a database including 2505 data points for 11 fluids and tubes with diameters from 0.27 to 6.05 mm. They have found that the Chen correlation [37] and those methods developed based of the Chen correlations are not suitable for microchannels, which actually means that the heat transfer mechanisms in macrochannel cannot explain the flow boiling heat transfer behaviors in microchannels. Furthermore, they have found that the Lazarek and Black [47] correlation and the Kew and Cornwell [45] correlation worked better for their database than other methods used in their study, which actually supports the nucleate dominant boiling mechanisms in microchannel flow boiling. They proposed a new modified correlation on the basis of the correlations of the Lazareck and Black and the Kew and Cornwell.

Li and Wu [55] examined the existing correlations for the saturated flow boiling heat transfer in microchannels with more than 3700 data points covering a wide range of working fluids, operational conditions and different microchannel dimensions. They have found that none of the existing correlations could predict the data sets precisely. They proposed a new flow boiling heat transfer with a function of the boiling number, the Bond number and the Reynolds number.

Tibirica and Ribatski [57] conducted experimental study on flow boiling of R134a and R245fa in a horizontal 2.3-mm-ID stainless steel tube with heating length of 464 mm. Their results were obtained under the conditions: mass fluxes from 50 to 700 kg/m²s, heat fluxes from 5 to 55 kW/m², saturation temperatures of 22, 31 and 41 °C, and vapor qualities from 0.05 to 0.99. They have found that the flow heat transfer coefficients are strongly affected by the heat flux, the mass flux and the vapor quality. In general, the heat transfer coefficient increases with increasing heat flux, mass flux, and saturation temperature, which can be reasonably explained by the macrochannel mechanisms. However, their results are different from those by Huo et al. [70] showing decreasing trends of the heat transfer coefficient with increasing vapor quality under similar test conditions. It should noted that a 2-mm-ID tube may still fall into the macrochannel range and the heat transfer behaviors should be similar to those in macrochannels.

Lin et al. [73] performed an experimental study of two-phase flow of refrigerant R141b in circular tubes of diameter of 1.8, 2.8, and 3.6 mm and one square tube of 2×2 mm^2 with different mass fluxes ranging from 50 to 3500 kg/m^2 s. Both nucleate and convective boiling mechanisms were said to occur in the tubes and the local heat transfer coefficient was found to be a weak function of mass flux while the mean heat transfer coefficient was independent of mass flux. They showed that the transition from nucleate to convective boiling at high heat fluxes occurred at higher qualities in their study. In the absence of dry-out in the saturation boiling region, they have found that the mean heat transfer coefficient varied only slightly with the tube diameter and was mainly a function of heat flux.

Chen and Garimella [75] investigated flow boiling of a dielectric fluid in a silicon chip-integrated microchannel heat sink. Twenty-four microchannels, each 389×389 μm in cross-section, were fabricated into the 12.7×12.7-mm silicon substrate. They also observed the flow regimes corresponding to the heat transfer behaviors. At low heat fluxes, bubbly flow is dominant, with the bubbles coalescing to form vapor slugs as the heat flux is increased. At high heat fluxes, the flow regimes in the downstream portion of the microchannels are characteristic of alternating wispy-annular flow and churn flow, while flow reversal is observed in the upstream region near the microchannel inlet. Local heat transfer measurements for three different flow rates show that at lower heat fluxes, the heat transfer coefficient increases with increasing heat flux. The heat transfer coefficient in fully developed boiling is seen to be independent of flow rate in this range. At higher heat fluxes, this trend is reversed, and the heat transfer coefficient decreases with further increases in heat flux due to partial dry-out in some of the microchannels. Heat fluxes at which fully developed boiling is achieved depend on the flow rate.

Liu and Garimella [76] carried out an experimental study of flow boiling heat transfer in microchannels (275×636 μm and 406×1063 μm) cut into a copper block, using DI water with mass fluxes ranging from 221 to 1283 kg/m^{2}s. Chen and Garimella [77] performed an experimental study of flow boiling of FC-77 in a copper rectangular microchannel (10 channels, 504 μm $\times$ 2.5 mm) heat sink at flow rates ranging from 80 to 133 kg/m^{2}s. Similar to the results by Chen and Garimella [75]. All these studies showed that beyond the onset of nucleate boiling, the boiling curves collapsed on a single curve for all mass flow rates; as the heat flux was increased further, the curves diverged and became dependent on mass flux. It is very important to understand the heat transfer trends and mechanisms by observing the flow phenomena and flow regimes [74]. These studies have provided good examples but systematic studies in this aspect are needed to better understand the heat transfer mechanisms.

The channel size has a significant effect on the two-phase flow and heat transfer behaviors in microchannel. So far, few have systematically investigated the effect of microchannel dimensions on the flow boiling heat transfer behaviors. A few studies have considered the effect of microchannel size on flow boiling patterns and the transition between different flow patterns as described in Chapter 3 while the effect of microchannel size on heat transfer coefficient has been largely unexplored.

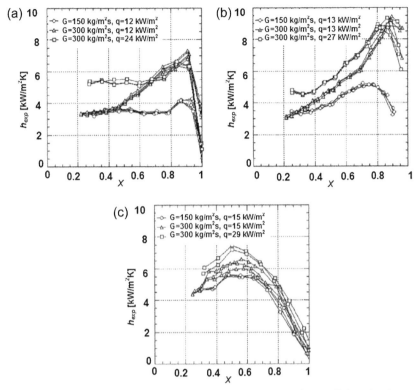

Figure 4.7 Effect of heat flux (q) and mass flux (G) on heat transfer coefficient for three different diameter tubes: (a) 3.1-mm-ID; (b) 1.12-mm-ID, and (c) 0.51-mm-ID [59].

Saitoh et al. [59] performed flow boiling experiments for R134a in single horizontal channels with internal diameters of 0.51, 1.12, and 3.10 mm. Figure 4.7(a–c) shows the effect of heat flux q and mass flux G on the flow boiling heat transfer coefficients for the three different diameter tubes. For the 3.1-mm-ID tube results as shown in Fig. 4.7(a), in the low vapor quality less than 0.6, when the mass flux is fixed at 300 kg/m²s, the heat transfer coefficient for the heat flux of 24 kW/m² is higher than that for the heat flux of 12 kW/m², whereas when the heat flux is fixed at 12 kW/m², the heat transfer coefficient for the mass flux of 150 kg/m²s is similar to that for the mass flux of 300 kg/m²s. In contrast, in the high vapor quality large than 0.6, when the heat flux is fixed at 12 kW/m², the heat transfer coefficient for the mass flux of 300 kg/m²s is higher than that for the mass flux of 150 kg/m²s, whereas when the mass flux is fixed at 300 kg/m²s, the heat transfer coefficient for the heat flux of 12 kW/m² is similar to that for the heat flux of 24 kW/m². These experimental results suggest that in the low vapor quality region, the heat transfer mechanism is nucleate boiling dominant and in the high vapor quality region, the heat transfer mechanism is forced convective evaporation dominant. For the 1.12-mm-ID tube results shown in Fig. 4.7(b), the heat transfer coefficients are are slightly higher than those for the 3.1- mm-ID tube shown Fig. 4.7(a), and the effects of heat flux and mass on the heat transfer coefficients are similar for both tubes. These results are in consistent with those by Tibirica and Ribatski [57] but again contradictory to those by Huo et al. [70]. For the 0.51-mm-ID tube results shown in Fig. 4.7(c), the heat transfer coefficient increases with increasing the heat flux but is not significantly affected by the mass flux. Among the three tubes, the heat transfer coefficients for the 0.51-mm-ID tube are the highest when the vapor quality is less than 0.5. As the tube diameter is decreased, the heat transfer coefficient starts to decrease at lower vapor quality. The vapor quality at which the heat transfer coefficient starts to decrease is 0.9 for the 3.1-mm-ID tube, between 0.8 and 0.9 for the 1.12-mm-ID tube, and between 0.5 and 0.6 for the 0.51-mm-ID tube. From their results, all three channels have showed strong effects of heat flux, signifying to them the dominance of nucleate boiling. On the issue of forced convective boiling, they observed that the 3.10-mm-diameter channel exhibited the strongest effect of mass flux on the heat transfer coefficient as compared with the 1.10-mm channel while the least effect on mass flux was observed for the 0.51-mm-diameter channel. The onset of dry-out was also observed to occur earlier in the lower vapor quality region with decreasing channel diameter. They have concluded that convective boiling dominance effectively decreases with decreasing channel diameter. It is very interesting to see the different heat transfer behaviors affected by the channel diameters. For both 3.1- and 1.12-mm-diameter tubes, the heat transfer behaviors are similar to the macrochannel heat transfer behaviors while for the 0.51-mm-diameter tube, the heat transfer behaviors have changed. As mentioned in Chapter 3, the distinction between macro- and microchannels may be classified according to the heat transfer behaviors or mechanisms. This study is a good example for this but further systematic studies are needed to well document the criteria for such a distinction.

Harirchian and Garimella [63] investigated the effect of mass flow rate and channel size on the flow boiling heat transfer behaviors of FC-77 in microchannel heat sinks with with a constant channel depth of 400 μm and different channel widths ranging from 100 to 5850 μm. Four mass flux values ranging from 250 to 1600 kg/m^{2}s were investigated for each test piece to identify the mass flux effect on the flow boiling heat transfer behavior in Fig. 4.8. The heat transfer coefficient increases with increasing the mass flux in the single-phase region for a fixed wall heat flux while the heat transfer coefficient becomes independent of mass flux and increases with heat flux in the boiling region. At high levels of wall heat flux, as the contribution from convective heat transfer begins to dominate that of nucleate boiling, the heat transfer coefficient becomes a function of mass flux and increases with increasing mass flux. Flow visualizations performed as part of this work show that the plots of heat transfer coefficient diverge from each other at the heat flux where the bubble nucleation was observed to be suppressed at the walls. Other microchannel sizes tested yielded similar trends for the dependence of heat transfer coefficient on mass flux. These results regarding the dependence of heat transfer coefficient on flow rate are also consistent with the findings of Chen and Garimella [75]. At high heat fluxes, a decrease in heat transfer coefficient is detected. Flow visualization has confirmed that this is attributed to a partial wall dry-out as explained by Chen and Garimella

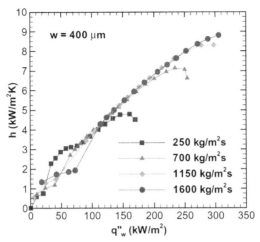

Figure 4.8 The effect of mass flux on the local heat transfer coefficients in microchannel [63].

[75]. Figure 4.9 shows the effect of microchannel width on local heat transfer coefficient as a function of (a) wall heat flux and (b) base heat flux. The heat transfer coefficient is independent of microchannel width for channels of width 400 μm and larger, and even for smaller microchannels has only a weak dependence on channel size. At the 250-μm width, the heat transfer coefficients are slightly lower than for the larger sizes at any wall heat flux as shown in Fig. 4.9(a). For the 100-μm-wide microchannels, the behavior is markedly different, with the heat transfer coefficient being relatively higher at the lower heat fluxes. As the heat flux increases, the curve crosses over and is lower than for the larger microchannels. This is attributed to the high exit vapor quality for the 100-μm microchannels. As a result of the high vapor qualities in the 100-μm microchannels, annular flow commences at lower heat fluxes than in the larger channel sizes, leading to higher heat transfer coefficients at the lower heat fluxes. Figure 4.9 (a) illustrates that increasing the width of the microchannels beyond 400 μm (for a fixed channel depth) does not affect the heat transfer coefficient for a fixed wall heat flux. However, from a design point of view, the dependence of heat transfer coefficient on microchannel size should be considered in terms of a given amount of heat dissipation from the chip, i.e., a fixed value of base heat flux. Figure 4.9 (b) shows the heat transfer coefficient (as a function of base heat flux) for different microchannel sizes. It is evident from Fig. 4.9 (b) that for a given heat dissipation from the chip, the heat transfer coefficient increases as the microchannels width increases. However, the maximum amount of heat that can be removed from the chip increases as the microchannels become smaller due to the larger surface enhancement with the smaller microchannels.

Qu and Mudawar [58] investigated flow boiling in a water-cooled microchannel heat sink. Their heat sink was made on a copper substrate and contained 21 parallel channels, each having a $231 \times 713 \ \mu m^2$ rectangular cross-section. The inlet Reynolds number ranged from 60 to 300. The inlet temperature was maintained constant at either

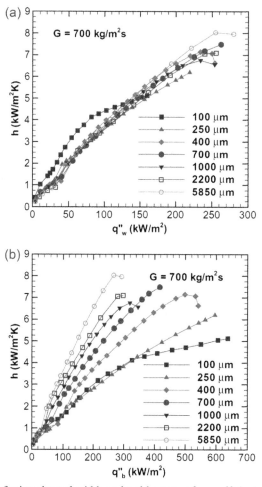

Figure 4.9 Effect of microchannel width on local heat transfer coefficient as a function of (a) wall heat flux and (b) base heat flux [63].

30 °C or 60 °C by appropriately adjusting the mass flux and heat flux, in the range of 135−402 kg/m^{2}s and 20−135 W/cm^2, respectively. Figure 4.10 shows their measured saturated flow boiling heat transfer coefficient versus thermodynamics equilibrium quality for two inlet temperatures of 30 °C and 60 °C. The heat transfer coefficient are be in the range of 22−44 kW/m^{2}K. It seems that the inlet temperature has some effect on the heat transfer coefficient but the knowledge in this aspect is very little in the literature. Contrary to behavior observed in macrochannels, the heat transfer coefficient decreased with increasing vapor quality. They attributed the unexpected heat transfer trends to appreciable droplet entrainment at the onset of the annular flow regime. Similar dependence of heat transfer coefficient on vapor quality was reported by Hetsroni et al. [93] for flow boiling of Vertrel XF in a heat sink that

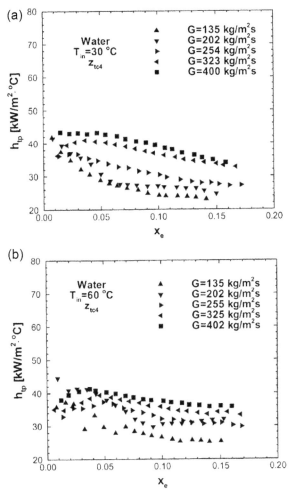

Figure 4.10 Saturated flow boiling heat transfer coefficient versus thermodynamics equilibrium quality for (a) $T_{in} = 30\,°C$ and (b) $T_{in} = 60\,°C$ [58].

had 21 parallel triangular microchannels each of hydraulic diameter 129 μm. Yen et al. [92] studied flow boiling of water and refrigerant HCFC123 and FC72 in microchannels, respectively. They also observed that the heat transfer coefficient decreases with an increase in vapor quality, in agreement with the results of Qu and Mudawar [58] and Hetsroni et al. [93].

More recently, Charnay et al. [86] conducted experimental study on flow boiling in a 3 mm diameter horizontal circular tube. Figure 4.11 shows the effect of heat flux on the heat transfer coefficients at mass velocity of 700 kg/m²s and two different saturation temperatures (60 and 80 °C). At 60 °C, the heat transfer coefficients are strongly affected by the heat fluxes at low vapor qualities where the flow regime is intermittent flow. The heat flux heat transfer coefficient curves at low and high heat fluxes converge

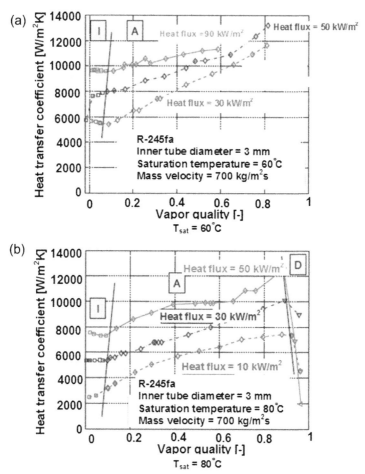

Figure 4.11 Influence of heat flux on heat transfer coefficient for R-245fa at 60 °C and 80 °C with a mass velocity of 700 kg/m²s (I: intermittent, A: annular, and D: dry-out) [86].

at high vapor qualities where the flow regime is annular flow. At 80 °C, the heat flux strongly affects the heat transfer coefficients at all vapor qualities before the dry-out. The heat transfer coefficient increases with increasing the vapor quality in annular flow. This is attributed to the enhanced contribution of convective boiling heat transfer with increasing the vapor quality. Considering the heat transfer mechanisms, nucleate boiling is dominant in the intermittent flow regime while nucleate and convective boiling are important in the annular flow regime. They also investigated the mass flux effect on the heat transfer coefficients as shown in Fig. 4.12. Two different trends have been identified as distinguished: Figure 4.12(a) shows that the higher the mass velocity, the larger the heat transfer coefficient for 60 °C and Fig. 4.12(b) shows the higher the mass flux, the smaller the heat transfer coefficient at lower vapor qualities while the higher the mass velocity, the larger the heat transfer coefficient at higher

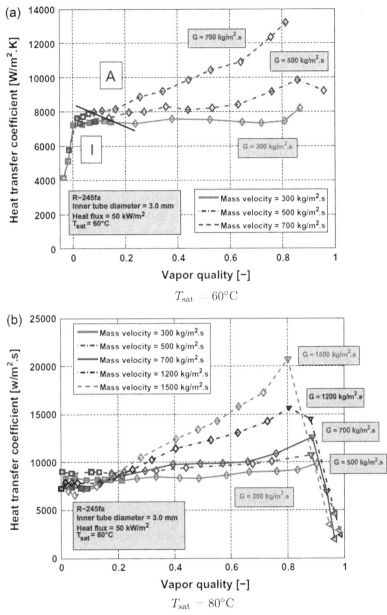

Figure 4.12 Influence of mass velocity on heat transfer coefficient for R-245fa at 60°C and 80°C with a heat flux of 50 kW/m² (I: intermittent flow and A: annular flow) [86].

vapor qualities 80 °C. It can be seen that the heat transfer coefficients are independent of the vapor qualities in the intermittent flow regime and the mass flux effect is less important because the nucleate boiling is the dominant heat transfer mechanism. In the annular flow regime, the heat transfer coefficient increases with increasing the vapor quality except for a mass velocity of 300 kg/m^2 s where the heat transfer coefficient is almost constant over the whole range of vapor quality. Convective boiling is the dominant heat transfer mechanism in this region. At large mass fluxes, the smaller bubbles tend to coalesce to form bigger bubbles. Thus, the bubbles frequency reduction could explain the results.

It is very important to relate the flow regimes and bubble dynamics to the measured heat transfer coefficient behavior because they are intrinsically related each other. Especially for microchannel, well designed test facility and accurate measurement system are essential for accurate experimental data. The different heat transfer trends may be explained by various mechanisms and they may also due to many other affecting factors such as the measurement accuracy, channel size and shape, fluid types and physical properties, the test parameter ranges, the data analysis, and so on. Different heating methods such as fluid heating and electrical heating, measurement methods and different data reduction methods could results in big contradictory results. For instance, local and average heat transfer coefficients may be different, depending on how the wall and local saturation temperatures are determined. Furthermore, careful energy balance and validation of the measurement system should be done before performing the flow boiling experiments. Experimental results from the literature can hardly be compared against one or another since there is no accepted benchmark. Furthermore, the issue of the dominant heat transfer mechanisms is not well explored and explained according to the experimental data. In other words, researchers are in somewhat of a 'dilemma' to conclude which is the dominant heat transfer mechanism in their studies as sometimes their proposed heat transfer mechanisms actually do not describe their heat transfer coefficient trends properly.

There is still a debate about the possible governing heat transfer mechanism for microchannel flow boiling (i.e., nucleate boiling dominant or forced convective boiling dominant). Generally, in the nucleate boiling dominant region, the heat transfer coefficient is mainly a function of heat flux and system pressure but is less independent of vapor quality and mass flux. In convective boiling dominant region the heat transfer coefficient largely depends on vapor quality and mass flux but is not a function of heat flux. On this basis, many researchers have addressed their experimental flow boiling results in microchannel as governed by the nucleate boiling or the convective boiling mechanism, depending on the heat transfer coefficient trend as a function of thermal—hydraulic parameters only. In fact, the current research results are diverse. Some anomaly heat transfer trends cannot be explained according to the relevant heat transfer mechanisms as described in the foregoing. For flow boiling in microchannels, understanding the fundamental heat transfer mechanisms should be relevant to the relevant flow regimes and bubble dynamics in microchannels, considering the channel size effect on the bubble growth and flow patterns [78,94—104]. Unfortunately, in most available publications, the flow patterns and flow boiling heat transfer behaviors have been separately investigated. In fact, the flow regimes and heat transfer behaviors are intrinsically related to each other. Both should be observed simultaneously in experiments for a better understanding the microchannel flow boiling phenomena and mechanisms. There are several such studies available as described earlier. For another example,

Jacob and Thome [101], Thome et al. [102] and Dupont et al. [103] proposed a heat transfer model based on elongated bubble flow in micro-scale channels, with the hypothesis that thin film evaporation is the dominant heat transfer mechanism as opposed to prior interpretations that conventional macro-scale nucleate boiling dominance. Their assumption is that micro-scale flow is reached when the bubble growth diameter reaches the tube internal diameter followed by subsequent detachment from the wall surface. In their heat transfer model, the critical bubble radius is evaluated using the effective nucleation wall superheat. However, systematic knowledge in this aspect has not yet completely achieved by relating heat transfer model, mechanism and flow patterns. Therefore, effort should be made to advance well documented knowledge in microchannel flow boiling heat transfer and the heat transfer mechanisms governing the flow boiling processes through well designed experiments on both flow boiling and flow patterns bu considering the channel size effect on them in future.

In summary, from the available studies on micro-scale flow boiling, flow boiling heat transfer mechanisms have been classified by into four different categories: (1) Nucleate boiling dominant because the heat transfer data are heat flux dependent; (2) Convective boiling dominant when the heat transfer coefficient depends on mass flux and vapor quality but not heat flux; (3) Both nucleate and convective boiling dominant and (4) Thin film evaporation of the liquid film around elongated bubble flows as the dominant heat transfer mechanism, which is heat flux dependent via bubble frequency and conduction across the film.

4.3 Correlations and Models of Flow Boiling Heat Transfer in Microchannels

4.3.1 Classification of Flow Boiling Heat Transfer Models

There are a large number of correlations and models available in the literature for flow boiling of saturated liquids in conventional channels. A number of researchers have developed correlations or models for microchannel flow boiling on the basis of these for macrochannel flow boiling. Most of these consider the contribution of two flow boiling heat transfer mechanisms: nucleate boiling dominant and convective boiling dominant. The heat transfer coefficient correlations can generally be divided into three groups:

1. The summation correlations: The heat transfer coefficient is considered to be the addition of the nucleate and convective boiling contribution such as the Chen correlation [37] in Eq. (4.2).
2. The asymptotic model: The heat transfer coefficient is assumed as one of the two mechanisms to be dominant such as the Steiner and Taborek model [101]:

$$h_{tp} = \left(h_l + h_{pool}\right)^{\frac{1}{n}}$$ (4.1)

where $n > 1$, the two-phase flow boiling heat transfer h_{tp} asymptotically approaches to the nucleate boiling heat transfer h_{pool} or convective boiling heat transfer h_l.

3. The flow pattern dependent model: This model consists of a flow pattern map and flow pattern specific models and correlation for the heat transfer such as the prediction methods by Kattan et al. [27–29], Cheng et al. [30–32], and Wojtan et al. [33,34] which are based on the asymptotic model and the relevant flow pattern maps.

Chen [37] proposed the first general flow boiling heat transfer correlation by using addition of two components corresponding to the two heat transfer mechanisms with a nucleate boiling suppression factor S and a convective enhancement factor E as:

$$h_{tp} = Eh_l + Sh_{pool} \tag{4.2}$$

The liquid phase heat transfer coefficient h_l is evaluated with the Dittus and Boelter equation [102]:

$$h_l = 0.023\text{Re}_l^{0.8}\text{Pr}_l^{0.4}\left[\frac{k_l}{D_h}\right] \tag{4.3}$$

$$\text{Re}_l = \frac{G(1-x)D_h}{\mu_l} \tag{4.4}$$

$$\text{Pr}_l = \frac{\mu_l c_{pl}}{k_l} \tag{4.5}$$

The pool boiling heat transfer coefficient h_{pool} is evaluated by the Forster and Zuber correlation [103]:

$$h_{pool} = 0.00122\frac{k_l^{0.79}c_{pl}^{0.45}\rho_l^{0.49}}{\sigma^{0.5}\mu_l^{0.29}i_{fg}^{0.24}\rho_g^{0.24}}(T_w - T_{sat})^{0.24}\left[p_{sat}(T_w) - p_l\right]^{0.75} \tag{4.6}$$

$$S = \frac{1}{1 + 2.35 \times 10^{-6}\text{Re}_{tp}^{1.17}} \tag{4.7}$$

$$\text{Re}_{tp} = \frac{G(1-x)D_h}{\mu_l}E^{1.25} \tag{4.8}$$

$$\text{For } \frac{1}{X_{tt}} \le 0.1 \quad E = 1 \tag{4.9}$$

$$\text{For } \frac{1}{X_{tt}} > 0.1 \quad E = 2.35\left(\frac{1}{X_{tt}} + 0.213\right)^{0.736} \tag{4.10}$$

S is the nucleate boiling suppression factor that takes into account steeper temperature gradient near the wall due to the fluid motion which tends to suppress the number of bubble active sites. The convective enhancement factor, E, takes into account the increment of convective effects relative to that of the single-phase flow of the liquid, promoted by the flow acceleration due to the evaporation process.

Gungor and Winterton [38] developed a general flow boiling heat transfer correlation which also considers the two heat transfer mechanisms as in the Chen correlation:

$$h_{tp} = Eh_l + Sh_{pool} \tag{4.11}$$

$$E = 1 + 24000Bo^{1.16} + 1.23\left(\frac{1}{X_{tt}}\right)^{0.86} \tag{4.12}$$

$$S = \frac{1}{1 + 1.15 \times 10^{-6}E^2Re_l^{1.17}} \tag{4.13}$$

The cooper pool boiling correlation [85,104] is used in Eq. (4.11):

$$h_{pool} = 55p_r^{0.12}\left(-\log p_r\right)^{-0.55}M^{-0.5}q^{0.67} \tag{4.14}$$

The heat transfer coefficient for the liquid phase h_l is evaluated with Eq. (4.3). If the tube is horizontal and the Froude number is less than 0.05 then E should be multiplied by:

$$E_2 = Fr^{(0.1-2Fr)} \tag{4.15}$$

and S should be multiplied by:

$$S_2 = \sqrt{Fr} \tag{4.16}$$

$$Fr\frac{G^2}{\rho_l^2 gD_h} \tag{4.17}$$

Liu and Winterton [39] developed a general flow boiling heat transfer method in order to predict saturated and subcooled flow boiling heat transfer based on the two flow boiling heat transfer mechanisms as in the Chen correlation. In their method, the nucleate boiling and convective boiling contributions are added up according to a power-type asymptotic approach with an exponent of 2 as:

$$h_{tp} = \sqrt{\left(Eh_{lo}\right)^2 + \left(Sh_{pool}\right)^2} \tag{4.18}$$

$$E = \left[1 + x\text{Pr}_l\left(\frac{\rho_l}{\rho_g} - 1\right)\right]^{0.35} \tag{4.19}$$

$$S = \frac{1}{1 + 0.055E^{0.1}Re_{lo}^{0.16}} \tag{4.20}$$

where h_{lo} is evaluated with the Dittus-Boelter [102] correlation taking total gas liquid two-phase flow as single-phase liquid flow and nucleate pool boiling heat transfer coefficient h_{pool} is evaluated with the Cooper nucleate pool boiling correlation Eq. (4.14) [104].

Besides the three flow boiling heat transfer correlations for macrochannel flow boiling listed here, there are also many correlations for macrochannel flow boiling heat transfer such as the Kandlikar correlation [40,41], the flow boiling heat transfer models of Kattan et al. [27—29], Cheng et al. [30—32], and Wojtan et al. [33,34]. It should be mentioned that the Gungor and Winterton correlation also covers small diameter tubes and the Cheng et al. model [30—32] for CO_2 flow boiling covers microchannels as well. The models of Kattan et al. [27—29], Cheng et al. [30—32], and Wojtan et al. [33,34] are flow patterned based methods, considering the heat transfer behavior and mechanisms. As pointed out by Cheng et al., such models including flow pattern information should be developed for microchannel flow boiling but not yet well developed [13].

4.3.2 Prediction Methods for Flow Boiling Heat Transfer in Microchannels

Over the past years, a number of correlations and models have been developed for microchannel flow boiling heat transfer. These heat transfer prediction methods are either by modifying the Chen correlation or correlating their own experimental data according to dimensionless numbers by a number of researchers. It should be pointed out that may correlations are only based on limited test fluids and conditions. As mentioned in Section 4.2, although various flow boiling heat transfer mechanisms have been proposed in different studies, their corresponding flow patterns and bubble evolution processes have not been observed by using flow visualization technology. Furthermore, these correlations actually based on the two heat transfer mechanisms for conventional channels do not really represent the very complex heat transfer mechanisms in microchannel flow boiling, which have not yet been well understood as pointed out in Section 4.2. Therefore, these correlations can only work for some specific fluids and conditions. In some cases, they cannot predict the heat transfer properly and even give wrong predictions. This is mainly due to the lack of understanding of the physical mechanisms, the limited applicable parameter ranges and inaccurate measured heat transfer data or completely wrong heat transfer data caused by the unreasonable test facility, measurement system and improper data reduction methods. Although it is difficult to judge the data in each individual study, it is still not difficult to observe unreasonably large heat transfer coefficients or abnormal heat transfer trends that cannot be explained by the physical processes and mechanisms in flow boiling if careful analysis can be done.

Table 4.2 summarizes the selected flow boiling heat transfer correlations in microscale channels. Lazarek and Black [47] proposed a new type of correlation in terms of dimensionless Nusselt, Reynolds, and boiling numbers. The strong dependence of the local saturated boiling heat transfer coefficient on heat flux with negligible influence of

Table 4.2 Selected Flow Boiling Heat Transfer Correlations for Microchannels in the Literature

Lazarek and Black [47]	$h_{tp} = 30\text{Re}_{lo}^{0.857} Bo^{0.714} \frac{k_l}{D_h}$	(4.21)
Kew and Cornwell [45]	$h_{tp} = 30\text{Re}_{lo}^{0.857} Bo^{0.714}(1-x)^{-0.143} \frac{k_l}{D_h}$	(4.22)
Tran et al. [68]	$h_{tp} = 8.4 \times 10^{-5} Bo^{0.6} We_l^{0.3} \left(\frac{\rho_l}{\rho_g}\right)^{-0.4}$	(4.23)
	$We_l = \frac{G^2 D_h}{\sigma \rho_l}$	(4.23a)
Warrier et al. [49]	$h_{tp} = \left(1 + 6Bo^{1/16} + f(Bo)x^{0.65}\right) h_{lo}$	(4.24)
	$f(Bo) = -5.3(1 - 855Bo)$	(4.24a)
Zhang et al. [50]	$h_{tp} = Sh_{pool} + Fh_{sp}$	(4.25)
	h_{sp} may be referred to the report	
	$S = \dfrac{1}{1 + 2.53 \times 10^{-6}\text{Re}_l^{1.17}}$	(4.25a)
	$E = MAX(E', 1)$	(4.25b)
	$E' = 0.64\left(1 + \frac{C}{X_{tt}} + \frac{1}{X_{tt}^2}\right)^{0.5}$	(4.25c)
Pamitran et al. [52]	$h_{tp} = Sh_{pool} + Fh_l$	(4.26)
	$S = 9.4626(\phi^2)^{-0.2747} Bo^{0.1285}$	(4.26a)
	$E = 0.062\phi^2 + 0.938$	(4.26b)
Saitoh et al. [59]	$h_{tp} = Sh_{pool} + Eh_l$	(4.27)
	$E = 1 + \dfrac{X_{tt}^{-1.05}}{1 + We_g^{-0.4}}$	(4.27a)
	$We_g = \frac{G^2 x^2 D_h}{\sigma \rho_g}$	(4.27b)
	$S = \dfrac{1}{1 + 0.4(\text{Re}_{tp} \times 10^{-4})^{1.4}}$	(4.27c)
	$h_{pool} = 207 \frac{k_l}{d_b}\left(\frac{qd_b}{k_l T_l}\right)^{0.745}\left(\frac{\rho_g}{\rho_l}\right)^{0.581} \text{Pr}_l^{0.533}$	(4.27d)
	$d_b = 0.51\left[\dfrac{2\sigma}{g\left(\rho_l - \rho_g\right)}\right]^{0.5}$	(4.27e)
Sun and Mishima [55]	$h_{tp} = \dfrac{6\text{Re}_{lo}^{1.05} Bo^{0.54}}{We_l^{0.191}(\rho_l/\rho_g)^{0.142}} \frac{k_l}{D_h}$	(4.28)
Li and Wu [56]	$h_{tp} = 334Bo^{0.3}\left(Bd\text{Re}_l^{0.36}\right)^{0.4} \frac{k_l}{D_h}$	(4.29)

vapor quality in their study suggested that the wall heat transfer process is controlled by nucleate boiling.

Kew and Cornwell [45] found that heat transfer and flow characteristics in micro-scale channels were significantly different from those observed in macro-scale channels. They modified the Lazarek and Black [47] correlation in terms of vapor quality x.

Tran et al. [68] recognized that the dominant heat transfer mechanism is nucleate boiling rather than convective boiling in small diameter channels. They proposed a heat transfer correlation with the Reynolds number being replaced with the liquid Weber number We_l to eliminate the viscous effect in favor of the surface tension effect. Warrier et al. [49] developed a different saturated flow boiling heat transfer correlation which is only dependent on the boiling number Bo and vapor quality x.

Zhang et al. [50] modified the Chen [37] correlation to predict flow boiling heat transfer in small diameter tubes. The poop boiling correlation by Foster and Zuber [103] is kept unchanged for the nucleate boiling heat transfer component. The boiling suppression factor S in the Chen correlation is also kept unchanged. To determine the convection evaporation enhancement factor E and the single-phase heat transfer coefficient, vapour, and liquid flow conditions (laminar or turbulent) are taken into account in their method.

Pamitran et al. [52] developed a heat transfer correlation for micro-scale channels based on the Chen [37] correlation. They proposed new calculation methods for nucleate boiling suppression factor S and convection evaporation enhancement factor E according to their experimental data.

Saitoh et al. [59] proposed a prediction method which covers both macro- and micro-scale flow boiling heat transfer by modifying the Chen correlation [37]. They considered the tube diameter effect on the heat transfer coefficient. Because the fluid flow conditions more strongly affect convective evaporation than nucleate boiling, the convective boiling enhancement factor S is expressed as a function of gas phase Weber number We_g and Matinelli number X_{tt}. The Stephen and Abdelsalam [105] nucleate pool boiling correlation is used in their method.

Sun and Mishima [55] found that the heat transfer coefficient is more dependent on the Weber number We_l than the vapor quality x, which is similar to the results obtained by Lazarek and Black [47]. They proposed a new flow boiling heat transfer correlation by modifying the Lazarek and Black correlation, taking the effect of Weber number into account in their correlation.

Li and Wu [56] found that the two-phase Nusselt number is greatly dependent on the Bond number Bd. Thus, they proposed a new flow boiling heat transfer correlation combining the effects of the Bond number Bd and liquid phase Reynolds number Re_l.

To evaluate these correlations for flow boiling heat transfer coefficients in micro-scale channels, a database has been established by collecting experimental data from the selected studies in the literature. Table 4.3 summarizes the selected studies for the microchannel flow boiling heat transfer database. The description of the database is given as follows:

Pamitran et al. [52] performed convective boiling heat transfer experiments in horizontal minichannels with inner diameters of 1.5 and 3.0 mm for binary mixture refrigerant R-410A. 147 data points were obtained with constant saturation temperature 10 °C, mass fluxes ranging from 300 to 600 kg/m^{2}s and heat fluxes from 10 to 30 kW/m^2.

Table 4.3 Summary of the Microchannel Flow Boiling Heat Transfer Database in the Literature

References	Fluid, Parameter Ranges: G (kg/m²s), q (kW/m²), T (°C) or P (kPa)	Channel Size D_h (mm), Horizontally (Unless Otherwise Stated)	No. of Data
Pamitran et al. [52]	R-410A, $G = 300-600$, $q = 10-30$, $T = 10$	$D_h = 1.5, 3.0$	147
Tibirica and Ribatski [57]	R134a, R245fa, $G = 100-700$, $q = 5-35$, $T = 22, 31, 41$	$D_h = 2.3$	136
Vlasie et al. [87]	R134a, R12, R141b, $G = 63.1-212$, $q = 5-90$, $T = 5-32$	$D_h = 2.0, 2.46, 3.69$	166
Agostini and Bontemps [88]	R134a, $G = 219$, $q = 29.049$, $P = 452, 557, 608$	$D_h = 2.01$, vertically	66
Yun et al. [89]	R410A, $G = 200-400$, $q = 10-20$, $T = 0, 5, 10$	$D_h = 1.36, 1.44$	113
Choi et al. [67]	R-22, R134a, $G = 200-600$, $q = 10-40$, $T = 10$	$D_h = 1.5\ 3.0$	335
Qi et al. [90]	N_2, $G = 448.4-1471.2$, $q = 82.5-133.5$, $P = 188-735.9$	$D_h = 0.531, 0.834, 1.042, 1.931$	284
Bertsch et al. [60]	R134a, R245fa, $G = 20-350$, $q = 0-220$, $T = 8-30$	$D_h = 0.54, 1.09$	183
In and Jeong [61]	R123, R134a, $G = 314, 392, 470$, $q = 10-20$, $T = 158-1100$	$D_h = 0.19$	217
Shiferaw et al. [91]	R134a, $G = 100-600$, $q = 16-150$, $P = 600-1200$	$D_h = 1.1$	324
Wang et al. [62]	R134a, $G = 310-860$, $q = 21-50$, $P = 650-750$	$D_h = 1.3$	365

Tibirica and Ribatski [57] studied flow boiling heat transfer coefficient of two fluids R134a and R245fa, flowing through a horizontal 2.3 mm inner diameter stainless steel tube. 136 data points were obtained over the range of mass fluxes from 100 to 700 kg/m^{2}s, heat fluxes from 5 to 35 kW/m^2 and saturation temperatures of 22, 31 and 41 °C.

Vlasie et al. [87] conducted experiments on two-phase flow heat transfer in microchannels with inner diameters of 2, 2.46 and 3.69 mm R134a, R12 and R141b were taken over a range of mass fluxes (63.1−212 kg/m^2 s^1), heat fluxes (5−90 kW/m^2) and saturation temperature (5−32 °C). 166 data points are obtained from their study.

Agostini and Bontemps [88] conducted an experimental investigation of flow boiling vertically for R134a in minichannels with hydraulic diameter of 2.01 mm. 64 data points were obtained from constant heat flux 29.049 kW/m^2 and constant mass fluxes 219 kg/m^{2}s with varying saturation pressures 452, 557 and 608 kPa.

Yun et al. [89] studied convective boiling heat transfer coefficients and pressure drop of R410A in rectangular microchannels with hydraulic diameter of 1.36 and 1.44 mm. The mass flux varied from 200 to 400 kg/m^2 s, heat flux from 10 to 20 kW/m^2 and the saturated temperature at 0, 5 and 10 °C. 113 data points were obtained from their study.

Choi et al. [67] conducted an experimental investigation of R-22 and R134a flowing in horizontal smooth minichannels with inner diameters of 1.5 and 3.0 mm. The local heat transfer coefficients were obtained for heat fluxes ranging from 10 to 40 kW/m^2, mass fluxes ranging from 200 to 600 kg/m^2 s and a saturation temperature of 10 °C. 335 data were collected from their study.

Qi et al. [90] studied the flow boiling of liquid nitrogen in the micro-scale channels with the diameters of 0.531, 0.834, 1.042 and 1.931 mm. 284 data points were collected from these different diameters with other parameters ranges: mass fluxes from 448.4 to 1417.2 kg/m^2 s, heat fluxes from 82.5 to 133.5 kW/m^2 and saturation pressures from 188 to 735.9 kPa.

Bertsch et al. [60] performed a study on flow boiling heat transfer with the refrigerants R-134a and R-245fa in copper micro-scale channels of hydraulic diameter 1.09 and 0.54 mm. The heat transfer coefficient is measured at saturation temperatures ranging from 8 to 30 °C, heat flux from 0 to 22 W/cm^2 and mass flux from 20 to 350 kg/m^{2}s. 183 data points under the condition of saturated pressure 125 and 550 kPa with R-245fa and R-134a respectively were collected in the test range.

In and Jeong [61] experimentally investigated flow boiling heat transfer coefficients for R123 and R134a in a single circular microchannel with inner diameter of 0.19 mm. The parameter ranges examined are: heat fluxes 10, 15 and 20 kW/m^2, mass fluxes 314, 392 and 470 kg/m^{2}s and different saturation pressures 158, 208 kPa for R123; 900, 1100 kPa for R134a; 217 data points were collected.

Shiferaw et al. [57] conducted experiments of flow boiling heat transfer in a stainless steel tube of 1.1 mm inner diameter with R134a. The test parameters are: mass fluxes ranging from 100 to 600 kg/m^{2}s, heat fluxes from 16 to 150 kW/m^2 and the pressure from 6 to 12 bar. 324 data points are obtained from one fluid R134a with varying other parameters.

Table 4.4 Test Fluids and Parameter Ranges of the Database

Data points number	2336
Test fluids	R410A, R141b, R134a, R245fa, R12, R123, R22 and N_2
Channel diameter	0.19−3.69 mm
Mass flux	20−1471.2 kg/m^2 s
Heat flux	5−150 kW/m^2
Orientation	Horizontal and vertical

Wang et al. [62] experimentally investigated flow boiling heat transfer characteristics of R134a in a horizontal mini-tube with inner diameter of 1.3 mm and provided a 365 data points over a range of vapor qualities up to 0.8, mass fluxes from 310 to 860 kg/m^2 s, heat fluxes from 21 to 50 kW/m^2 and saturation pressures from 0.65 to 0.75 MPa.

A heat transfer database including 2336 data points extracted from the selected 11 published reports has been compiled. The database covers a wide range of test fluids and experimental parameter ranges listed in Table 4.4. The database was compared to the 12 flow boiling heat transfer correlation including the correlations of Chen [37], Gungor and Winterton [38] and Liu and Winterton [39] for conventional channels and the nine selected correlations for microchannels in Table 4.2. The statistical analysis is based on relative error ξ_i (the percentage of predicted points within ±30%):

$$\xi_i = \frac{\text{Predicted} - \text{Measured}}{\text{Measured}} \tag{4.30}$$

The heat transfer coefficients of the eight test fluids were compared with the predicted results by the 12 correlations. Table 4.5 shows the statistical analysis of the predicted results with the selected flow boiling correlations for each individual fluid and all fluids using the relative error ξ_i within ±30%. According to the proportions of data falling within the ranges of ±30% error band, comparative results of the two best correlations for each fluid are shown in Figs 4.13−4.20.

Overall, the Saitoh et al. [59] and the Li and Wu [56] correlations provide better predictions than other correlations for the whole database. However, the Saitoh et al. heat transfer correlation only predicted 54.9% of the whole database and the Li and Wu correlation only predicts 56.7% of the whole database. Although some correlations may work well for some fluids, e.g., the Saitoh et al. correlation predicts well the R12, R22, and R141b data, they do not work for other fluids, e.g., the Saitoh et al. correlation only predicts 3.1% of R410A data. It is interesting that the Chen correlation predicts well theR245fa data better than all other correlations except the Li and Wu

Table 4.5 Statistical Analysis of the Predicted Results With the Selected Heat Transfer Correlations for Each Individual Fluid and All Fluids (Relative Error ξ_i within ±30%)

Correlation	R134a	R245fa	R12	R22	R123	R141b	R410A	N$_2$	Total
Chen [37]	40.5%	72.7%	57.4%	41.6%	36.6%	35.9%	4.6%	15.9%	35.2%
Lazark and Black [47]	48.6%	0.8%	92.6%	69.0%	6.3%	66.7%	52.5%	2.5%	40.9%
Kew and Cornwell [45]	55.5%	0.8%	92.6%	83.2%	16.1%	59.0%	56.8%	2.5%	46.3%
Gungor-Winterton [38]	64.0%	71.1%	61.1%	25.7%	4.5%	7.7%	10.8%	5.6%	44.8%
Liu-Winterton [39]	28.5%	39.7%	48.2%	70.8%	33.9%	71.8%	5.8%	44.7%	32.2%
Tran et al. [68]	18.2%	0.8%	55.6%	24.8%	0.0%	76.9%	21.6%	45.8%	22.5%
Warrier et al. [49]	12.8%	5.8%	0.0%	31.9%	24.1%	48.7%	6.6%	39.4%	16.9%
Zhang et al. [50]	41.5%	66.1%	57.4%	42.5%	55.4%	43.6%	3.9%	14.8%	36.2%
Pamitran et al. [52]	48.9%	43.8%	35.2%	0.0%	7.1%	0.0%	48.3%	0.7%	36.7%
Saitoh et al. [59]	63.0%	65.3%	98.2%	87.6%	52.7%	87.2%	3.1%	37.0%	54.9%
Sun and Mishima [55]	67.4%	0.8%	94.4%	76.1%	18.8%	79.5%	27.0%	4.6%	50.0%
Li and Wu [56]	66.0%	76.0%	24.1%	35.4%	85.7%	7.7%	1.2%	68.3%	56.7%

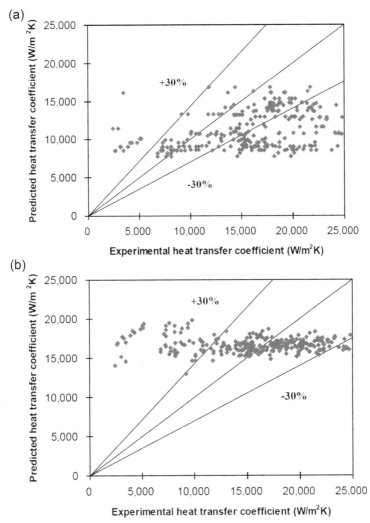

Figure 4.13 Comparison of N_2 heat transfer coefficient data with the best predictive correlation for (a) Tran et al. [68] and (b) Li and Wu [56].

correlation. This might be by chance to capture the data for that fluid, or some other reasons such as the R245fa might be still in macro-scale.

In general, the correlations for flow boiling conventional channels do not work for microchannel flow boiling. It can be concluded that none of the existing correlations could be a general prediction method for flow boiling heat transfer micro-scale channels as shown in Table 4.5. No correlations can satisfactorily predict all the experimental data for the tested eight fluids in the database. It is impossible to reach a conclusion from this comparisons due to the big discrepancies among the experimental data from different researches as illustrated in Section 4.2. Furthermore, another reason

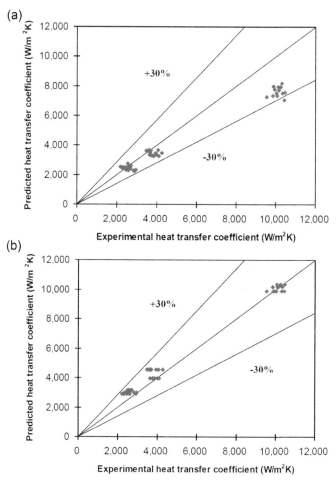

Figure 4.14 Comparison of R12 heat transfer coefficient data with the best predictive correlation for (a) Saitoh et al. [59] and (b) Sun and Mishima [55].

is due to the flow boiling mechanisms in microchannels. In fact, simply modifying the heat transfer correlation for conventional channels does not representing the actual heat transfer mechanisms governing the microchannel flow boiling phenomena. Contradictory heat transfer trends are inconsistent with the said heat transfer mechanisms in some published reports. Simply referring to the two different heat transfer mechanisms for microchannel flow boiling, which can in fact simultaneously occur in microchannels, does not properly representing the actual mechanisms occurred in the microchannels. Furthermore, a large number of experiments were conducted in the presence of two-phase flow instabilities, which may drastically affect the heat transfer trends [95]. Moreover, erroneous data regression procedure is frequently adopted to develop a new correlation. Inherent difficulties verified in conventional flow

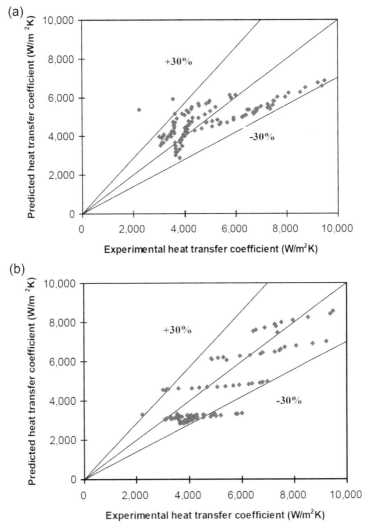

Figure 4.15 Comparison of R22 heat transfer coefficient data with the best predictive correlation for (a) Saitoh et al. [59] and (b) Kew and Cornwell [45].

boiling heat transfer measurements are incremented in the case of microchannels due to the reduced scales involved. Energy balances using single-phase flow heat transfer measurements to validate the experimental apparatus are not a common practice in the available studies. The different data reduction methods may be another big factor which affects the experimental results significantly, and thus further affects the prediction method developed based on these data. It should be pointed out that even when using state-of-the-art instrumentation and calibration procedures, measured flow boiling heat transfer coefficient errors are higher than ±30% or even more, which

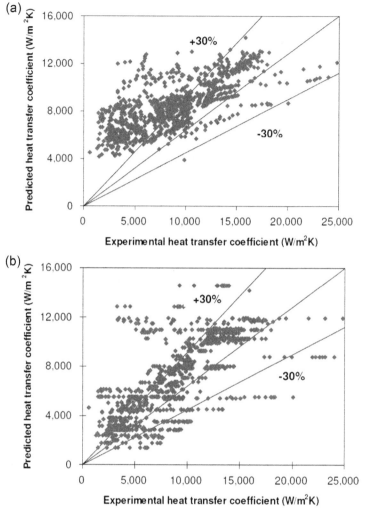

Figure 4.16 Comparative results of R134a data to the best two correlations for (a) Li and Wu [56] and (b) Sun and Mishima [55].

are normal due to the nature of complex two-phase flow boiling. However, complete different data could be produced due to other reasons such as improper test system and measurement methods etc. Therefore, it is essential to design a proper test facility, use the measurement instrument and sensors, use proper data reduction method, design reasonable test runs and do all necessary calibration and validation before performing the experiments to provide reliable and reasonable data for further developing generalized prediction methods for microchannels. Mechanistic prediction methods based on flow pattern information and well developed heat transfer mechanisms should be target in future.

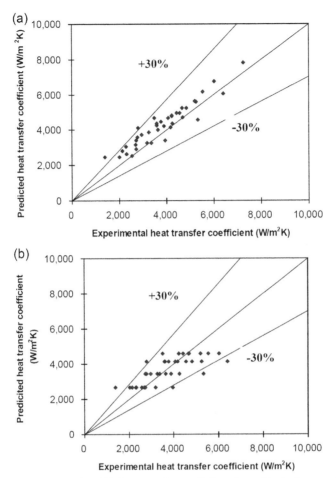

Figure 4.17 Comparison of R141b heat transfer coefficient data with the best predictive correlation for (a) Saitoch et al. [59] and (b) Sun and Mishima [55].

4.4 Models of Flow Boiling Heat Transfer for Specific Flow Patterns in Microchannels

Analytical models for evaporation heat transfer models for specific flow patterns are needed in understanding the theoretical basis of the flow boiling heat transfer but such models have not well developed so far. There are several such models for elongated bubble flow and annular flow in the literature [107–114].

Thome et al. [108] developed a micro-scale model that describes the heat transfer processes during the cyclic passage of elongated bubbles in a micro-scale channel. In this model, bubbles are assumed to nucleate and quickly grow to the channel size upstream such that successive elongated bubbles are formed that are confined radially

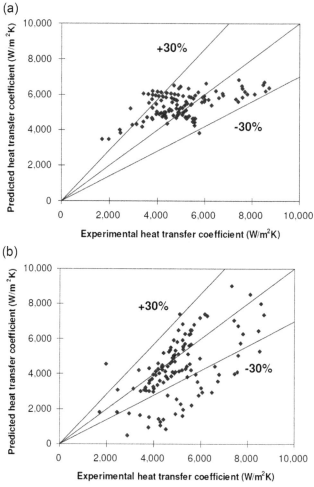

Figure 4.18 Comparative results of R245fa data to the best two correlations for (a) Li and Wu [56] and (b) Chen [37].

by the tube wall and grow in length, trapping a thin film of liquid between the bubble and the inner tube wall. The thickness of this film plays an important role in heat transfer. As shown, in Fig. 4.21, at a fixed location, the process proceeds as follows: (1) a liquid slug passes (without any entrained vapor bubbles, contrary to macro-scale flows), (2) an elongated bubble passes (whose liquid film is formed from liquid removed from the liquid slug), and (3) if the thin evaporating film of the bubble dries out before the arrival of the next liquid slug, then a vapor slug passes. A time-averaged local heat transfer coefficient is obtained during the cyclic passage. Dupont et al. [109] compared the time-averaged local heat transfer coefficient to the experimental data taken from seven independent studies covering seven fluids including R-11, R-12, R-113, R-123, R-134a, R-141b, and CO_2, covering tube diameters from 0.77 to 3.1 mm,

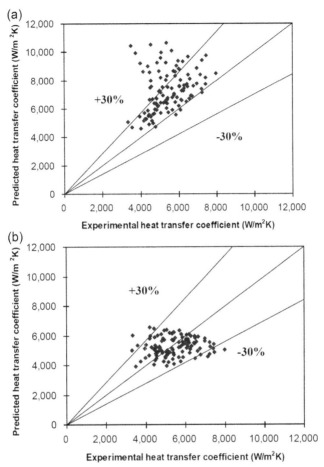

Figure 4.19 Comparison of R123 heat transfer coefficient data with the best predictive correlation for (a) Zhang et al. [50] and (b) Li and Wu [56].

mass velocities from 50 to 564 kg/m²s, saturation pressures from 124 to 5766 kPa, heat fluxes from 5 to 178 kW/m², and vapor qualities from 0.01 to 0.99. Their new three zone model predicts 67% of the database to within ±30%. The new model illustrates the importance of the strong cyclic variation in the heat transfer coefficient and the strong dependency of heat transfer on the bubble frequency, the minimum liquid film thickness at dry-out and the liquid film formation thickness. It should be mentioned here that different flow patterns such as bubbly flow, intermittent flow and annular flow etc. may be relevant to their database. Only based on elongated bubble flow considering the liquid film heat transfer does not really represent the actual mechanisms in the database. In fact, their model predicted an increase in heat transfer coefficient with a decrease in diameter for low values of vapor quality and a decrease in heat transfer coefficient for large vapor qualities. This may be the

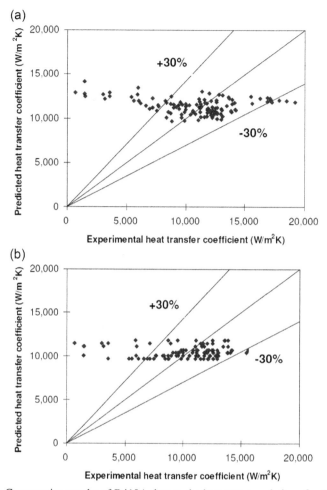

Figure 4.20 Comparative results of R410A data to the best two correlations for (a) Kew and Cornwell [45] and (b) Lazarek and Black [47].

reason why their model predicts 67% of the database, which is actually low. However, this is a very good start to develop theoretical model for microchannel flow boiling.

Consolini and Thome [110] developed a one-dimensional model of confined coalescing bubble flow for the prediction of microchannel convective boiling heat transfer. In their model, two or more bubbles bond under the action of inertia and surface tension, the passage frequency of the bubble−liquid slug pair declines, with a redistribution of liquid among the remaining flow structures. They made assumption that heat transfer occurs only by conduction through the thin evaporating liquid film trapped between the bubbles and the channel wall. Thus, their model includes a simpli-fied description of the dynamics of the formation and flow of the liquid film and the thin film evaporation process, taking into account the added mass transfer by breakup

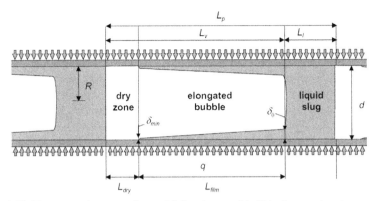

Figure 4.21 Three-zone heat transfer model for elongated bubble flow regime in microchannels: diagram illustrating a triplet composed of a liquid slug, an elongated bubble, and a vapor slug [108].

of the bridging liquid slugs. Their new model has been confronted against experimental data taken within the coalescing bubble flow mode that have been identified by a diabatic micro-scale flow pattern map [115]. The comparisons for three different fluids (R-134a, R-236fa, and R-245fa) gave encouraging results with 83% of the database predicted within a ±30% error band. Furthermore, they have found that their new model is able to predict a "nucleate boiling curve" with an exponent of 0.74 typical of numerous microchannel flow boiling studies, thus suggesting film evaporation as the controlling heat transfer mechanism rather than nucleate boiling. They suggested film evaporation as the controlling heat transfer mechanism rather than nucleate boiling in microchannel flow boiling. However, the experimental data have shown complex heat transfer mechanisms as mentioned in Section 4.2. Furthermore, accurate flow pattern map for microchannel flow boiling is needed to use their model. It is actually a big challenge as no universal flow pattern maps are available to predict the flow patterns precisely as described in Chapter 3.

Annular flow is a common flow pattern observed in microchannel flow boiling. Analytical model for annular flow may be developed from the mass, momentum and energy conservation with the aid of some assumptions and empirical correlations for wall and interfacial shear stress and droplets entrainment and deposition rates such as the study by Cheng [111]. Similarly, annular flow models for microchannel flow boiling have also been investigated by Qu and Mudawar [112] and Kim and Midawaer [113]. These models have some theoretical basis. However, in order to validate these models, a well-documented universal flow pattern map is needed to segment the heat transfer data but a generalized flow pattern map is not yet available.

Analytical models for flow boiling in microchannels are very complex, and the required assumptions to solve these models restrict the ability to capture the real physics of boiling mechanisms. Also, most empirical correlations are not able to predict other experimental data, even under a similar range of operating conditions where the correlations were obtained. The complex nature of flow boiling in microchannels such as liquid—vapor interactions, bubble growth in the flow as well as in the thin liuqid film make analytical or empirical modeling of the two-phase flow a very difficult task. A serious need was felt to conduct a comprehensive study of phase change phenomena in microchannels to understand the fundamental mechanisms involved in the boiling process before attempting any modeling as pointed out by Magnini and Thome [114]. More accurate models for the heat transfer coefficient will be obtained if the modeling efforts are concentrated on each particular flow pattern. Therefore, flow pattern maps with well developed flow patterns and transition lines may facilitate the modeling efforts but universal flow maps are not available so far. Therefore, it is essential to further conduct systematic experimental research in microchannel flow boiling to obtain the heat transfer and flow pattern data simultaneously to provided good database. Furthermore, new heat transfer mechanisms should be developed based on both heat transfer behaviors and the relevant flow patterns. Effort should be made to develop universal flow boiling heat transfer models in future [106,114].

4.5 Concluding Remarks

A large number of studies on microchannel flow boiling heat transfer have been conduced over the past years. Various heat transfer behaviors trends are available, which have been mostly explained with the two flow boiling heat transfer mechanisms by different researchers. However, the actual heat transfer mechanisms are much more complex than the two mechanisms and should be well understood. Furthermore, a number of heat transfer models and correlations have also been developed based on microchannel flow boiling experimental data. From the analysis and comparison of the experimental data from different studies, it can be concluded that well performed and documented experimental studies on microchannel flow boiling are still needed. The channel size effect on the flow boiling heat transfer behaviors and mechanisms have not yet well be understood. In general, the available heat transfer correlations and models poorly predict the experimental database collected from the literature. No universal prediction methods are available for microchannel flow boiling heat transfer so far. Furthermore, the available heat transfer correlations and models lack the heat transfer mechanism basis. The below main issues have been identified in the available studies as:

1. Big discrepancies among experimental results from different studies at similar conditions have been found. Different heat transfer trends have been identified.
2. The two main flow boiling heat transfer mechanisms in conventional channels are generally used to explain the experimental results in microchannels but some anormal trends cannot be reasonably explained.
3. The existing correlations and models poorly predict the independent microchannel flow boiling heat transfer data. The current prediction methods lack the heat transfer mechanism basis. Reliable universal prediction methods are not available so far.
4. Mechanistic prediction methods and models are not well developed.

Therefore, as a priority, well-performed and -documented experimental studies on microchannel flow boiling are needed in future. Systematic experiments on two-phase flow and flow boiling at a wide range of conditions should be conducted by considering the effect of channel size on both heat transfer coefficients and the relevant flow patterns. As already mentioned, flow pattern visualization and transition criteria of two-phase flow and flow boiling of microchannels should be systematically investigated, which should be related to the flow boiling heat transfer behaviors and mechanisms. Generalized heat transfer models should be targeted by incorporating the flow boiling heat transfer mechanisms, flow patterns, channel sizes and fluid properties etc. Development of the theory of two-phase flow and flow boiling of microchannels should also be focused on by analyzing the heat transfer model for specific flow patterns. This should include a detailed study of the mechanisms of mass, momentum and heat transfer under conditions of interaction of hydrodynamic and thermal effects and phase changes in microchannels.

It is envisaged that a systematic research on two-phase flow and flow boiling in microchannels will bring advancement of knowledge and new theory of two-phase flow and flow boiling in microchannels and meet the practical requirements in various applications. However, it is still a challenge to in microchannel flow boiling field due to the complexity and difficulty of two-phase flow and flow boiling phenomena microchannels. Efforts should be made to contribute to both experimental and theoretical studies in the future.

Nomenclature

Bd	Bond number, $[g(\rho_l - \rho_g)D_h^2/\sigma]$
Bo	Boiling number, $[q/(Gi_{fg})]$
C	Constant
c_p	Specific heat at constant pressure, J/kg K
D_h	Internal tube hydraulic diameter, m
d_b	Bubble diameter, m
E	Convective boiling heat transfer enhancement factor
E'	Convective boiling heat transfer enhancement factor
E_2	Correct factor defined by Eq. (4.15)
Fr	Froude number, $[G^2/(\rho_L^2 g D_h)]$
G	Total gas and liquid two-phase mass velocity, kg/m^2 s
g	Gravitational acceleration, 9.81 m/s^2
h	Heat transfer coefficient, W/m^2 K
i	Data point number
i_{fg}	Latent heat of evaporation, J/kg

Continued

k	Thermal conductivity, W/mK
M	Molecular weight
p	Pressure, N/m^2
p_r	Reduced pressure $[p/p_{crit}]$
Pr_l	Prandtl number, $[\mu_l c_{pl}/k_l]$
q	Heat flux, W/m^2
Re_{lo}	Reynolds number considering the total gas–liquid two-phase flow as liquid flow $[GD_h/(\mu_L)]$
Re_l	Reynolds number considering only liquid phase flow $[G(1-x)D_h/(\mu_L)]$
Re_{tp}	Two-phase Reynolds number defined by Eq. (4.8)
S	Nucleate boiling suppression factor
$S2$	Correction factor defined by Eq. (4.16)
T	Temperature, K
T_{sat}	Saturation temperature, K
We_l	Liquid Weber number considering the total vapor–liquid flow as liquid flow $[G^2 D_h/(\rho_l \sigma)]$
We_g	Gas phase Weber number $[G^2 x^2 D_h/(\rho_g \, \sigma)]$
X_{tt}	Martinelli number, $\{[(1-x)/x)]^{0.9}[\rho_g/\rho_l]^{0.5}[\mu_l/\mu_g]^{0.1}\}$
x	Vapor quality
Greek symbols	
ϕ^2	Two-phase friction multiplier
μ	Dynamic viscosity, Ns/m^2
ρ	Density, kg/m^3
σ	Surface tension, N/m
ξ_i	Relative error defined by Eq. (4.21)
Subscripts	
b	Bubble
cb	Convective boiling
$crit$	Critical
fg	Latent
g	Gas phase
l	Liquid phase
lo	Considering the total gas–liquid two-phase flow as liquid flow
nb	Nucleate boiling

p	Constant pressure
pool	Pool boiling
sat	Saturation
sp	Single phase
tp	Two phase
w	Wall

References

[1] L. Cheng, Microscale and nanoscale thermal and fluid transport phenomena: Rapidly developing research fields, Int. J. Microscale Nanoscale Therm. Fluid Transp. Phenom. 1 (2010) 3−6.

[2] G.P. Celata, Microscale heat transfer in single- and two-phase flows: Scaling, stability and transition, Int. J. Microscale Nanoscale Therm. Fluid Transp. Phenom. 1 (2010) 7−36.

[3] S.K. Suha, G. Zummo, G.P. Celata, Review on flow boiling in microchannels, Int. J. Microscale Nanoscale Therm. Fluid Transp. Phenom. 1 (2010) 111−178.

[4] J.R. Thome, The new frontier in heat transfer: microscale and nanoscale technologies, Heat Transfer Eng. 27 (9) (2006) 1−3.

[5] S.G. Kandlikar, Fundamental issues related to flow boiling in minichannels and microchannels, Exp. Therm. Fluid Sci. 26 (2002) 389−407.

[6] J.R. Thome, State-of-the art overview of boiling and two-phase flows in microschannels, Heat Transfer Eng. 27 (9) (2006) 4−19.

[7] L. Cheng, E.P. Bandarra Filho, J.R. Thome, Nanofluid two-phase flow and thermal physics: a new research frontier of nanotechnology and its challenges, J. Nanosci. Nanotech. 8 (2008) 3315−3332.

[8] L. Cheng, L. Liu, Boiling and two phase flow phenomena of refrigerant-based nanofluids: fundamentals, applications and challenges, Int. J. Refrigeration 36 (2013) 421−446.

[9] L. Cheng, Nanofluid heat transfer technologies, Recent Patents on Eng. 3 (1) (2009) 1−7.

[10] L. Cheng, Fundamental issues of critical heat flux phenomena during flow boiling in microscale-channels and nucleate pool boiling in confined spaces, Heat Transfer Eng. 34 (2013) 1011−1043.

[11] J.R. Thome, Boiling in microchannels: a review of experiment and theory, Int. J. Heat Fluid flow 25 (2004) 128−139.

[12] G. Ribatski, A critical overview on the recent literature concerning flow boiling and two-phase flows inside micro-scale channels, Exp. Heat Transfer 26 (2013) 198−246.

[13] L. Cheng, G. Ribatski, J.R. Thome, Gas-liquid two-phase flow patterns and flow pattern maps: fundamentals and applications, ASME Appl. Mech. Rev 61 (2008) 050802.

[14] L. Cheng, D. Mewes, Review of two-phase flow and flow boiling of mixtures in small and mini channels, Int. J. Multiphase Flow 32 (2006) 183−207.

[15] G.P. Celata, S.K. Suha, G. Zummeo, Heat transfer characteristics of flow boiling in a single horizontal microchannel, Int. J. Therm. Sci 49 (2010) 1086−1094.

[16] L. Cheng, Critical heat flux in microscale channels and confined Spaces: a review on experimental studies and prediction methods, Russ. J. General Chem. 82 (12) (2012) 2116–2131.

[17] L. Cheng, J.R. Thome, Cooling of microprocessors using flow boiling of CO_2 in a micro-evaporator: preliminary analysis and performance comparison, Appl. Therm. Eng. 29 (2009) 2426–2432.

[18] C.B. Tibiriçá, G. Ribatski, An experimental study on flow boiling heat transfer of R134a in a 2.3 mm Tube, Int. J. Microscale Nanoscale Therm. Fluid Transp. Phenom. 1 (2010) 37–58.

[19] L. Cheng, L. Liu, Analysis and evaluation of gas-liquid two-phase frictional pressure drop prediction methods for microscale channels, Int. J. Microscale Nanoscale Therm. Fluid Transp. Phenom. 2 (2011) 259–280.

[20] L. Cheng, H. Zou, Evaluation of flow boiling heat transfer correlations with experimental data of R134a, R22, R410A and R245fa in microscale channels, Int. J. Microscale Nanoscale Therm. Fluid Transp. Phenom. 1 (2010) 363–380.

[21] L. Cheng, D. Mewes, A. Luke, Boiling phenomena with surfactants and polymeric additives: a state-of-the-art review, Int. J. Heat Mass Transfer 50 (2007) 2744–2771.

[22] G. Ribatski, L. Wojtan, J.R. Thome, An analysis of experimental data and prediction methods for two-phase frictional pressure drop and flow boiling heat transfer in microscale channels, Exp. Therm. Fluid Sci. 31 (2006) 1–19.

[23] J. Moreno Quibén, L. Cheng, R.J. da Silva Lima, J.R. Thome, Flow boiling in horizontal flattened tubes: part I — two-phase frictional pressure drop results and model, Int. J. Heat Mass Transfer 52 (2009) 3634–3644.

[24] J. Moreno Quibén, L. Cheng, R.J. da Silva Lima, J.R. Thome, Flow boiling in horizontal flattened tubes: part II — flow boiling heat transfer results and model, Int. J. Heat Mass Transfer 52 (2009) 3645–3653.

[25] J.G. Collier, J.R. Thome, Convective Boiling and Condensation, Oxford University Press Inc, New York, 1994.

[26] V.P. Carey, Liquid Vapor Phase Change Phenomena, Hemisphere, Washington D.C, 1992.

[27] N. Kattan, J.R. Thome, D. Favrat, Flow boiling in horizontal tubes. Part 1: development of a diabatic two-phase flow pattern map, J. Heat Transfer 120 (1998) 140–147.

[28] N. Kattan, J.R. Thome, D. Favrat, Flow boiling in horizontal tubes: Part 2-New heat transfer data for five refrigerants, J. Heat Transfer 120 (1998) 148–155.

[29] N. Kattan, J.R. Thome, D. Favrat, Flow boiling in horizontal tubes: Part-3: development of a new heat transfer model based on flow patterns, J. Heat Transfer 120 (1998) 156–165.

[30] L. Cheng, G. Ribatski, L. Wojtan, J.R. Thome, New flow boiling heat transfer model and flow pattern map for carbon dioxide evaporating inside horizontal tubes, Int. J. Heat Mass Transfer 49 (2006) 4082–4094.

[31] L. Cheng, G. Ribatski, J. Quibén Moreno, J.R. Thome, New prediction methods for CO_2 evaporation inside tubes: Part I — a two-phase flow pattern map and a flow pattern based phenomenological model for two-phase flow frictional pressure drops, Int. J. Heat Mass Transfer 51 (2008) 111–124.

[32] L. Cheng, G. Ribatski, J.R. Thome, New prediction methods for CO_2 evaporation inside tubes: Part II — an updated general flow boiling heat transfer model based on flow patterns, Int. J. Heat Mass Transfer 51 (2008) 125–135.

[33]   L. Wojtan, T. Ursenbacher, J.R. Thome, Investigation of flow boiling in horizontal tubes: Part I — a new diabatic two-phase flow pattern map, Int. J. Heat Mass Transfer 48 (2005) 2955—2969.

[34]   L. Wojtan, T. Ursenbacher, J.R. Thome, Investigation of flow boiling in horizontal tubes: Part II — development of a new heat transfer model for stratified-wavy, dryout and mist flow regimes, Int. J. Heat Mass Transfer 48 (2005) 2970—2985.

[35]   L. Cheng, T. Chen, Flow boiling heat transfer in a vertical spirally internally ribbed tube, Heat Mass Transfer 37 (2001) 229—236.

[36]   L. Cheng, T. Chen, Study of flow boiling heat transfer in a tube with axial micro-grooves, Exp. Heat Transfer 14 (1) (2001) 59—73.

[37]   J.C. Chen, Correlation for boiling heat transfer to saturated fluids in convective flow, Industrial and Eng. Chem. — Process Design and Development 5 (3) (1966) 322—329.

[38]   K.E. Gungor, R.H.S. Winterton, A general correlation for flow boiling in tubes and annuli, Int. J. Heat Mass Transfer 29 (3) (1986) 351—358.

[39]   Z. Liu, R.H.S. Winterton, A general correlation for saturated and subcooled flow boiling in tubes and annuli, based on a nucleate pool boiling equation, Int. J. Heat Mass Transfer 34 (11) (1991) 2759—2766.

[40]   S.G. Kandlikar, A general correlation for two-phase flow boiling heat transfer coefficient inside horizontal and vertical tubes, J. Heat Transfer 102 (1990) 219—228.

[41]   S.G. Kandlikar, A model for predicting the two-phase flow boiling heat transfer coefficient in augmented tube and compact heat exchanger geometries, J. Heat Transfer 113 (1991) 966—972.

[42]   L. Cheng, T. Chen, Comparison of six typical correlations for upward flow boiling heat transfer with kerosene in a vertical smooth tube, Heat Transfer Eng. 21 (5) (2000) 27—34.

[43]   S.-M. Kim, I. Mudawar, Review of databases and predictive methods for heat transfer in condensing and boiling mini/micro-channel flows, Int. J. Heat and Mass Transfer 77 (2014) 627—652.

[44]   B. Agostini, J.R. Thome, Comparison of an extended database of flow boiling heat transfer coefficient in multi-microchannel elements with the three-zone model, in: ECI International Conference on Heat Transfer and Fluid Flow in Microscale, Castelvecchio Pascoli, Italy, 2005.

[45]   P.A. Kew, K. Cornwell, Correlations for the prediction of boiling heat transfer in small-diameter channels, Appl. Therm. Eng. 17 (1997) 705—715.

[46]   T. Harirchian, S.V. Garimella, A comprehensive flow regime map for microchannel flow boiling with quantitative transition criteria, Int. J. Heat Mass Transfer 53 (2010) 2694—2702.

[47]   G.M. Lazarek, S.H. Black, Evaporative heat transfer, pressure drop and critical heat flux in a small vertical tube with R-113, Int. J. Heat Mass Transfer 25 (7) (1982) 945—960.

[48]   Z.Y. Bao, D.F. Fletcher, B.S. Haynes, Flow boiling heat transfer of freon R11 and HCFC123 in narrow passages, Int. J. Heat Mass Transfer 43 (2000) 3347—3358.

[49]   G.R. Warrier, V.K. Dhir, L.A. Momoda, Heat transfer and pressure drop in marrow rectangular channels, Exp. Therm. Fluid Sci. 26 (2002) 53—64.

[50]   W. Zhang, T. Hibiki, K. Mishima, Correlation for flow boiling heat transfer in mini-channels, Int. J. Heat Mass Transfer 47 (2004) 5749—5763.

[51]   J. Lee, I. Mudawar, Two-phase flow in high-heat-flux micro-channel heat sink for refrigeration cooling applications: Part II-heat transfer characteristics, Int. J. Heat Mass Transfer 48 (2005) 941—955.

[52] A.S. Pamitran, K. Choi, J. Oh, H. Oh, Forced convective boiling heat transfer of R-410A in horizontal minichannels, Int. J. Refrigeration 30 (2007) 155—165.

[53] S.S. Bertsch, E.A. Groll, S.V. Garimella, A composite heat transfer correlation for saturated flow boiling in small channels, Int. J. Heat Mass Transfer 52 (2009) 2110—2118.

[54] C.L. Ong, J.R. Thome, Flow boiling heat transfer of R134a, R236a and R245fa in a horizontal 1.030 mm circular channel, Exp. Therm. Fluid Sci. 33 (2009) 651—663.

[55] L. Sun, K. Mishima, An evaluation of prediction methods for saturated methods for saturated flow boiling heat transfer, Int. J. Heat Mass Transfer 52 (2009) 5323—5329.

[56] W. Li, Z. Wu, A general correlation for evaporative heat transfer in micro/mini-channels, Int. J. Heat Mass Transfer 53 (2010) 1778—1787.

[57] C.B. Tibirica, G. Ribatski, Flow boiling heat transfer of R134a and R245fa in a 2.3 mm tube, Int. J. Heat Mass Transfer 53 (2010) 2459—2468.

[58] W. Qu, I. Mudawar, Flow boiling heat transfer in two-phase micro-channels heat sinks — I. Experimental investigation and assessment of correlation methods, Int. J. Heat Mass Transfer 46 (2003) 2755—2771.

[59] S. Saitoh, H. Daiguji, E. Hihara, Effect of tube diameter on boiling heat transfer of R-134a in horizontal small-diameter tubes, Int. J. Heat Mass Transfer 48 (2005) 4973—4984.

[60] S.S. Bertsch, E.A. Groll, S.V. Garimella, Refrigerant flow boiling heat transfer in parallel microchannels as a function of local vapor quality, Int. J. Heat Mass Transfer 51 (2008) 3724—3735.

[61] S. In, S. Jeong, Flow boiling heat transfer of R134a, R236fa and R245fa in a horizontal 1.03mm circular channel, Exp. Therm. Fluid Sci. 33 (2009) 651—663.

[62] L. Wang, M. Chen, M. Groll, Flow boiling heat transfer characteristics of R134a in a horizontal mini tube, J. Chem. Eng. Data 54 (2009) 2638—2645.

[63] T. Harirchian, S.V. Garimella, Microchannel size effects on local flow boiling heat transfer to a dielectric fluid, Int. J. Heat Mass Transfer 51 (2008) 3724—3735.

[64] K.H. Bang, W.H. Choo, Flow boiling in minichannels of copper, brass, and aluminum round tubes, in: Proceedings of 2nd International Conference on Microchannels and Minichannels, Rochester, USA, 2004, pp. 559—564.

[65] N.H. Kim, Y.S. Sim, C.K. Min, Convective boiling of R22 in a flat extruded aluminum multi-port tube, in: Proceedings of 2nd International Conference on Microchannels and Minichannels, Rochester, USA, 2004, pp. 507—514.

[66] R. Yun, J. Heo, Y. Kim, J.T. Chung, Convective boiling heat transfer characteristics of R410A in microchannels, in: Proceedings of 10th International Refrigeration and Air Conditioning Conference at Purdue, West Lafayette, USA, 2004.

[67] K. Choi, A.S. Pamitran, C.Y. Oh, J.T. Oh, Boiling heat transfer of R-22, R-134a, and CO_2 in horizontal smooth minichannels, Int. J. Refrigeration 30 (2007) 1336—1346.

[68] T.N. Tran, M.W. Wambsganss, D.M. France, Small circular and rectangular channel boiling with two refrigerants, Int. J. Multiphase Flow 22 (1996) 485—498.

[69] W. Owhaib, C. Martin-Callizo, B. Palm, Evaporative heat transfer in vertical circular microchannels, Appl. Therm. Eng. 24 (2004) 1241—1253.

[70] X. Huo, L. Chen, Y.S. Tian, T.G. Karayiannis, Flow boiling and flow regimes in small diameter tubes, Appl. Therm. Eng. 24 (2004) 1225—1239.

[71] Y.Y. Yan, T.F. Lin, Evaporation heat transfer and pressure drop of refrigerant R134a in a small pipe, Int. J. Heat Mass Transfer 41 (1998) 4183—4194.

[72] C. Huh, M.H. Kim, Two-phase pressure drop and boiling heat transfer in a single horizontal microchannel, in: Proceedings of the 4th Int. Conf. Nanochannels, Microchannels and Minichannels, ICNMM2006, 2006, pp. 1097−1104. B.

[73] S. Lin, P.A. Kew, K. Cornwell, Flow boiling of refrigerant R141b in small tubes, Chem. Eng. Res. Des. 79 (4) (2001) 417−424.

[74] T. Harirchian, S.V. Garimella, Effects of channel dimension, heat flux, and mass flux on flow boiling regimes in microchannels, Int. J. Multiphase Flow 35 (2009) 349−362.

[75] T. Chen, S.V. Garimella, Measurements and high-speed visualization of flow boiling of a dielectric fluid in a silicon microchannel heat sink, Int. J. Multiphase Flow 32 (8) (2006) 957−971.

[76] D. Liu, S.V. Garimella, Flow boiling heat transfer in microchannels, J. Heat Transfer 129 (10) (2007) 1321−1332.

[77] T. Chen, S.V. Garimella, Flow boiling heat transfer to a dielectric coolant in a microchannel heat sink, IEEE Trans. Compon. Packag. Technol. 30 (1) (2006) 24−31.

[78] L. Zhang, E.N. Wang, K.E. Goodson, T.W. Kenny, Phase change phenomena in silicon microchannels, Int. J. Heat Mass Transfer 48 (2005) 1572−1582.

[79] A.E. Bergles, J.H. Lienhard, G.E. Kendall, P. Griffith, Boiling and evaporation in small diameter channels, Heat Transfer Eng. 24 (1) (2003) 18−40.

[80] M. Lee, L.S.L. Cheung, Y. Lee, Y. Zohar, Height effect on nucleation-site activity and size-dependent bubble dynamics in microchannel convective boiling, J. Micromech. and Microeng. 15 (2005) 2121−2129.

[81] M. Lee, Y.Y. Wong, M. Wong, Y. Zohar, Size and shape effects on two-phase flow patterns in microchannel forced convection boiling, Journal of Micromech. and Microeng. 13 (2003) 155−164.

[82] V. Dupont, R. Thome, Evaporation in microchannels: influence of the channel diameter on heat transfer, Microfluidics and Nanofluidics 1 (2) (2005) 119−127.

[83] T. Chen, S.V. Garimella, Effect of dissolved air on subcooled flow boiling of a dielectric coolant in a microchannel heat sink, J. Electronic Packaging 128 (4) (2006) 398−404.

[84] P.-S. Lee, S.V. Garimella, Saturated flow boiling heat transfer and pressure drop in silicon microchannel arrays, International J. of Heat Mass Transfer 51 (3−4) (2008) 789−806.

[85] M.G. Cooper, Heat flow rates in saturated nucleate pool boiling − a wideranging examination using reduced properties, Adv. in Heat Transfer 16 (1984) 157−239.

[86] R. Charnay, R. Revellin, J. Bonjour, Flow boiling characteristics of R-245fa in a minichannel at medium saturation temperatures, Exp. Therm. Fluid Sci. 59 (2014) 184−194.

[87] C. Vlasie, H. Macchi, J. Guilpart, B. Agostini, Flow boiling in small diameter channels, Int. J. Refrigeration 27 (2004) 191−201.

[88] B. Agostini, A. Bontemps, Vertical flow boiling of refrigerant R134a in small channels, Int. J. Heat and Fluid Flow 26 (2005) 296−306.

[89] R. Yun, J.H. Heo, Y. Kim, Evaporative heat transfer and pressure drop of R410A in microchannels, Int. J. Refrigeration 29 (2006) 92−100.

[90] S.L. Qi, P. Zhang, L.X. Xu, Flow boiling of liquid nitrogen in micro-tubes: Part II − heat transfer characteristics and critical heat flux, Int. J. Heat Mass Transfer 50 (2007) 5017−5030.

[91] D. Shiferaw, T.G. Karayiannis, D.B.R. Kenning, Flow boiling in a 1.1 mm tube with R134a: experimental results and comparison with model, Int. J. Therm. Sci 48 (2009) 331−341.

[92]   T.-H. Yen, M. Shoji, F. Takemura, Y. Suzuki, N. Kasagi, Visualization of convective boiling heat transfer in single microchannels with different shaped cross-sections, Int. J. Heat Mass Transfer 49 (2006) 3884−3894.

[93]   G. Hetsroni, A. Mosyak, Z. Segal, G. Ziskind, A uniform temperature heat sink for cooling of electronic devices, Int. J. Heat Mass Transfer 45 (2002) 3275−3286.

[94]   J. Pettersen, Flow vaporization of CO_2 in microchannel tubes, Exp. Therm. Fluid Sci. 28 (2004) 111−121.

[95]   G. Wang, P. Cheng, A.E. Bergles, Effects of inlet/outlet configurations on flow boiling instability in parallel microchannels, Int. J. Heat and Mass Transfer 51 (2008) 2267−2281.

[96]   H.Y. Zhang, D. Pinjala, T.N. Wong, Experimental characterization of flow boiling heat dissipation in a microchannel heat sink with different orientations, in: Proceeding of 7th Electronics Packaging Technology Conference, EPTC, 2, 2005, pp. 670−676.

[97]   L. Jiang, M. Wong, Y. Zohar, Phase change in microchannel heat sinks with integrated temperature sensors, J. Microelectromech. Syst 8 (1999) 358−365.

[98]   S. Saisorn, J. Kaew-On, S. Wongwises, An experimental investigation of flow boiling heat transfer of R-134a in horizontal and vertical mini-channels, Exp. Therm. Fluid Sci. 46 (2013) 232−244.

[99]   W. Yu, D.M. France, M.W. Wambsganss, J.R. Hull, Two phase pressure drop, boiling heat transfer, and critical heat flux to water in a small-diameter horizontal tube, Int. J. Multiphase Flow 28 (2002) 927−941.

[100] G. Hetsroni, A. Mosyak, E. Pogrebnyak, Z. Segal, Explosive boiling of water in parallel micro-channels, Int. J. Multiphase Flow 31 (2005) 371−392.

[101] D. Steiner, J. Taborek, Flow boiling heat transfer in vertical tubes correlated by an asymptotic model, Heat Transfer Eng. 13 (1992) 43−69.

[102] F.W. Dittus, L.M.K. Boelter, Heat transfer in automobile radiators of the tubular type, in: Pub. Eng., vol. 2, University of California, Berkeley, 1930, p. 443.

[103] H.K. Foster, N. Zuber, Bubble dynamics and boiling heat transfer, AIChE J. 1 (1955) 531−535.

[104] M.G. Cooper, Saturation nucleate pool boiling: a simple correlation, Int. Chem. Eng. Symp. Ser. 86 (1984) 785−793.

[105] K. Stephen, M. Abdelsalam, Heat-transfer correlation for natural convection boiling, Int. J. Heat Mass Transfer 23 (1980) 73−87.

[106] S.G. Kandlikar, Scale effects on flow boiling heat transfer in microchannels: a fundamental perspective, Int. J. Therm. Sci. 49 (2010) 1073−1085.

[107] A.M. Jacobi, J.R. Thome, Heat transfer model for evaporation of elongated bubble flows in microchannels, J. Heat Transfer 124 (2002) 1131−1136.

[108] J.R. Thome, V. Dupont, A.M. Jacobi, Heat transfer model for evaporation in microchannels. Part I: Presentation of the model, Int. J. Heat Mass Transfer 47 (2004) 3375−3385.

[109] V. Dupont, J.R. Thome, A.M. Jacobi, Heat transfer model for evaporation in microchannels. Part II: comparison with the database, Int. J. Heat Mass Transfer 47 (2004) 3387−3401.

[110] L. Consolini, J.R. Thome, A heat transfer model for evaporation of coalescing bubble in micro-channel flow, Int. J. Heat Fluid Flow 31 (2010) 115−125.

[111] L. Cheng, Modelling of heat transfer of upward annular flow in vertical tubes, Chem. Eng. Comm. 194 (2007) 975−993.

[112] W. Qu, I. Mudawar, Flow boiling heat transfer in two-phase micro-channel heat sinks—II. Annular two-phase flow model, Int. J. Heat Mass Transfer 46 (2003) 2773−2784.

[113] S.-M. Kim, I. Mudawar, Theoretical model for local heat transfer coefficient for annular flow boiling in circular mini/micro-channels, Int. J. Heat Mass Transfer 73 (2014) 731−742.

[114] S.M. Magnini, J.R. Thome, Proposed models, ongoing experiments, and latest numerical simulations of microchannel two-phase flow boiling, Int. J. Multiphase Flow 59 (2014) 84−101.

[115] R. Revellin, J.R. Thome, A new type of diabatic flow pattern map for boiling heat transfer in microchannels, J. Micromech. Microeng. 17 (2006) 788−796.

Pressure Drop

Sujoy K. Saha[1], Gian P. Celata[2]
[1]Department of Mechanical Engineering, Indian Institute of Engineering Science
and Technology, Shibpur, Howrah, West Bengal, India; [2]Energy Technology Department,
ENEA Casaccia Research Centre, S. M. Galeria, Rome, Italy

5.1 Introduction

Internal flow configurations are found to be most convenient for heating and cooling
and hence have been widely used for this purpose. With heat being a form of useful
energy, its optimum utilization is highly desirable from industrial economy and environmental protection points of view. Researchers and scientists are striving to decrease
the operating cost by identifying sources of thermal resistance and subsequently
reducing them through active and passive techniques [1]. The transport sector (road,
aviation, and marine) has further motivated to reduce the weight and size of heat
exchangers due to space and their carrying capacity limitations. The electronics industry is observing exponential growth in developing high-speed processing computers,
multifeature mobile units, and surgical and biomedical instruments, but this has led
to increased component density per unit area. As a result, cooling of electronic components has become a serious concern, since ineffective cooling will lead to malfunc-
tion or early failure. No doubt this challenge and opportunity of cooling of electronic
components have evolved in the concept of application of mini/microchannels. Tuckerman and Pease in 1981 [2] demonstrated the first time the solution to the electronic
cooling problem by performing experiments on silicon-based microchannel heat sink.
Since then, many research findings, through analytical and experimental
investigations, have been reported. The early (1991−2002) studies can be grouped
as analytical [3−9], numerical [10,11], or experimental [12−21]. To differentiate
microchannels from conventional heat exchangers, different criteria for classifications
have been proposed. Mehendale et al.'s [22] classification was based simply on the
dimensions (hydraulic diameter) of the channels, as given in Table 5.1.

Table 5.1 Classification of Channels [22]

Class of Heat Exchanger	Range of Hydraulic Diameter (D_h)
Conventional	$D_h \geq 6$ mm
Compact	1 mm $\leq D_h \leq 6$ mm
Macro or mini	100 μm $\leq D_h \leq 1$ mm
Micro	1 μm $\leq D_h \leq 100$ μm

Microchannel Phase Change Transport Phenomena. http://dx.doi.org/10.1016/B978-0-12-804318-9.00005-4

On the other hand, classification provided by Kandlikar [23] is based on flow consideration and is suggested as

Conventional heat exchangers	$D_h \geq 3$ mm
Macro or mini heat exchangers	$200\ \mu m \leq D_h \leq 3$ mm
Micro heat exchangers	$10\ \mu m \leq D_h \leq 200\ \mu m$

Many reviews [24–37] have been periodically written encompassing and highlighting new findings, advances, challenges, and roadmaps to erase the gray areas in microchannel design and development. These periodic updates play vital roles in identifying new research problems or domains. Steinke and Kandlikar [38] have discussed the possibility of using conventional heat transfer enhancement techniques in mini-/microchannels. The summary of techniques is given in Table 5.2, and Figs. 5.1 and 5.2 depict the potential candidates that can be used in mini/microchannels for heat transfer enhancement in single-phase flow.

5.2 Studies on Flow Characteristics of Water in Microtubes

For practical application of mini/microchannels, understanding of fluid and heat transfer characteristics is of utmost importance. In this section, flow characteristics in microchannels are discussed and, in addition, views and experimental findings of researchers regarding the possibilities of applicability of conventional theories and correlations are collated. Tuckerman and Pease [2], who were the first to propose electronic cooling using reduced-size channels, reported compliance of classic laminar flow theory while fluid flowed in microchannels. Choi et al. [39] reported significant difference between experimental results of friction factor and heat transfer coefficient with that calculated from classic correlations.

Li and Mala [40] conducted experiments with deionized water in order to investigate its flow characteristics as it flows through stainless steel (SS) and fused silica (FS) microtubes having internal diameters ranging from 50 to 254 μm. The details of individual microtubes are given in Table 5.3.

During the experiment, the flow rate was kept constant, and then pressure difference required to force the liquid through the microtube was measured. The pressure drop was also calculated using classic correlation, which is given by the Poiseuille flow equation,

$$Q = \frac{\pi R^4}{8\mu l}\, \Delta p$$

Figure 5.3 demonstrates the comparison of the experimentally measured value of the pressure gradient with the predicted values calculated using classic correlations (shown by dotted curve) for both SS and FS microtubes.

Table 5.2 Summary of Enhancement Techniques for Use in Micro and Minichannels [38]

Enhancement Technique	Conventional $D_h > 3$ mm	Minichannel 3 mm $\geq D_h > 200$ µm	Microchannel 200 µm $\geq D_h > 10$ µm
		Passive Techniques	
Surface roughness	Roughness structure remains in boundary layer, provides early transition to turbulence	Use different surface treatments, roughness structures can remain in boundary layer and protrude into bulk flow	Can achieve with various etches; roughness structures may greatly influence flow field
Flow disruption	Using twisted tape, coiled wires, offset strip fins, fairly effective	Can extend conventional methods here; offset strip fins, some twisted tapes, small gauge wire	Can use sidewall or in channel; optimize geometry for minimal impact on flow
Channel curvature	Not practical due to large radius of curvature; have been demonstrated $D_h = 3.33$ mm	More possible than convectional; incorporate return bends for compact heat exchangers	Most practical; achievable radius of curvature; large no serpentine channels
Reentrant obstructions	Effect not as prevalent; bulk flow reaches fully developed flow quickly; harder to return flow to developing state	Can incorporate structures to interrupt flow; header design could contribute to pre-existing turbulence	Short paths make for dominate behavior; can incorporate opportunities to maintain developing flows
Secondary flows	Flows obstructions can generate secondary flows; combination of inserts and obstructions	Could use jets to aid in second flow generation; combination of inserts and obstructions	Can fabricate geometries to promote mixing of fluid in channel
Out-of-plane mixing	Not very effective; space requirements prohibitive	Possible use; three-dimensional mixing may not be that effective	Greatest potential; fabricate complex 3D geometries very difficult
Fluid additives	Phase change material (PCMs) dominate	PCMs possible; fluid additives possible	Fluid additives; microparticles and nanoparticles possible

Continued

Table 5.2 Summary of Enhancement Techniques for Use in Micro and Minichannels [38]—cont'd

Enhancement Technique	Conventional D_h > 3 mm	Minichannel 3 mm ≥ D_h > 200 μm	Microchannel 200 μm ≥ D_h > 10 μm
		Active Techniques	
Vibrations	Surface and fluid vibration used currently	Possible to implement; can use in compact heat exchanger	External power is a problem; integrate piezoelectric actuators
Electrostatic Fields	Electrohydrodynamic forces currently used; integrated electrodes	Could be easier to integrate into compact heat exchanger; external power not be problematic	Can integrate electrodes into channel walls; power consumption problematic
Flow pulsation	Established work showing enhancement	Can implement in compact heat exchangers fluid delivery	Possible to implement, could make fluid delivery simpler
Variable roughness structures	Difficult to integrate very small variable structures into a convectional channel	Difficult to integrate into compact heat exchangers	Possible to integrate; piezoelectric actuators change roughness structures

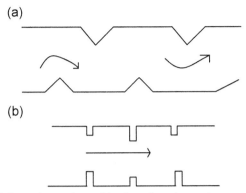

Figure 5.1 Sidewall flow obstructions: (a) triangular obstructions and (b) square obstructions [38].

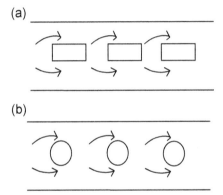

Figure 5.2 Flow obstructions in the channels: (a) simple rectangular geometry and (b) circular profile [38].

From the above plot, the following observations were reported.

1. Considerable agreement between experimentally measured pressure gradient and predicted values from classic correlation were noticed for lower Reynolds number, while for higher Reynolds number, significant differences between the two values were observed.
2. The difference between experimentally measured pressure gradient and predicted values from correlation was further magnified with decrease in microtube diameter.

The reasons for this deviation were attributed to early transition of flow from laminar to turbulent and/or surface roughness.

The term "friction constant ratio" was introduced to predict the intensity of deviation and was defined as,

$$C^* = \frac{f_{exp}\text{Re}}{f_{th}\text{Re}}$$

Table 5.3 Individual Microtubes [40]

Dimensions of SS and FS Microtubes			
Diameter D ± 2 μm	Length of Tubes (cm) ± 0.005		
	Shorter l_1	Longer l_2	Difference Δl
SS			
63.50	3.0	5.50	2.50
101.6	4.0	6.10	2.10
130.0	4.2	8.15	3.95
152.0	4.8	8.00	3.20
203.0	3.5	6.10	2.60
254.0	5.0	8.80	3.80
FS			
50.0	2.85	5.30	2.45
76.0	3.20	5.90	2.70
80.0	3.7	6.30	2.60
101.0	3.30	6.20	2.90
150.0	3.40	6.45	3.05
205.0	3.20	6.10	2.90
250.0	3.10	6.20	3.10

Mean surface roughness of both the SS and FS microtubes ±1.75 μm.

In case of fully developed laminar flow through conventional heat exchanger, product (f·Re) remains constant but for flow in microtubes this product no longer remains constant. The same is found experimentally and shown in Fig. 5.4. Obviously, the desired value of C^* is 1, but it is found that the majority of data fall between $0.6 \leq C \leq 1.4$ [24].

5.2.1 Effect of Surface Condition on Flow Characteristics

Cheng and Wu [41] reported experimental findings on pressure drop of deionized water in silicon trapezoidal microchannels with different surface conditions. Trapezoidal microchannels were fabricated by the wet etching method, and a total of 13 microchannels, having different surface roughness and geometrical parameters, were fabricated by varying concentrations (of KOH solution), temperatures, and times in the wet etching process. Figure 5.5 shows a typical trapezoidal cross section of the microchannel, while Table 5.4 describes the geometric parameters and dimensionless surface roughness as the ratio of absolute surface roughness (k) to the hydraulic diameter (D_h) of the 13 microchannels used during the experiment.

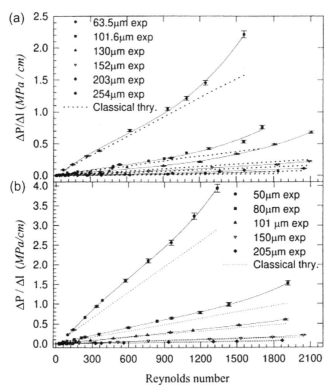

Figure 5.3 Experimentally measured pressure gradient ΔP/Δ*l* versus Re for (a) SS and (b) FS microtubes, and comparison with the classic theory [40].

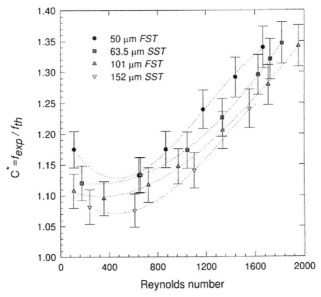

Figure 5.4 Plot of friction constant ratio versus Re for SS and FS microtubes [40].

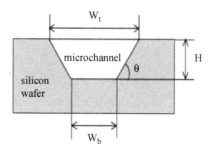

Figure 5.5 Cross section of microchannels [41]. W_t, Top width; W_b, Bottom width; H, Height of channel; θ, Channel angle.

Researchers have used the term "apparent friction factor" to evaluate the total pressure drop of the deionized water flowing through the microchannels and worked out apparent friction constant as,

$$f_{app} \, \text{Re} \, = \, \frac{\Delta p \cdot N \cdot A_c \cdot D_h^{\,2}}{2vML}$$

where Δp is the pressure drop, N is total number of microchannels, A_c is the cross-sectional area of a microchannel, D_h and L are the hydraulic diameter and length of the microchannels, v is the kinetic viscosity of water, and M is the mass flow rate of water.

The investigations into the effects of the surface roughness and geometric parameters are illustrated here.

5.2.2 Impact of Surface Roughness

As the size of the channel reduces to the order of few microns, significant effects of surface roughness on fluid flow and heat transfer characteristics are observed. Figure 5.6 depicts the influence of surface roughness where a pair of trapezoidal channels (microchannels #7 and #9) having the same geometric parameters but different relative surface roughness is investigated. Further, a pair of triangular cross sections (microchannels #8 and #10) was also investigated keeping the same geometric parameters but different relative surface roughness. It was observed that at the same Reynolds number, in the case of the trapezoidal channel, the one that was having higher surface roughness (microchannels #9) exhibited higher values of apparent friction constant. Similar trends were observed for triangular microchannels. Further, it was noticed that the friction constant of high roughness microchannels increases faster than those of low roughness microchannels with increasing Reynolds number.

5.2.3 Effect of Geometric Parameters

To predict the effect of geometrical parameters (height-to-top width ratio $[H/W_t]$, bottom-to-top width ratio $[W_b/W_t]$, and length-to-diameter ratio $[L/D_h]$), trapezoidal

Table 5.4 Geometric Parameters and Surface Properties of Microchannels [41]

Channel No.	W_t (10^{-6} m)	W_b (10^{-6} m)	H (10^{-6} m)	W_b/W_t	H/W_t	k/D_h	L/D_h	Surface Material
#1	1473.08	1375.86	56.22	0.934	0.0382	9.85×10^{-5}	285.41	Si
#2	770.48	672.63	56.34	0.873	0.0731	7.59×10^{-5}	298.67	Si
#3	549.83	454.71	56.33	0.827	0.1024	5.69×10^{-5}	310.46	Si
#4	423.2	327.4	56.13	0.774	0.1326	8.63×10^{-5}	325.11	Si
#5	248.83	153.03	56.24	0.615	0.2260	7.64×10^{-5}	370.85	Si
#6	157.99	61.62	56.28	0.390	0.3562	4.30×10^{-5}	453.79	Si
#7	437.21	270.19	110.7	0.618	0.2532	3.26×10^{-5}	191.77	Si
#8	171.7	0	110.8	0	0.6453	3.62×10^{-5}	362.35	Si
#9	429.99	262.30	109.1	0.610	0.2537	5.87×10^{-3}	195.34	Si
#10	168.03	0	108.9	0	0.6481	1.09×10^{-2}	369.29	Si
#11	555.00	459.54	56.77	0.828	0.1023	9.76×10^{-5}	307.90	SiO_2
#12	251.5	155.7	56.50	0.619	0.2247	5.71×10^{-5}	368.11	SiO_2
#13	158.12	62.30	56.49	0.394	0.3573	6.94×10^{-5}	451.40	SiO_2

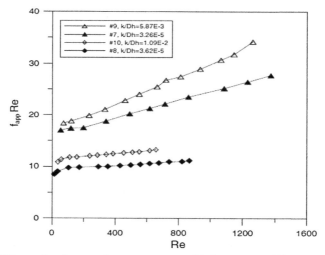

Figure 5.6 Effects of surface roughness on apparent friction constant [41].

microchannels #1 to #6 were prepared to have approximately same surface roughness while having different geometrical parameters. Figure 5.7 exhibits the influence of geometrical parameters on apparent friction constant. One can clearly see from the plot the differences in the apparent friction constant that is because of different cross-sectional aspect ratios, indicating a significant influence of geometrical parameters on friction characteristics.

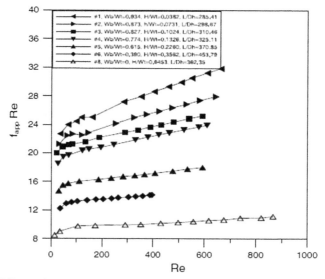

Figure 5.7 Effects of geometric parameters on apparent friction constant [41].

5.3 Effect of Header Shapes on Fluid Flow Characteristics

Anbumeenakshi and Thansekhar [42] experimentally investigated the effect of coating and shapes of header on pressure drop in microchannels, using deionized water as the coolant. For this purpose, 25 parallel rectangular microchannels, made of aluminum, were copper coated by using electro codeposition technique. They compared pressure drop of bare and copper-coated microchannels for three different header combinations: triangular, trapezoidal, and rectangular. The microchannel test section used during the experiment is shown in Fig. 5.8, while headers of different shapes are shown in Fig. 5.9(a–c).

5.3.1 Effect of Coating and Header Combination

During the experiment, mass flow rate was varied in the range of 50–120 kg/h, while Reynolds number ranged from 200 to 600. Figures 5.10–5.12 depict the experimental data plots of pressure drop against the Reynolds number for all permutation and combinations of bare and coated microchannels with different headers. The experimental findings reveal that as the Reynolds number increases, the pressure drop increases for both the bare surface and the coated surface microchannel, but an increase in pressure drop in coated microchannels is found more often compared with the bare one. On the other hand, among the various header combinations, trapezoidal headers had the lowest pumping power.

5.4 Pressure Loss Investigation in Rectangular Channels with Large Aspect Ratio

Hassan et al. [43] experimentally verified that the conventional laws and correlations describing the flow and convective heat transfer in ducts of large dimension are also applicable in microchannels of hydraulic diameter greater than or equal to 100 μm.

Figure 5.8 Microchannel test section [42].

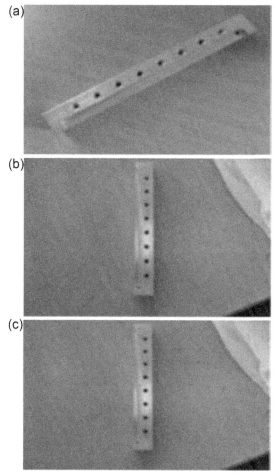

Figure 5.9 Headers of different shapes: (a) rectangular header [42]. (b) triangular header [42] and (c) trapezoidal header [42].

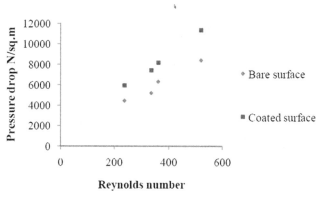

Figure 5.10 Rectangular header combination [42].

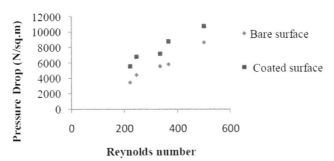

Figure 5.11 Triangular header combination [42].

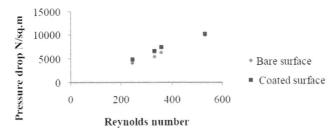

Figure 5.12 Trapezoidal header combination [42].

An experiment was conducted using tap water as a working fluid in a straight microchannel of rectangular cross section and large aspect (width to height) ratio. During the investigation, the Reynolds number had been varied from 100 to 5000 while channel heights (e) were varied from 50 to 500 μm, keeping a constant width. Figure 5.13 shows a schematic diagram of the experimental setup.

5.4.1 Pressure Drop Observations

Poiseuille number, being an indicator of pressure loss, has been estimated from experimental data for all microchannel heights covering laminar and turbulent regimes.

The Poiseuille number (P_o) has been obtained from the product of

$$\text{Re} = \frac{uD_h}{v}$$

$$f = \frac{\Delta P}{L} \frac{2D_h}{\rho u^2}$$

The experimentally measured Poiseuille number is then compared with the theoretical Poiseuille number values ($P_{o\ th}$) that are being calculated using the correlation proposed by Shah and London [44] for laminar flow regime, and for turbulent flow

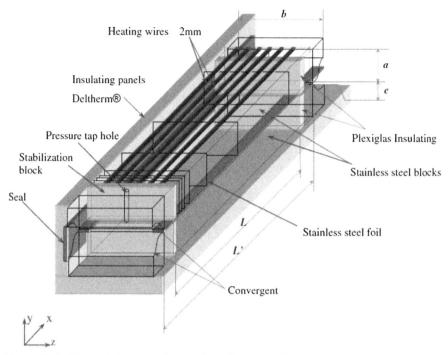

Figure 5.13 Schematic diagram of microchannel setup [43].

regime, the comparison has been made with values calculated from the correlations suggested by Blasius [45] and that of Kakaç et al. [46].

$$P_{o\ th} = 24\left[\,1 - 1.3553\alpha^* + 1.9467\alpha^{*2} - 1.7012\alpha^{*3} + 0.9564\alpha^{*4}\right.$$
$$\left. - 0.2537\alpha^{*5}\right]$$

$$f = 0.0791\mathrm{Re}^{-0.25}\ for\ 5000 \leq \mathrm{Re} \leq 3 \times 10^4$$

$$f = (1.0875 - 0.1125\alpha^*)f_c\ for\ 4 \times 10^3 \leq \mathrm{Re} \leq 10^7$$

where $f_c = 0.00128 + 0.1143\mathrm{Re}^{-0.311}$ _and_ $\alpha^* = \frac{e}{b}$, where e is the channel height and b is width.

The comparison of variation of experimental and theoretical values of Poiseuille number against the Reynolds number corresponding to different microchannel heights (e) is demonstrated in Fig. 5.14(a–d).

From these comparison plots, it can be seen that in the laminar regime the experimental Poiseuille numbers are closely following the theoretical values obtained from correlation [44]. On the other hand, in the turbulent regime, the experimental values lie between the Poiseuille numbers obtained from the Blasius [45] correlation and the one calculated using the Kakaç et al. [46] correlation.

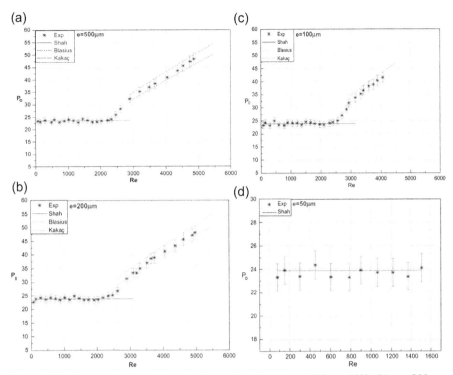

Figure 5.14 Pressure drop variation for channel height: (a) $e = 500$ μm [43], (b) $e = 200$ μm [43], (c) $e = 100$ μm [43] and (d) $e = 50$ μm [43].

5.5   Effect of Shape and Geometrical Parameters on Pressure Drop

Gunnasegaran et al. [47] investigated the influence of geometrical parameters and the shapes (rectangular, trapezoidal, and triangular) of microchannels on the pressure drop, considering water as a working fluid. They adopted a numerical approach. The diagram of the microchannels' heat sink and their different shapes subjected to investigation are given in Fig. 5.15. It is important to mention that in this pressure drop analysis, minor losses are not accounted for, which otherwise has weight in real-time design of heat sink.

To investigate the effect of geometrical parameters of each shape, each dimension is assigned three different cases, as given in Tables 5.5−5.7 [47].

Figures 5.16−5.18 depict the variation of pressure drop against the Reynolds number for different shapes of microchannels. It was observed that pressure drop had an increasing trend with Reynolds number, and this nature of variation was found to be common irrespective of the shape of the microchannels under study. Further, it was observed that hydraulic diameter had significant effect on pressure drop. Maximum pressure drop was observed corresponding to minimum hydraulic diameter, while

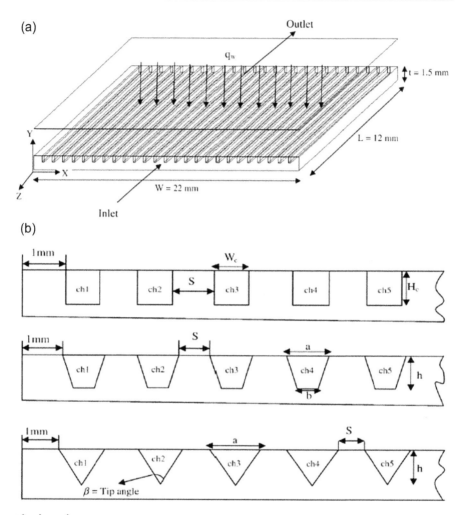

ch=channel

Figure 5.15 (a) Schematic diagram of the microchannels' heat sink. (b) Section of different microchannel shapes with dimensions [47].

minimum pressure drop was corresponding to maximum hydraulic diameter at a particular Reynolds number; this trend was common irrespective of the shape of the microchannels under study. The cause of increased pressure drop with reduction in hydraulic diameter is obvious because as hydraulic diameter reduces, free flow passage also reduces, resulting in an increased pressure drop.

5.5.1 Effect of Heat Flux on Pressure Drop

In Fig. 5.19, the variations in pressure drop against the Reynolds number corresponding to three different values of heat flux ($q = 100$, 500, and 1000 W/m^2) have been

Table 5.5 Parameters for Three Different Sets of Rectangular Microchannels [47]

	Case 1	Case 2	Case 3
H_c (μm)	460	430	390
W_c (μm)	180	280	380
L_c (μm)	10,000	10,000	10,000
S (μm)	596	500	404
D_h (μm)	259	339	385
Number of channels	25	25	25

Table 5.6 Parameters for Three Different Sets of Trapezoidal Microchannels [47]

	Case 1	Case 2	Case 3
a (μm)	180	280	380
b (μm)	125	225	325
h_c (μm)	460	430	390
L_c (μm)	10,000	10,000	10,000
S (μm)	596	500	404
D_h (μm)	229	318	370
Number of channels	25	25	25

Table 5.7 Parameters for Three Different Sets of Triangular Microchannels [47]

	Case 1	Case 2	Case 3
a (μm)	180	280	380
h_c (μm)	460	430	390
L_c (μm)	10,000	10,000	10,000
S (μm)	596	500	404
D_h (μm)	148	203	238
β	22.14°	36.07°	51.95°
Number of channels	25	25	25

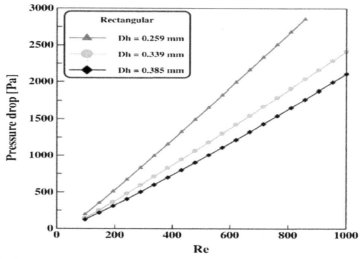

Figure 5.16 Pressure drop variations versus Reynolds number for rectangular microchannels [47].

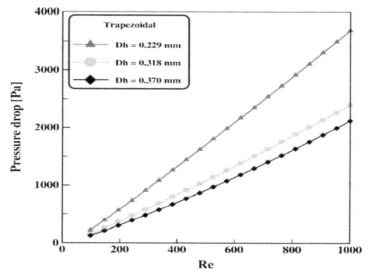

Figure 5.17 Pressure drop variations versus Reynolds number for trapezoidal microchannels [47].

shown. From this plot, it is clear that at a particular Reynolds number, the pressure drop is maximum corresponding to the lowest heat flux (i.e., 100 W/m²), while it is minimum corresponding to maximum heat flux (i.e., 1000 W/m²). The reason for such behavior is attributed to the fact that as the liquid temperature increases, its viscosity decreases and since the experiment is conducted under three different values of

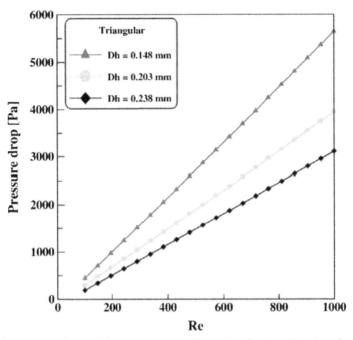

Figure 5.18 Pressure drop variations versus Reynolds number for triangular microchannels [47].

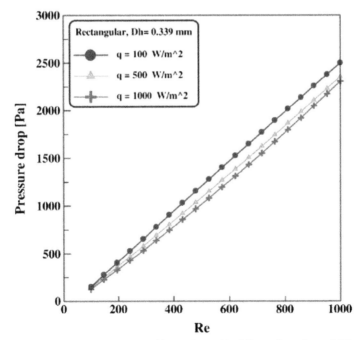

Figure 5.19 Pressure drop versus Reynolds number with different heat fluxes [47].

constant heat flux conditions, viscosity of the liquid will be minimal, corresponding to the highest heat flux value. Due to this low viscosity, the pressure drop will also be lower. Qu and Mudawar [48] had also reported similar variation of pressure drop for different heat flux intensities.

5.5.2 Effect of Aspect Ratio on Pressure Drop

Dimensionless parameter, Poiseuille number, which is a product of friction factor and Reynolds number, has been used as an indicator of pressure drop.

The friction factor is calculated using the Darcy equation [49] and is given by

$$f = \frac{2 D_h \Delta p}{\rho u^2 L_c}$$

where D_h is the hydraulic diameter, Δp is the pressure drop, ρ is the density of fluid, u is the velocity of fluid, and L_c is the length of channel.

The effect of aspect ratio of rectangular microchannel on Poiseuille number is demonstrated in Fig. 5.20. Pressure drop is observed to increase with increase in aspect ratio. The same observation was reported by Kandlikar et al. [50] for rectangular

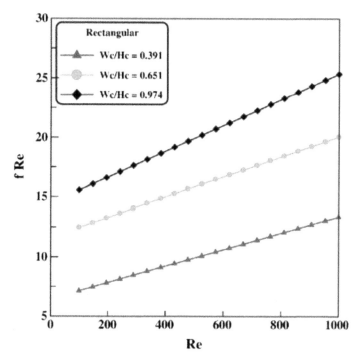

Figure 5.20 Poiseuille number at different width−height (W_c/H_c) ratios for rectangular microchannels [47].

channels. The increase in pressure drop is attributed to vortices that are induced due to reduced free flow passage with increase in aspect ratio.

Closure

This article essentially focuses on the pressure drop studies in microchannels. The discussion is started with motivation and need for mini/microchannels, and subsequently the potential of applicability of conventional theory and correlations was presented in condensed form. Although no consensus in the findings (experimental/analytical) have been observed regarding the validity of conventional theory and correlations, this information will become potential fuels for startups and brainstorming, and this, in turn, will encourage researchers and scientists to even the gap in understanding thermohydraulics of microchannels. Last, there is an urgent need and challenge to develop instruments, methods, and techniques for accurate measurement of parameters at the micro level (e.g., surface roughness, etc.), because many parameters have negligible influence in conventional heat exchanger thermohydraulic performance and, as the size of the channel reduces, they strongly influence the accuracy of microchannel analysis, if neglected.

Nomenclature

Q	Volume flow rate
R	Radius
μ	Dynamic viscosity
Δp	Pressure drop
C^*	Friction constant ratio
Re	Reynolds number
N	Total number of microchannels
D_h	Hydraulic diameter
v	Kinematic viscosity
M	Mass flow rate of water
L	Length
A_c	Channel cross-sectional area
f_{app}	Apparent friction factor
f_{exp}	Friction factor from experimental
f_{th}	Friction factor from correlation
ρ	Density

$P_{o\ th}$	Poiseuille number from correlation
α^*	Aspect ratio
f	Friction factor
e	Channel height
b	Width of channel
L_c	Channel length

References

[1]   S. Liu, M. Sakr, A comprehensive review on passive heat transfer enhancements in pipe exchangers, Renewable Sustainable Energy Rev. 19 (2013) 64−81.

[2]   D.B. Tuckerman, R.F. Pease, High performance heat sinking for VLSI, IEEE Electron Device Lett. EDL-2 (1981) 126−129.

[3]   R.W. Keyes, Heat transfer in forced convection through fins, IEEE Trans. Electron Devices ED-31 (1984) 1218−1221.

[4]   V.K. Samalam, Convective heat transfer in micro-channels, J. Electron. Mater. 18 (1989) 611−617.

[5]   R.J. Phillips, Micro-channel heat sinks, in: A. Bar-Cohen, A.D. Kraus (Eds.), Advances in Thermal Modeling of Electronic Components, vol. 2, ASME Press, New York, 1990, pp. 109−184.

[6]   R.W. Knight, J.S. Goodling, D.J. Hall, Optimal thermal design of forced convection heat sinks − analytical, ASME J. Electron. Packag. 113 (1991) 313−321.

[7]   R.W. Knight, D.J. Hall, J.S. Goodling, R.C. Jaeger, Heat sink optimization with application to micro-channels, IEEE Trans. Compon. Hybrids Manuf. Technol. 15 (1992) 832−842.

[8]   A. Bejan, A.M. Morega, Optimal arrays of pin fins and plate fins in laminar forced convection, ASME J. Heat Transfer 115 (1993) 75−81.

[9]   D.Y. Lee, K. Vafai, Comparative analysis of jet impingement and micro-channel cooling for high heat flux applications, Int. J. Heat Mass Transfer 42 (1999) 1555−1568.

[10] A. Weisberg, H.H. Bau, J.N. Zemel, Analysis of micro-channels for integrated cooling, Int. J. Heat Mass Transfer 35 (1992) 2465−2474.

[11] A.G. Fedorov, R. Viskanta, Three-dimensional conjugate heat transfer in the micro-channel heat sink for electronic packaging, Int. J. Heat Mass Transfer 43 (2000) 399−415.

[12] T. Kishimito, T. Ohsaki, VLSI packaging technique using liquid-cooled channels, IEEE Trans. Compon. Hybrids Manuf. Technol. CHMT-9 (1986) 328−335.

[13] S. Sasaki, T. Kishimito, Optimal structure for micro-grooved cooling fin for high-power LSI devices, Electron. Lett. 22 (1986) 1332−1334.

[14] D. Nayak, L.T. Hwang, I. Turlik, A. Reisman, A high performance thermal module for computer packaging, J. Electron. Mater. 16 (1987) 357−364.

[15] M.M. Rahman, F. Gui, Experimental measurements of fluid flow and heat transfer in micro-channel cooling passages in a chip substrate, Adv. Electron. Packag. ASME EEP-4 (1993) 495−506.

[16]   M.M. Rahman, F. Gui, Design, fabrication, and testing of micro-channel heat sinks for aircraft avionics cooling, in: Proceedings of the 28th Intersociety Energy Conversion Engineering Conference, vol. 1, 1993, pp. 1–6.

[17]   T.S. Ravigururajan, J. Cuta, C.E. McDonald, M.K. Drost, Single-phase flow thermal performance characteristics of a parallel micro-channel heat exchanger, in: National Heat Transfer Conference, vol. 7, ASME HTD-329, 1996, pp. 157–166.

[18]   T.M. Harms, M.J. Kazmierczak, F.M. Cerner, A. Holke, H.T. Henderson, J. Pilchowski, et al., Experimental investigation of heat transfer and pressure drop through deep micro-channels in a (1 0 0) silicon substarte, in: Proceedings of the ASME Heat Transfer Division, HTD-vol. 351, 1997, pp. 347–357.

[19]   K. Kawano, K. Minakami, H. Iwasaki, M. Ishizuka, Micro channel heat exchanger for cooling electrical equipment, Appl. Heat Transfer Equip. Syst. Educ. ASME HTD-361-3/PID-3 (1998) 173–180.

[20]   T.M. Harms, M.J. Kazmierczak, F.M. Cerner, Developing convective heat transfer in deep rectangular micro-channels, Int. J. Heat Fluid Flow 20 (1999) 149–157.

[21]   M.M. Rahman, Measurements of heat transfer in micro-channel heat sinks, Int. Commun. Heat Mass Transfer 27 (2000) 495–506.

[22]   S.S. Mehendale, A.M. Jacobi, R.K. Ahah, Fluid flow and heat transfer at micro and meso-scales with application to heat exchanger design, Appl. Mech. Rev. 53 (2000) 175–193.

[23]   S.G. Kandlikar, W.J. Grande, Evolution of micro-channel flow passages thermohydraulic performance and fabrication technology, Heat Transfer Eng. 24 (2003) 3–17.

[24]   M. Asadi, G. Xie, B. Sunden, A review of heat transfer and pressure drop characteristics of single and two-phase micro-channels, Int. J. Heat Mass Transfer 79 (2014) 34–53.

[25]   Satish G. Kandlikar, Y. Stephane Colin, S. Peles, R. Garimella, Juergen j. Fabian Pease, D.B. Brandner, Tuckerman, Heat transfer in micro-channels—2012 Status and Research Needs, ASME J. Heat Transfer 16 (2013) 091001.

[26]   I. Mudawar, Assessment of high-heat-flux thermal management schemes, IEEE Trans. Compon. Packag. Technol. 24 (2) (2001) 122–141.

[27]   R. Savino, D. Paterna, Y. Abe, Recent developments in heat pipes: an overview, Recent Pat. Eng. 1 (2) (2007) 153–161.

[28]   C.B. Sobhan, R.L. Raq, G.P. Peterson, A review and comparative study of the investigations on micro heat pipes, Int. J. Energy Res. 31 (6–7) (2007) 664–688.

[29]   S. Kakac, A. Pramuanjaroenkij, Review of convective heat transfer enhancement with nanofluids, Int. J. Heat Mass Transfer 52 (2009) 3187–3196.

[30]   T.A. Ameel, R.O. Warrington, R.S. Wegeng, M.K. Drost, Miniaturization technologies applied to energy systems, Energy Convers. Manage. 38 (1997) 969–982.

[31]   M. Gad-el-Hak (Ed.), The MEMS Handbook, CRC Press, New York, 2001.

[32]   N.T. Nguyen, Micromachined flow sensors—a review, Flow Meas. Instrum. 8 (1997) 7–16.

[33]   A.R. Abramson, C.L. Tien, Recent developments in microscale thermophysical engineering, Microscale Thermophys. Eng. 3 (1999) 229–244.

[34]   A.B. Duncan, G.P. Peterson, Review of microscale heat transfer, ASME Appl. Mech. Rev. 47 (1994) 397–428.

[35]   D.K. Bailey, T.A. Ameel, R.O. Warrington, T.I. Savoie, Single phase forced convection heat transfer in microgeometries—a review, in: IECEC Conference ASME-FL, Orlando, USA, 1995. ES-396.

[36]   B. Palm, Heat transfer in micro-channels, Microscale Thermophys. Eng. 5 (2001) 155–175.

[37] W.J. Bowman, D. Maynes, A review of micro-heat exchanger flow physics, fabrication methods and application, in: Proceedings of ASME IMECE 2001, New York, USA, 2001. HTD-24280.

[38] Mark E. Steinke, S.G. Kandlikar, Review of single phase heat transfer enhancement techniques for application in micro-channels, minichannels and microdevices, Heat Technol. 22 (2) (2004).

[39] S.B. Choi, R.R. Baron, R.O. Warrington, Fluid flow and heat transfer in microtubes, ASME DSC 40 (1991) 89−93.

[40] Gh Mohiuddin Mala, D. Li, Flow characteristics of water in microtubes, Int. J. Heat Fluid Flow 20 (1999) 142−148.

[41] P. Cheng, H.Y. Wu, An experimental study of convective heat transfer in silicon micro-channels with different surface conditions, Int. J. Heat Mass Transfer 46 (2003) 2547−2556.

[42] C. Anbumeenakshi, M.R. Thansekhar, Experimental investigation of the combined effect of coating and header combination in micro-channels, Procedia Technol. 14 (2014) 520−527.

[43] O. Mokrani, B. Bourouga, C. Castelain, H. Peerhossaini, Fluid flow and convective heat transfer in flat micro-channels, Int. J. Heat Mass Transfer 52 (2009) 1337−1352.

[44] R.K. Shah, A.L. London, Laminar Flow Forced Convection in Ducts, Academic Press, 1978.

[45] H. Blasius, Das Ähnlichkeitsgesetz bei Reibungvorgängen in Flüssigkeiten, Forchg. Arb. Ing.-Wes. 131 (Berlin) (1913).

[46] S. Kakaç, R.K. Shah, W. Aung, Handbook of Single-Phase Convective Heat Transfer, Wiley-Interscience, 1987.

[47] P. Gunnasegaran, H.A. Mohammed, N.H. Shuaib, R. Saidur, The effect of geometrical parameters on heat transfer characteristics of micro-channels heat sink with different shapes, Int. Commun. Heat Mass Transfer 37 (2010) 1078−1086.

[48] W. Qu, I. Mudawar, Experimental and numerical study of pressure drop and heat transfer in a single-phase micro-channel heat sink, Int. J. Heat Mass Transfer 45 (2002) 2549−2565.

[49] L.F. Moody, Friction factors for pipe flow, J. Heat Transfer 66 (8) (1944) 671−684.

[50] S. Kandlikar, S. Garimella, D. Li, S. Colin, M.R. King, Heat Transfer and Fluid Flow in Minichannels and Microchannels, Elsevier, USA, 2005.

Critical Heat Flux for Boiling in Microchannels

P.K. Das[1], A.K. Das[2]
[1]Department of Mechanical Engineering, Indian Institute of Technology Kharagpur, Kharagpur, West Bengal, India; [2]Department of Mechanical and Industrial Engineering, Indian Institute of Technology Roorkee, Roorkee, Uttarakhand, India

6.1 Introduction

The intense drive for miniaturization in electronic industry over the last several decades posed a continuous demand for the efficient dissipation of heat through smaller and smaller surface area. Such demands are also common in gas turbine systems, high performance computing systems and data centers, conventional automobiles, as well as cars of the next generation using electricity or hydrogen as the power source, and so on [48]. Thermal management is very crucial in a variety of cutting edge technologies like cooling of rocket engine nozzle, heat transfer from fusion reactor blankets, heat dissipation from satellite electronics etc. Such a tremendous demand for efficient heat dissipation gave birth to different techniques, a variety of art for augmenting heat transfer and finally the devices with microchannel cooling passages. Microchannel heat exchangers and heat sinks with a liquid as the coolant has almost become the state of the art in cooling of all the cutting edge applications.

Through microchannels are often used for single-phase coolants, it is needless to say that boiling in microchannel can enhance the rate of heat transfer by orders of magnitude compared to its single-phase counterpart due to the contribution by latent heat transfer. Further, boiling heat transfer produces more uniform temperature distribution in the heat sink or heat exchanger and can operate satisfactorily with a much lesser amount of coolant inventory requiring a smaller prime mover. It is, therefore, not surprising that many heat exchangers are operated in the boiling heat transfer mode to have the benefit of extracting a very high heat flux at a relatively low temperature difference.

Out of different modes of boiling, nucleate boiling is the most effective in transferring heat. This mode of boiling is characterized by the formation of vapor nuclei on the solid substrate. These vapor nuclei grows into bubbles as they get a continuous feeding of vapor due to heat transfer. Big enough bubbles leave the substrate, taking away the thermal energy from the surface. With the increase of heat flux, the frequency of bubble departure increases along with the activation of new nucleation sites. This is also accompanied by a rise in surface temperature. With the continuation of this process almost the entire surface takes part in the generation of bubbles and the largest possible rate of heat transfer is noted in nucleate boiling. This very important state of boiling is known as critical heat flux (CHF) condition. If the heat flux is increased beyond this,

Microchannel Phase Change Transport Phenomena. http://dx.doi.org/10.1016/B978-0-12-804318-9.00006-6

there will be a shift to a less effective mode of heat transfer called film boiling, many times with a phenomenal increase in temperature. This has a potential risk of damaging the heating surface. Consequently, all the practical applications try to restrict the heat transfer below the CHF point. It is therefore of utmost importance to know the CHF limit for boiling through microchannels, to understand its mechanism and to have pre-dictive methods for defining a safe operating limit for this newer generation of heat exchangers and heat sinks.

The present article aims at a comprehensive review of the CHF during flow boiling through microchannels based on published literature. Due to its importance, boiling in microchannels has been covered in the literature, including some. Some of the inter-esting reviews are by Saha et al. [85], Saha and Celata [80,86], Kim and Mudawar [48], Thome [88]. On CHF, Kandlikar [39,40] provided a very elaborate discussion and Das et al. provided a state-of-the-art review in 2012. In the present article, the review of Das et al. [18] was elaborated on by incorporating information reported after 2012. At this juncture, it may be noted that there is no universally accepted definition of microchannel. Saha et al. [85] reviewed the available opinions and suggestions regarding the definition of microchannels. One may appreciate the controversy in demarcating microchannel behavior in two-phase flow from this review. In the present discussion, results from the investigations on millimeter-sized channels, or milli-channels, are also included.

6.2   CHF in Pool Boiling and Flow Boiling in Macrochannels—Present State of Understanding

Over the past decades, CHF under various conditions like pool boiling, flow boiling through "macro"channels, annulus, rod bundles, and over tube bank has been studied extensively. Despite an overwhelming volume of literature on this topic, consisting of experimental data, mechanistic models, and numerical simulations, understanding of CHF is only partial.

The basic concept of CHF may be appreciated from the most fundamental studies on boiling heat transfer by Nukiyama [63] and his proposition of the pool boiling curve. Back in 1934, he demonstrated three distinct regimes in pool boiling through an innovative experiment. The regimes of nucleate boiling, transition boiling, and film boiling, as shown in Fig. 6.1, demonstrate the three distinct modes of heat transfer of a temperature-controlled heater during boiling. With the increase in temperature, the heat flux increases rapidly in the nucleate boiling regime (B to C) until the CHF (C) is reached. Beyond that, transition boiling (C to D), characterized by the partial blanket-ing of the heater surface by vapor patches, occurs, and heat flux decreases with the increase of temperature.

This is not only an unstable region of boiling but also has to be avoided in most of the industrial applications for the safety of the equipment. In the film boiling region (D to E), increasing heat flux is also associated with an increase in temperature, but this is not a preferred region of operation as the high temperature may cause a damage

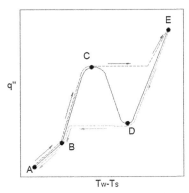

Figure 6.1 Pool boiling curve. Temperature controlled experiments; A-B-C-D- Eheat flux controlled experiment; B-C-E during heating, E-D-B during cooling.

of the heater. On the other hand, if the heat flux of the heater is varied independently, which is the case in the most practical situations, there could be a sudden rise in temperature due to a small change in heat flux at the CHF point (C). The boiling regime abruptly jumps from the nucleate boiling to the film boiling. This is, again, detrimental to the heater and is to be avoided at any cost. Therefore, the importance of CHF and the necessity of its reliable prediction need not be overexaggerated. This also justifies the enormous volume of literature on this topic. Although the above description has been made in the context of pool boiling, the same holds good for flow boiling except for some typical cases [44].

CHF is an outcome of a complex interplay of fluid dynamics, thermodynamics, and heat transfer. It is influenced by a large number of process and system variables and is not fully understood until now [14]. It would therefore be prudent to have a brief overview of CHF in the most simple case (i.e., pool boiling) before discussing the issue in connection with microchannel boiling, where not only the geometrical confinement but also the flow velocity has a profound effect.

For pool boiling, mainly three different mechanisms of CHF have been postulated. At high heat flux, the vapor generated on the heating surface leaves in the form of circular jets arranged in regular array over the surface. Zuber [98] and later, Lienhard and Dhir [57] postulated that the hydrodynamic instability of these jets is responsible for CHF as that can disturb the replenishment of liquid to the surface. It has been postulated that several classic instabilities like Kelvin-Helmholtz and Rayleigh-Taylor instabilities are involved in this phenomenon. Figure 6.2 depicts the scenario during CHF as postulated by the hydrodynamic instability theory and considered as one of prominent mechanisms of CHF [44].

On the other hand, Haramura and Katto [27] explained the phenomenon from the near wall fluid behavior. They postulated that a growing bubble is fed by the vapor evaporated from a thin microlayer at its bottom and CHF occurs due to the dry-out of the microlayer as can be seen from Fig. 6.3. The basic mechanism can also be extended for flow boiling and for boiling with various other configurations and

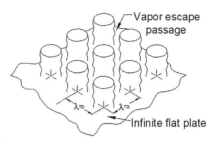

Figure 6.2 Instability of vapor jets emerging from a flat horizontal surface.

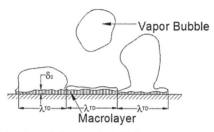

Figure 6.3 Dry-out of microlayer below vapor bubbles.

conditions. It is also interesting to note that though the basis of the above two models are different; both of them predict similar parametric variation of CHF.

Dhir and Liaw [17] proposed the existence of a thermal boundary layer adjacent to the heater surface during nucleate and transition boiling. It was further postulated that depending on the wettability and the temperature of the surface, the vapor stems undergo changes. Ultimately, the merging of the vapor stems triggers CHF as depicted in Fig. 6.4. The idealized model is given by Liaw and Dhir [55] and discussed by Katto [44].

Hydrodynamic instability and microlayer dry-out appear to be the two main competitors as far as the mechanism of CHF in pool boiling is concerned. Bergles [6] provided a comparison of these two models. The effort for improvement on the basic models on CHF still continues. In this context, one may refer to the recent model of Kandlikar [34]. In this model, Kandlikar included the effect of contact angle, surface orientation, and subcooling. The above models can be termed as mechanistic models and are examples of simple physics-based approach to describe a very complex phenomenon.

There have also been attempts to identify the basic mechanism of CHF so that a single model can be applicable for pool and flow boiling. In this regard, importance has been given to the microlayer evaporation. The model of Ha and No [24,25] may be mentioned. In the basic model, it is assumed that a dry spot is formed when there are a critical number of bubbles surrounding a single bubble.

CHF in subcooled boiling is immensely important in industry for energy transfer and safety consideration. The effect of bulk fluid movement and the lateral movement of the bubbles on CHF in such situations are very prominent. Kandlikar [35] provided a comprehensive review of the state of the understanding of CHF during subcooled

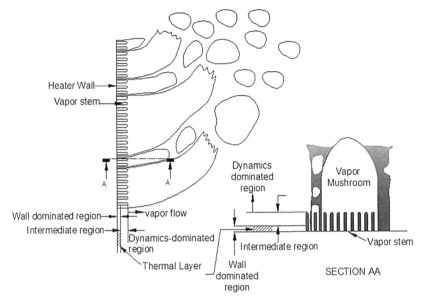

Figure 6.4 Vapor stems connecting a bubble with the heater surface.

flow boiling. In this connection one may refer to the work of Celata and Mariani [13] where the models based on the following mechanisms have been identified.

- Boundary layer ejection model.
- Critical enthalpy in bubble layer model
- Liquid flow blockage model
- Vapor removal limit and near wall bubble crowding model
- Liquid sublayer dry-out model

It needs to be mentioned that none of the mechanisms can be considered for the modeling of CHF in microchannel flow boiling without substantial modifications. However, the insight provided by these models helps understanding the probable mechanism of CHF in microchannels. Particularly, thin film dry-out could be a viable mechanism even in case of microchannel. But the dynamic situation during flow boiling cannot be ignored. The importance of the receding contact angle during evaporation in a moving fluid should be considered as a mechanistic representation. Further, the effect of the confinement on bubble growth and coalescence on CHF should be critically judged.

6.3   Some General Observations on Boiling in Microchannels and Associated CHF

During flow boiling, for effective heat transfer, one can have two distinct modes before dry-out occurs. These modes are also observed in microchannels as briefed by Kim and

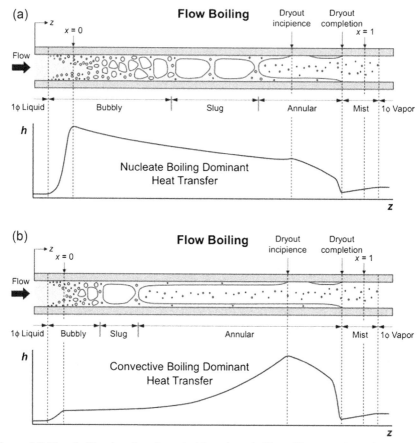

Figure 6.5 Flow boiling in microchannels (a) nucleate boiling, (b) convective boiling.

Mudawar [48] in their recent review. In one of these modes, heat transfer is primarily by nucleate boiling and in the other by convective boiling. Figure 6.5 shows these two modes of boiling and the associated variation of wall heat flux as depicted by Kim and Mudawar [48].

In the nucleate boiling dominant case, the bubbly and slug flow regimes occupy significant length of the channel and have a brief annular region upstream of them. The heat transfer coefficient (h) decreases monotonically from the saturated boiling condition (x = 0) until the completion of dry-out condition (x = 1). On the other hand, for the convective boiling case, after a brief region of bubbly and slug flow the most of the tube length is occupied with annular flow regime. The heat transfer coefficient increases until the inception of dry-out (local dry-out). Then it rapidly fall as complete dry-out occurs. Due to the unique hydrodynamics the connotation of CHF is slightly different for flow through microchannels. As Kim and Mudawar [48] explain, CHF refers to the heat transfer limit denoted by the substantial reduction in local heat transfer coefficient as the heated wall is grossly deprived of the liquid access. If the system is heat flux controlled this may give rise to a sudden shoot up in wall temperature initiating a catastrophic damage or failure of the system. They have indicated three mechanisms for CHF in microchannels, namely departure from nucleate boiling (DNB), dry-out, and premature CHF. Lee and Mudawar [101] demonstrated a typical case of DNB for a microchannel heat sink of $D_H = 334.1$ μm with HFE-7100 as working fluid. The visualization shows formation of a continuous vapor blanket at the wall due to bubble coalescence while there is no dearth liquid at the core. This is a clear demonstration of CHF from DNB and is favored by inlet subcooling, high mass velocity, and small length-to-diameter ratio. Dry out is generally associated with saturated inlet conditions, low mass velocity, and large length-to-diameter ratio. This mode of CHF is prevalent in annular flow regime as the liquid film at the wall thins down due to continuous evaporation and finally exposes the wall to the core. The observation by Lee and Mudawar [53] is depicted in Fig. 6.6 for a liquid R134a in a heat sink containing rectangular microchannels. In this case, CHF may occur locally due to incipient dry-out prior to the complete dry-out.

Lee and Mudawar [101] also observed premature CHF in a microchannel heat sink using HFE-7100 where the channel restriction forces the generated vapor mass to experience a backflow and to migrate to the inlet plenum. This phenomenon is associated with a lot of oscillation and flow fluctuation and occurs due to the low liquid momentum which cannot push through the flow overcoming the excessive pressure drop.

6.4 Experimental Investigations of CHF

The last decade has witnessed a very large number of experimental investigations on flow boiling through microchannels. Experimental results covering a large variation of geometry, fluid and working conditions have been obtained from different laboratories all over the globe [45,59]. Some of these works are fundamental in nature while many of them are targeted to determine the performance of typical heat sinks or heat exchangers relevant to the products of the sponsors. Though Bergles and Kandlikar [7] reported the lack of information in single microchannel, a good number of experiments have also been conducted in single channels in the millimetric and submillimetric ranges.

That the boiling behavior could be radically different in a narrow confinement from that on an external surface or through a conventional tube was apprehended by a number of investigators [3,4,46,47]. The heat transfer coefficient, the flow regime, and the CHF change depending on the geometry of the confinement, gap size, and orientation of the heater surface (horizontal—up or down, inclined or vertical—up or down). Direct comparisons between the results from different sources are not possible due to the variation of orientation and geometry. However, all the investigations indicate a strong influence of flow regimes on CHF. Decrease of CHF due to counter current flow limitation (CCFL) was also observed. CCFL is reverse flow of the liquid phase as a result of vapor flow with a high velocity in a counter current direction. An analogue to this phenomenon is observed in microchannel geometry. Reduction of

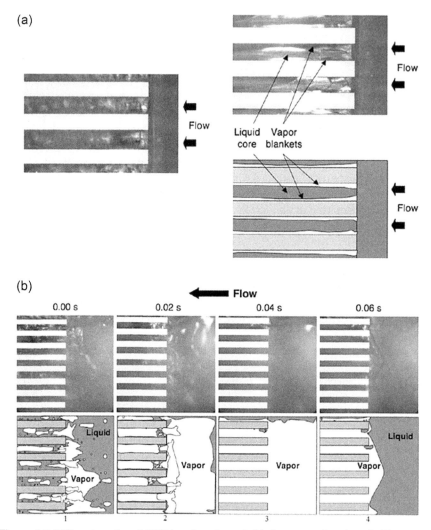

Figure 6.6 Different modes of CHF in microchannel: (a) departure of nucleate boiling, (b) premature CHF.

CHF occurs as liquid flows in the reverse direction forced by the expanding vapor bubbles. Due to the hindrance of heat removal, CHF is favored.

Experiments on microchannels cover a wide range of geometry, physical dimensions, working fluids, and operating parameters. The expanse of the parametric range and variation can be appreciated from the database provided in two earlier reviews [21,73]. The readers are referred to these two sources as they complement the present topic. It may be noted that the two tables have only a marginal overlap. In the present work, we provide some of the very recent and important experimental investigations in Table 6.1. The CHF data of the table also signify the current direction of research.

Table 6.1 Typical Experimental Investigations on CHF during Flow Boiling through Microchannel

Authors	Geometry	Channel	Fluid	Test Conditions	Remark
Mauro et al. [58]	199 μm wide and 756 μm deep (copper heat sink)	29	R134a, R236fa, R245fa	Mass fluxes = 250–1500 kg/m²s inlet subcooling from 25 to 5 K saturation temperatures = 20–50 °C	Effect of subcooling is negligible. Mass flux increases CHF increases. Effect of saturation temp ~ effect of system pressure.
Park and Thome [65]	(1) 467 μm wide and 4052 μm deep (2) 199 μm wide and 756 μm deep; total length = 30 mm, heated length = 20 mm (copper heat sink)	20 and 29	R134a, R236fa, R245fa	Mass flux = 100–400 kg/m²s; inlet subcooling from 20 to 3 K	For low mass velocity all the refrigerants showing same CHF. Depending on CHF and pressure drop, R134a is better coolant than R236fa and R245fa.
Roday and Jensen [78,79]	Microtube, inner diameter = 0.286–0.700 mm, total length = 121.96–138.65 mm, heated length = 21.66–90.84 mm (stainless steel)	Single	Water, R123	Mass flux = 320–1570 kg/m²s, exit pressure = 25.3–225 kPa, subcooled temperature = 2–80 °C	CHF dependent on mass flux, heated length, channel diameter, exit pressure as well as liquid subcooling.

Continued

Table 6.1 Typical Experimental Investigations on CHF during Flow Boiling through Microchannel—cont'd

Authors	Geometry	Channel	Fluid	Test Conditions	Remark
Agostini et al. [1,2]	Parallel channels, (223 μm wide, 680 μm high and 20 mm long with 80 mm thick fins separating the channels. (Silicon)	67	R236fa, R245fa	Mass flux = 276–992 kg/m²s, subcool temperature = 0.4–15 °C	CHF increases with mass flux. Effect of inlet subcooling and saturation temperature is negligible. Effect of exit pressure has not studied.
Wojtan et al. [93]	Microtube, inner diameter = 0.5, 0.8 mm, heated length = 20–70 mm (stainless steel)	Single	R134a, R245fa	Mass flux = 400–1600 kg/m²s, subcooled temperature = 4–12 °C	CHF dependent on mass flux, heated length, channel diameter but not on liquid subcooling, Qu–Mudawar correlation over predicts CHF for the working range, pressure effect has not studied.

Reference	Channel dimensions	Number of channels	Fluid	Operating conditions	Remarks
Hetsroni et al. [31]	Hydraulic diameter 220, 130, and 100 μm (parallel triangular microchannels) (silicon substrate) (silicon substrate).	13, 21, and 26	Water, ethanol	Mass flux 32–200 kg/m²s; heat flux 120–270 kW/m²; vapor quality 0.01–0.08	Liquid film thickness reaches minimum CHF occurs.
Lee and Mudawar [53]	231 μm wide by 713 μm deep (copper block)	53	R134a	Mass velocity = 127–654 kg/m²s	At exit quality = 0.55 CHF can be observed.
Qu et al. [100]	215 × 821 μm	21	Deionized water	Inlet temperature 30–60 °C; mass velocity 86–368 kg/m²s	With low mass flux no effect of inlet temperature.
Stoddard et al. [84]	Annuli, inner diameter = 6.45 mm, gap = 0.724–1.001 mm. Heated length = 185 mm	Single	Water	Mass flux = 100–380 kg/m²s; exit pressure = 0.344–1.034 Mpa; wall heat flux: 0.231–1.068 MW/m	Comparison of vertical and horizontal identical channel has been done.

Based on the available literature, one may note that experiments have been conducted in microtubes and microchannels of rectangular cross section. In the case of microtubes, both single [76–79] and multiple tubes [11] have been used. On the other hand, the use of a single microchannel is not very common for the study of CHF; data are available only from heat sink with parallel microchannels. Single microtubes are metallic, while parallel channel heat sinks are made of both silicon and metal (copper and stainless steel). Single microchannels in silicon substrates are also sometimes used.

The range of working fluids covers water, refrigerants (R134a, R245fa, R236fa, R123, R32, R113, etc.), CO_2, nitrogen, helium, ethanol, etc. Though the time is not right yet to propose the effect of fluid properties in definitive teRMS, CHF is significantly influenced by the type of fluid. Unless systematic experiments are planned to estimate CHF for different fluids using the same experimental facility, a comparison in the true sense is not possible. Such studies are limited in numbers. J.R. Thome et al. [58,65] have used R134a, R236fa, and R245fa in the same experimental setup and obtained a common trend of CHF for all the refrigerants. Nevertheless, the least value of CHF was recorded for R236fa, and R134a was recognized as the best cooling medium as far as pressure drop and CHF values are concerned. The investigators have also noted a negligible effect of saturation temperature on CHF within their range of experimental parameters. However, further investigations are required in this direction.

Of different geometrical parameters, the hydraulic diameter and length-to-diameter ratio have a profound effect on CHF. For subcooled boiling, CHF increases with the decrease of channel diameter. This trend (Fig. 6.7(a)) has consistently been observed by a number of researchers [91,5]. On the other hand, for $x_{eq} > 0$, CHF decreases with the decrease in diameter. Bergles [5] postulated that as the channel diameter decreases, there is also a decrease in the departure diameter of the vapor bubbles. Further, bubble velocity relative to the liquid increases and condensation at the tip of bubble becomes stronger.

The combined effect of these three facts is responsible for an increase in CHF with the decrease in diameter. Almost all the investigations report a monotonic increase in CHF with the increase in mass flux, which is not an unexpected result. CHF also increases approximately linearly with the increase in subcooling as shown in Fig. 6.7(b) [94]. However, the influence of subcooling is more pronounced at higher mass flow rate.

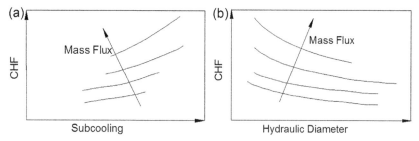

Figure 6.7 Variation of CHF: (a) with inlet subcooling, and (b) with hydraulic diameter.

6.5   Prediction of CHF through Correlations

The earlier section discusses some important experimental investigations on flow boiling through microchannels. It has been observed that some trends of the variation of CHF have already emerged. This has encouraged the researchers to try to predict CHF as a function of geometric and operational parameters as well as fluid properties. Such predictions can be made through correlations or modeling. In this section, the correlations for the CHF in microchannels have been examined while the effort for modeling CHF has been scrutinized in the next section.

In the endeavor of proposing correlations, often the researchers restrict themselves to limited data set generated from identical test conditions. This casts a doubt on the applicability of the correlation over a wider range. Availability of limited data is one of the reasons for this. Fortunately, in case of microchannel boiling, a huge volume of data has been generated within a small period of time and some of the correlations have been developed using a large data set. One can identify three different trends as far as the prediction of CHF in microchannels through correlations is concerned. Researchers tried to use correlations, already available for macro systems, to modify some available correlations to comply with the experimental results of microchannels, and to develop altogether new correlations. Table 6.2 provides a compilation of some important correlations relevant for the CHF in microchannel. It needs to be mentioned that this table is not meant to present an all-inclusive list. Nevertheless, it presents correlations very relevant for the present topic and those that have shown promise in predicting CHF in microchannels.

In Table 6.2, correlations have been listed chronologically, as they have been developed or proposed. For ease of ready cross-reference, numbers enclosed in brackets have been inserted after some references within the main body of the text. These numbers correspond to the chronological serial numbers against the correlation as entered in Table 6.2.

In this table, most of the correlations try to find out a logical relationship between Bl_{chf} or q_{chf} and the operating, geometric and property variables. Bl_{chf}, the nondimensional CHF, is defined as $Bl_{chf} = q_{chf}/Gh_{lv}$. From the very definition it recognizes a direct proportionality between the CHF and the mass velocity. That the CHF increases with the mass flow rate has been an accepted fact in macro-systems and has also been established for microchannels beyond contradiction. Furthermore, from the basic physics, the CHF should be directly proportional to the enthalpy needed for vaporization. This trend also supports almost all the correlations and theories developed from diverse assumptions and approaches.

Four of the correlations of Table 6.2 [41,67,69,93] share similar structure and some common groups of nondimensional numbers. In all these correlations, q_{chf} is proportional to a property group $(\rho_l/\rho_v)^\alpha$ where $\alpha > 0$. The density difference between the phases indicates the ease with which the vapor can get separated from the liquid phase from an evaporating surface. On the other hand, this also signifies the location of the operating point with respect to the critical point and the magnitude of the enthalpy of vaporization. In general, the greater the difference between the densities of saturated vapor and saturated liquid, the higher will be the value of CHF. Further, CHF is proportional to $(1/We_{lo})^\beta$ and $(L_n/d_n)^{-\gamma}$ where β and γ are positive entities and

Table 6.2 Correlations Used for the Prediction of CHF in Microchannels

Sl. No	Correlation	Author(s)	Details	Error	Remarks
1	$Bl_{chf} = 0.10(\rho_v/\rho_l)^{0.133}\,(1/We_{lo})^{0.333}\,\dfrac{1}{1+0.03\,(L_n/d_h)}$	Katto [41]	Circular, conv. channel	Percentage of data points within the 30% error band: 40.7%; MEA = 46%; SD: 69.1%	Developed for microchannels. However, different authors have compared it against microchannel experimental data and it has fared well. Wu et al. [95] has compiled the results from a microchannel database drawn from various sources
2	$q_{chf} = Gh_{lv}q_{Co}\left(1 + k\,\dfrac{\Delta h_{sub}}{h_{lv}}\right)$	Katto and Ohno [43]		**Aqueous:** No. of data points: 2427 Percent of data points within the 30% error band: 69.1%; MAE:27.9%; MRE: 3.2% **Non-aqueous:** No. of data points:569 Percent of data points within the 30%; error band: 69.1%; MAE: 26.3%; MRE: 0.2%;	
3	$q''_{hor} = k_{hor}q''_{tab}.$ $k_{hor} = \begin{cases} 0.0, G \le G_{\min}; \\ (G - G_{\min})/(G_{\max} - G), G_{\min} < G < G_{\max}; \\ 1.0, G \ge G_{\max}. \end{cases}$	Groneveld [23]	Correlations on flow through horizontal channels as this are scarce, especially at low mass fluxes. According to Yu et al. [97], this formula gives the correct trend of CHF in small tubes		

Continued

	Equation	Reference	Description	Statistics
4	$$G_{min} = \frac{\sqrt{gD\rho_g(\rho_l - \rho_g)}}{x}\left(\frac{1}{0.65 + 1.11x_{tt}^{0.6}}\right)^2,$$ $$G_{max} = \left[\frac{gD^{1.2}\rho_l(\rho_l - \rho_g)}{0.092(1 - x_{tt})^{1.8}\mu_l^{0.2}}\left\{-0.3470 + 0.092\ln(x_{tt})\right.\right.$$ $$\left.\left. -0.0556\ln^2(x_{tt})\right\}\right]^{0.056}.$$ $$q_{chf}/Gh_{lv} = 0.124\left(\frac{L}{d}\right)^{-0.89}\left(\frac{10^4}{Y_{shah}}\right)^n (1 - x_{in})$$ $$Y_{shah} = G^{1.8}d^{0.6}\left(\frac{C_p}{K_l\rho_l^{0.8}g^{0.4}}\right)\left(\frac{\mu_l}{\mu_v}\right)^{0.6}$$ If $Y_{shah} \leq 10^4$, $n = 0$. If $10^4 < Y_{shah} \leq 10^6$, $n = \left(\frac{d}{L}\right)^{0.54}$ If $Y_{shah} > 10^6$, $n = \frac{0.12}{(1 - x_{in})^{0.5}}$	Shah [81]	Derived from a database containing 23 different fluids and diameter range 0.315–37.5 mm	**Aqueous:** No. of data points: 2427 Percent of data points within the 30%; error band: 76.3%; MAE: 25.0%; MRE: 11.3% **Non-Aqueous:** No. of data points: 569 Percent of data points within the 30%; error band: 41.7%; MAE: 33.7; MRE: -14.4
5	$$\frac{C}{C_{Tong}} = 1 - \frac{52.3 + 80x_{eq,o} - 50x_{eq,o}^2}{60.5 + (10p_0)^{1.4}}$$	Inasaka and Nariai [102]	Can verify CHF data within ±20% accuracy for channels down to 2 mm diameter	Mean deviation: 30.5%
6	$$Bo = \frac{C}{\sqrt{Re}} = \frac{q_c}{Gh_{fg}}$$	Celeta et al. [103]	Can verify CHF data within RMS error of 21.2% for channels down to 0.3 mm diameter	Mean deviation: 30.1%
7	$$Bo = \frac{c_1 We_D^{c_2}(\rho_f/\rho_g)^{c_3}[1 - c_4(\rho_f/\rho_g)^{c_5}x_{eq,in}]}{1 + 4c_1c_4 We_D^{c_2}(\rho_f/\rho_g)^{c_3+c_5}(L/d_h)}$$ $$c_1 = 0.0722; \quad c_2 = -0.312; \quad c_3 = -0.644;$$ $$c_4 = 0.900; \quad c_5 = 0.724$$	Hall and Mudawar [26]		Mean deviation: 19.2%

Table 6.2 Correlations Used for the Prediction of CHF in Microchannels—cont'd

Sl. No	Correlation	Author(s)	Details	Error	Remarks
8	$Bl_{chf} = 33.43(\rho_v/\rho_l)^{1.11}(1/We_{lo})^{0.21}(L_h/d_h)^{-0.36}$	Qu and Mudawar [69]	Rectangular channel, hydraulic dia = 0.38–2.54 mm	Percent of data points within the 30%; error band: 37.6%; MAE: 44.9%; SD: 56.4%	Average of results obtained by 9 authors, compiled by Wu et al. [95]
				Aqueous: No. of data points: 2427 Percent of data points within the 30%; error band: 1.9%; MAE: 418.4%; MRE: 330.1%	Based on average of results obtained by 10 authors, compiled by Revellin et al. [70,74,75]
				Non-Aqueous: No. of data points: 569 Percent of data points within the 30%; error: 4.2%; MAE: 1883.9%; MRE: 1883.3%	Based on average of results obtained by 12 authors, compiled by Revellin et al. [74,75]
9	$Bl_{chf} = 0.437(\rho_v/\rho_l)^{0.073}(1/We_{lo})^{0.24}(L_h/d_h)^{-0.72}$	Wojtan et al. [93]	Circular channel, D_h = 0.5–0.8 mm; valid for R134a, R245fa, etc.	Percent of data points within the 30%; error band: 66.7%; MAE: 46.1%; SD: 105.1%	Based on the average of results obtained by 8 authors, compiled by Wu et al. [95]
				Aqueous: Percent of data points within the 30%; error band: 27.7%; MAE: 95.0%; MRE: 64.1%	Based on the average of results obtained by 10 authors, compiled by Revellin et al. [74,75]
				Non-aqueous: No. of data points: 569 Percent of data points within the 30%; error band: 64.9%; MAE: 25.1%; MRE: 15.8	Based on the average of results obtained by 12 authors, compiled by Revellin et al. [74,75]

10	$\dfrac{q_{chf}}{Gih_{lv}} = 0.0352\left[We_D + 0.0119\left(\dfrac{L}{d}\right)^{2.31}\left(\dfrac{\rho_v}{\rho_l}\right)^{0.361}\right]^{-0.311}$ $\times 2.05\left[\left(\dfrac{\rho_v}{\rho_l}\right)^{0.17} - x_{in}\right]$	Zhang et al. [99]		**Aqueous:** No. of data points: 2427 Percent of data points within the 30%; error band: 82.8%; MAE: 18.2%; MRE: 2.7% **Non-aqueous:** No. of data points: 569 Percentage of data points within the 30% error band: 63.6%; MAE: 31.7%; MRE: 1.6%	
11	$\dfrac{q_c}{Gih_{lv}} = 0.437\left(\dfrac{\rho_v}{\rho_l}\right)^{0.073} We^{-0.24}\left(\dfrac{L_H}{d}\right)^{-0.72}$	Katto and Ohno, modified by Wotzan et al. [104]		Percent of data points within the 30%; error band: 82.4%; MAE: 7.6%	
12	$Bl_{chf} = \{[0.0934(P_e/P_{Cr})^2 - 1.3 \times 10^{-4}]\,x_e^{0.59}\}^{1/1.08}$	Kosar and Peles [50]	Rectang. channel, $D_h \sim$ 0.228 mm, for R123	Percent of data points within the 30%; error band: 26.6%; MAE: 114.1%; SD: 146.7%	
13	$Bl_{chf} = (0.214 + 0.14Co)(\rho_v/\rho_l)^{0.133}(1/We_{lo})^{0.333}$ $\dfrac{1}{1 + 0.03(L_H/d_H)}$ $q_{chf} = q_{co}\left(1 + K'\dfrac{\Delta h_{sub}}{h_{lv}}\right)$	Qi et al. [67]	Circ. channel hyd. dia. = 0.53–1.93 mm	Percent of data points within the 30%; error band: 8.3%; MAE: 313.4%; SD: 312.4%	Based on the average of results obtained by 8 authors, compiled by Wu et al. [95]
14	If $\dfrac{1}{We_{lo}} < 3 \times 10^{-6}$ then $K' = 1.8\left(\dfrac{L}{130d}\right)^{\frac{-5\rho_v}{\rho_l}}$; else $K' = 0.075\left(\dfrac{L}{130d}\right)^{\frac{5\rho_v}{\rho_l}}\rho_l\,We_{lo}^{0.25}$	Qi et al. [67]		**Aqueous:** No. of data points: 2427 Percent of data points within the 30%; error band: 21.8%; MAE: 169.6%; MRE: 147.5 **Non-Aqueous:** No. of data points: 569 Percent of data points within the 30%; error band: 13.2%; MAE: 206%; MRE: 203.9%	Based on average of results of 10 authors, compiled by Revellin et al. [74,75]

Continued

Table 6.2 Correlations Used for the Prediction of CHF in Microchannels—cont'd

Sl. No	Correlation	Author(s)	Details	Error	Remarks
15	$\dfrac{q_{chf}}{Gh_{lv}} = 0.12\left(\dfrac{\rho_v}{\rho_l}\right)^{0.062} We^{-0.141}\left(\dfrac{L_{th}}{d}\right)^{-0.7}\left(\dfrac{\mu_l}{\mu_v}\right)^{0.183}\left(\dfrac{d}{d_h}\right)^{0.11}$	Ong and Thome [64]		No. of points: 21 Percent of data points within the 30%; error band: 94.4%; MAE: 13.6%	
16	$Bl_{chf} = q_{chf}/(Gh_{lv}) = 0.62(L_{th}/d_h)^{-1.19}x_e^{0.817}$	Wu et al. [95]	Valid for 5 fluids incl. water	This is based on 629 (133 aqueous, 469 nonaqueous) data points, covering 5 halogenated refrigerants, nitrogen and water. It predicts almost 97% of the nonaqueous data (except R12) and 94% of water data within the ±30% error data band. For nonaqueous data, the MAE and SD are 11.1% and 13.9% respectively. For water, they are 17.2% and 18.4% respectively	
17	$Bl_{chf} = 0.6(L_{th}/d_h)^{-1.2}x_e^{0.82}$	Wu and Li [96]		Percent of data points within the 30%; error band: 96.5%; MAE: 10.2%; SD: 13.7%; MRE: 2.4%	Region-I: (726 data points) $L_{th}/d_h \leq 150$, $BoRe_l^{0.5} \leq 200$
18	$Bl_{chf} = 1.16 \times 10^{-3}\left(We_m Ca_l^{0.8}\right)^{-0.16}$	Wu and Li [96]		Percent of data points within the 30%; error band: 82.0%; MAE: 16.1%; SD: 20.5%; MRE: 5.9%	Region-II: (133 points) $L_{th}/d_h \leq 150$, $BoRe_l^{0.5} \leq 200$.

MAE, mean absolute error; MRE, mean relative error; SD, standard deviation; RMS, root mean square.

$We_{lo} = G^2L/\sigma\rho$. The Weber number ($We_{lo}$) is the ratio of inertia to surface force. A dominance of surface force will enhance the CHF. Finally, for saturated flow boiling, an increase in D will increase the value of heat flux.

It may be noted that the first model [41] was not developed specifically for microchannels. Rather it is very general, as CHF was assumed to depend on a large number of parameters such as:

$$Bl_{chf} = f\left(\frac{\rho_l}{\rho_v}, \frac{\mu_l}{\mu_v}, \frac{G^2L}{\sigma\rho}, \frac{\mu_l}{GL}, \frac{g(\rho_l - \rho_v)}{G^2}\right) \tag{6.1}$$

Then the author has argued that the effect of viscosity is negligible in case of CHF and for forced convection, gravity plays a minor role. Further, the effect of gravity is not that prominent in microchannels. This reduces the above equation to

$$Bl_{chf} = f\left(\frac{\rho_l}{\rho_v}, \frac{G^2L}{\sigma\rho}\right) \tag{6.2}$$

Further, Katto [41] assumes an idealized representation of boiling in a uniformly heated vertical tube as shown in Fig. 6.8. For a limiting condition of $d \rightarrow 0$, through vector dimensional analysis, he postulates:

$$Bl_{chf}\frac{L}{d} = const.\left(\frac{\rho_l}{\rho_v}\right)^a\left(\frac{\sigma\rho}{G^2L}\right) \tag{6.3}$$

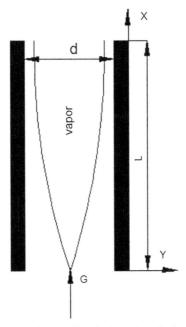

Figure 6.8 Idealization of boiling in a uniformly heated vertical tube.

This is the basis of the first correlation in Table 6.2. A few points may be noted regarding Katto's [41] work. First, the correlation was developed for uniformly heated vertical channels. In a vertical channel, stratification due to gravity is absent. In case of microchannels, even in horizontal orientation, the effect of gravity will not be prominent. Rather, the surface force will be important. Moreover, the above correlation was developed based on the assumption of a small d. Therefore, it is very logical that the basic form of the correlation also satisfies the CHF variation in microchannels. Wu et al. [95] used the Katto correlation for a large number of refrigerants, water and nitrogen, taking data from diverse sources and obtained a reasonable prediction.

The correlation by Katto and Ohno [43] has been used by a large number of researchers for predicting saturated CHF in a single channel. Originally, the correlation was developed for circular tubes and applied down to 3-mm-diameter tubes for normal refrigerants. Though the basic form of the correlation is similar to that of Katto [41], it is capable of accounting for the subcooling with some modifications. Based on the assumption that CHF increases linearly with subcooling, the form of the equation (Table 6.2) was suggested. The correlation gives a reasonable prediction as indicated in the table.

It may be noted that compared to the correlations of CHF for vertical flow, the correlations for flow through horizontal channels, are scarce, especially at low mass fluxes. Groeneveld [23] suggested a simple method of prediction for CHF in horizontal channel by multiplying the corresponding value in vertical tubes with a constant. According to Yu et al. [97], this formula gives the correct trend of CHF in small tubes. When the CHF values obtained from the Groeneveld formula [22] are plotted against mass quality, it is seen that CHF decreases with decreasing mass flux.

Shah [81] proposed a correlation for CHF in uniformly heated vertical channels from a database of 62 independent sources using 23 fluids consisting of water, cryogens, organics, and liquid metals for channel diameters varying from 0.315 to 37.5 mm and heated length-to-diameter ratios from 1.2 to 940. Shah's [81] final correlation basically contains three piecewise defined subcorrelations, as determined by a parameter Y. Of these three correlations, the "upstream conditions" correlation covers the situation where CHF at a location depends on the upstream conditions, e.g., inlet subcooling and distance from tube inlet. The "local condition correlation" relates CHF to the local quality.

Tong [89] derived a CHF correlation by applying boundary layer theory to the problem of a permeable plate with gas injection. His correlation contained a parameter C_{Tong}, which was given as a function of quality.

$$q_c = CG\frac{h_{fg}}{\text{Re}^{0.6}} \tag{6.4}$$

$$C = 1.70 - 7.43x_{eq.,o} + 12.22x_{eq.,o}^2 \tag{6.5}$$

Inasaka and Nariai [102] modified the parameter C_{Tong} as follows:

$$\frac{C}{C_{Tong}} = 1 - \frac{52.3 + 80x_{eq.,o} - 50x_{eq.,o}^2}{60.5 + (10p_0)^{1.4}} \tag{6.6}$$

This modified equation has been successfully applied to channels of diameters down to 2 mm.

Celeta et al. [103] also modified Tong's correlation, both by modifying the parameter C as well as by changing the exponent of the Reynolds number. This correlation agreed well with the data for channels down to 0.3 mm. These endeavors exhibit how an existing correlation can be modified to meet the requirement of microchannel boiling.

The next equation is due to Hall and Mudawar [26], who derived a statistical five-parameter correlation using a comprehensive database covering almost all the available literature. A total of 4860 data points were used and the RMS error was 14.3%. Data for diameters as low as 0.25 mm were successfully correlated.

Qu and Mudawar [69] constructed a correlation having a very low error. They originally compared the Katto-Ohno [43] correlation against experimental data for water in the diameter range $1-3$ mm. For circular single channels, the accuracy was fairly high. However, they found that as the CHF was approached, vapor backflow occurred in the upstream plenum as a result of flow instabilities. This resulted in CHF being more or less independent of inlet subcooling. The new correlation is aimed at addressing these issues. When compared against their data for water and R113, flowing through 21 parallel 215×821 μm channels, the error was about 4%. Such a correlation would be useful for parallel multichannel heat sink.

Wojtan et al. [93] derived a simple correlation from R134a and R245fa data for 0.509 and 0.790 mm ID microchannels. The heated length varied from 20 to 70 mm.

Zhang et al. [99] used an extensive database collated from 10 different laboratories and covering 2500 data points for water only, for a parametric trend analysis and derived the correlation with all fluid properties specified at the inlet. Diameters as low as 0.33 mm have been considered.

Wojtan et al. [93] used the data from their own experiments to compare the original Katto-Ohno [43] correlation and felt the need for certain modifications. The analysis of their CHF data in single, circular and uniformly heated microchannels led to a new fitting of the Katto-Ohno [43] curve. The Mean Absolute Error (MEA) is 7.6% and the recommended limit of applicability is $\rho_v/\rho_l \leq 0.15$ for a lack of data in the high reduced pressure range.

Kosar and Peles [50] concluded that CHF decreased with exit quality. However, this conclusion cannot be taken as a profound one because the effect of quality at a constant mass flux was not specified. The decrease in CHF was mainly resulted from the decrease in mass flux. At a constant heated length and specified mass flux, the CHF increased substantially with a reduction in tube diameter to 0.286 mm from 0.427 mm. A similar trend was also observed by Qi et al. [67]. However, the Kosar and Peles [50] correlation has poor predictive ability for both water and nonaqueous fluids. It overrates the experimental B_l values greatly with a large scatter for nonaqueous data, and under-predicts the experimental B_l values for water data. Qi et al. [67] proposed a correlation that has been verified for aqueous data only.

Their other correlation [67] was derived from flow-boiling data of nitrogen in channels of diameters $0.531-1.042$ mm. All properties are to be specified at the outlet.

Ong and Thome [64] has proposed an improved CHF correlation, based on the modification of the Wojtan et al. [93] CHF correlation. The proposed correlation aims to fit the current CHF experimental data, the experimental CHF data of Wojtan et al. [93] for circular channels with 0.51 and 0.79 mm internal diameters and square multichannels CHF data of Park for R134a, R236fa and R245fa with heated diameters $= 0.35$ and 0.88 mm. The exponents for the density ratio, Weber number and the length/diameter ratio have all been modified to fit the data. Two new teRMS, namely (d/d_h) and (μ_l/μ_v) have been introduced to account for the microscale confinement effect and the influence of viscosity on CHF. The new CHF correlation is valid for diameters between 0.35 and 3.04 mm.

Wu et al. [95] outline a complicated nonetheless powerful algorithm for the derivation of their correlation. Heat flux is nondimensionalized using mass flux and latent heat of vaporization in boiling number Bl. The heated equivalent diameter is used, reflecting the actual heating conditions. The following relationships were proposed:

$$Bl_{chf} = a_1(L_h/d_{he})^{a_2}$$
$$a_1 > 0 \tag{6.7}$$
$$a_2 < 0$$

The exit quality is a function of three parameters: mass flux, inlet subcooling, and CHF. Thus, the following functional form was proposed:

$$Bl_{chf} = a_3 x_e^{a_4},$$
$$a_3, a_4 > 0 \tag{6.8}$$

By regression analysis, the parameters were determined as:

$$a_1 = 0.364,$$
$$a_2 = 1.19 \tag{6.9}$$

A new nondimensional correlation is thus developed avoiding predictive discontinuities.

The final correlation of Table 6.2 [96] has also been derived using the logic along the similar lines as discussed above.

In this section we intend to discuss different correlations used for the prediction of CHF in micro and minichannels. We would also like to critically examine the suitability of the correlations and their range of applications. In this regard the excellent survey done by Revellin et al. [74,75] should be mentioned. They have considered a large database containing 2996 data points from 19 different laboratories covering different types of fluids. Irrespective of the fluids a few general trends were observed. For saturated boiling CHF shows the following trends when the other parameters are kept constant.

- CHF increases with mass flux.
- CHF increases with the increase in D. A reverse trend is observed for subcooled boiling.

- CHF increases with the reduction in heating length.
- CHF increases with subcooling.
- CHF increases with saturation pressure.

6.6 Physical Mechanism and Mechanistic Models

Despite the continued effort to reveal the physics of CHF in macrosystems, our understanding is only partial until now [15]. Even for the simplest case no stand alone theory
has been developed from the first principle. It goes without saying that the mechanism
of CHF in microchannel is more elusive and requires more time before any hypothesis
can be established with confidence. Nevertheless certain facts are emerging due to the
recent investigations.

One of the earliest efforts to consolidate the nature of CHF in microchannels is due
to Bergles and Kandlikar [7]. Based on the literature available [83] at that point of time
they have made a number of interesting observations which are discussed below.

Single-channel CHF data were very rare (at the time of their review). For parallel
multi-microchannels (rectangular cross section in most of the cases) the available
CHF data were taken under unstable conditions. The critical condition was associated
with the compressible volume instability upstream or with the parallel channel
Ledinegg instability. This reduced the CHF values compared to those achievable under
stable flow conditions.

The comment of Bergles and Kandlikar [7] was based on the pioneering work of Qu
and Mudawar [68]. In this work the investigators reported the flow boiling of deionized, deaerated water in a heat block consisting of twenty-one 215×821 μm channels.
The mass flux range was 86–368 kg/m^{2}s with a variation of inlet temperature from 30
to 60 °C, while the outlet pressure was 1.13 bar. The authors reported a unique phenomenon as the CHF was approached. There was a reverse flow of the vapor mass
from all the microchannels into the inlet plenum. Based on the description of the authors the phenomenon has been schematically reproduced in Fig. 6.9.

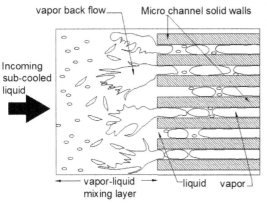

Figure 6.9 Back flow of vapor mass into the inlet plenum during boiling in a multiple microchannel block [68].

From the experimental data they have correlated the heat flux as

$$\frac{q''}{Gh_{fg}} = 33.43 \times \left(\frac{\tilde{n}_g}{\tilde{n}_f}\right)^{1.11} \left(\frac{G^2 L}{\sigma \tilde{n}_f}\right)^{-0.21} \left(\frac{L}{d_h}\right)^{-0.36} \tag{6.10}$$

It may be noted that the trend of the correlation is opposite to other common observation that CHF increases with decreasing channel diameter. Bergles and Kandlikar [7] suggested that this anomaly is due the back flow of vapor which certainly blocks the inflow and reduces the inlet subcooling.

When there are a number of parallel channels connected to an inlet plenum, two types of instabilities are possible. Presence of a large compressible volume at the upstream gives rise to constant volume instability (CVI) whose criteria are given by the following expression:

$$\left.\frac{\partial \Delta \rho}{\partial w}\right|_{CVI} = 0 \tag{6.11}$$

Another type of instability is the excursive or Ledinegg instability which occurs due to the flow through a number of parallel microchannels connected by common plenums. This could be understood from the supply-demand curves of a pump–pipeline combination. The typical nature of the demand curve of a microchannel with boiling is responsible for such a typical static instability [10]. Termed as the Ledinegg or excursive instability (EI), its criterion for instability is given by

$$\left.\frac{\partial \Delta \rho}{\partial w}\right|_{EI} = 0 \tag{6.12}$$

Another important aspect of multiple microchannel block is the conjugate effect. A comprehensive analysis of CHF in such systems can be done considering the entire heat sink as a single unit. Though Bergles and Kandlikar [7] commented on the scarcity of CHF data in single microchannels, many investigations have been conducted in narrow channels (<3 mm) over the years. In such channels, at relatively high mass flux, a unique phenomenon was observed [19,23,32].

The transition boiling curve has a positive slope contradicting the usual concept of CHF. Based on the sketches provided by Groeneveld [23], the probable boiling curves are shown in Fig. 6.10 for two cases, (1) variation of mass flux and (2) variation of quality. Though, most of the microchannel applications are limited to not-very-high mass flux, care must be taken in their design and operation as the CHF behavior can deviate substantially from the usual trend of boiling curve.

Any mechanistic model of flow boiling should take the prevailing flow regimes into cognition. Aided by high speed photographic techniques, several researches have critically examined the flow regimes during boiling through microchannels. In general, slug flow and annular flow are two common flow regimes through microchannels. Kandlikar [37] reported that during boiling, vapor phase may exist

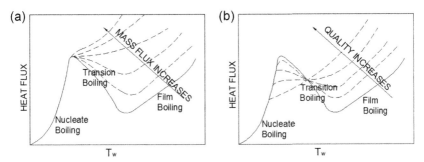

Figure 6.10 Flow boiling curve: (a) variation with mass flux, (b) variation with quality.

in the form of nucleating bubbles, dispersed bubbles, elongated bubbles and also in the core of annular flow. And liquid, on the other hand, can remain as bulk liquid and can appear as slugs, thin films on the channel wall or as dispersed droplets in the vapor stream. Accordingly, there could be a number of flow regimes. A unique phenomenon related to microchannel flow boiling is the presence of rapidly expanding vapor bubbles.

From high speed photography and from numerical simulation the existence of rapidly growing bubble in superheated liquid has been conceived. This causes a reverse flow of the liquid in the direction of the inlet manifold.

Harirchian and Garimella [28−30] reported five major flow regimes, namely bubbly, slug churn, wispy annular, annular and a post dry-out regime of inverted annular flow. It may be noted that the observed flow regimes are not much different from those observed in large size conduits. The authors have reported an elaborate account on the influence of channel dimension and mass flux on the flow regimes. Nevertheless, a dedicated study on the relation between the flow regimes and CHF or dry-out is still pending. Such a study should also consider the orientation of the channel and the direction of flow.

Kandlikar [38] proposed a scale analysis based theoretical model for CHF. The model was an extension of the theory of CHF available for microchannels but incorporated a force balance among the evaporation momentum, surface tension, inertia and viscous forces. The constants in the model were derived from the available experimental data.

Based on the available literature until date, it may be noted that the dry-out during flow boiling has been associated with the elongated bubbles [33] or annular flow regime. This is consistant with the fact that slug flow or annular flows are most common in small channels [9]. Kandlikar [39,40] depicted the local dry-patch formation (Fig. 6.11) in elongated bubbles and discussed the role of micro-layer evaporation and the advancing and receding meniscus of an expanding bubble. Revellin and Thome [71] suggested a mechanistic mode of CHF for flow boiling through heated microchannels. Based on the experimental observations it has been assumed that the CHF occurs in microchannel during annular flow due to film dry-out. There could be two possibilities. At low mass fluxes dry-out occurs when the liquid film is totally exhausted. This occurs at $x = 1.0$.

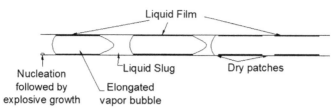

Figure 6.11 Dry patch formation in elongated bubble.

At high mass flow rates vapor shear may overcome surface tension forces and succeed in removing the liquid film from the wall. For dry-out to occur the average film thickness need not be zero, but the liquid film becomes wavy. The interfacial waves are large enough so that at places, the troughs of them touch the wall. In this scenario dry-out occurs at $x < 1$. These two situations are schematically depicted in Fig. 6.12. The authors have separately considered one dimensional conservation equations for mass and momentum for the two phases. Further, the pressure difference across the interface was evaluated by the Young-Laplace equation. The set of differential equations are solved with suitable boundary conditions and the closure equation for wall friction. Additionally, the heights of the interfacial waves are determined considering the shear stress at the interface and the occurrence of Kelvin-Helmholtz type instability. The final expression of the film thickness considering the above mechanisms is obtained as:

$$\delta = CR \left(\frac{u_v}{u_l}\right)^{j_1} \left(\frac{(\rho_l - \rho_v)}{\sigma}\right)^{k_1} \tag{6.13}$$

The values of C, j_1, and k_1 are taken from experiments. The simulation stops when the above equation is satisfied at the outlet of the microchannel. The model has been validated for three different refrigerants (R134a, R245fa, and R113) from two different laboratories. More than 96% of the data could be predicted with an error band of $\pm 20\%$, while the mean absolute error is 8%.

In a subsequent communication, Revellin et al. [72] extended the film dry-out model described above. During flow boiling in microchannels, the vapor dry-out quality (x_{do}) depicts two different variations with the mass flux, G. For laminar dry-out, x_{do} varies as G_j and for transition film dry-out, x_{do} varies with G_j, where j is a positive index. The second type of variation was observed only for CO_2 at high mass velocities. The authors have explained that this difference in trend using the basic model of

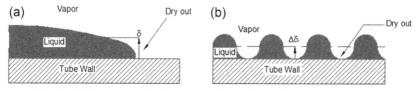

Figure 6.12 Liquid film distribution in annular flow boiling: (a) smooth film, (b) wavy film.

Revellin and Thome [71]. Finally, they commented that the behavior of CO_2 is not radically different from that of the other liquids. The apparent difference comes due to its higher reduced pressure used in many boiling experiments.

Revellin and Thome [73] further argued that though water exhibits a higher CHF compared to all the other fluids studied, it will not be a suitable candidate for the present application of electronic component cooling due to its very low saturation pressure at 30–40 °C. Using the model already developed by them, they have critically examined the suitability of four different refrigerants for the same application. According to the order of merit from best to worst, they can be listed as follows: R245fa, R134a, R236fa, and FC-72. However R134a and R236fa give almost identical CHF values and the decision should be made by the user depending on the application. Further the authors have proposed that a better design should conform to the following combination: (1) a shorter channel length, (2) a low saturation temperature, (3) high mass flux, (4) higher degree of subcooling, and (5) a larger dimension of the microchannel for the typical fluid.

Revellin et al. [74,75] carried out a theoretical study on the optimization of a constructal based tree shaped microchannel network in a disc shaped heat sink. They have used the theoretical CHF model of Revellin and Thome [71]. The coupling of pumping power with the base CHF leads to a different optimum design. They also demonstrated that the most complex design need not be the best design.

Kosar [49] developed a simple model of CHF for saturated flow boiling. He proposed that during saturated boiling, the flow regimes transform from bubbly to slug and then to spray annular flow pattern which is characterized by a thin annular liquid film at the wall and liquid droplets at the vapor core as shown in Fig. 6.13. It may be noted that Fig. 6.13 is much similar to the description of flow regimes during flow boiling in heated vertical channels of larger dimensions. Or in other words, in case of microchannels, stratification due to gravity is not prominent. The complete evaporation of the liquid film gives rise to dry-out. The model is based on simple algebraic relationships of mass and energy balance.

The model was compared for a range of mini and microchannels (0.223 mm $< d_h < 3.1$ mm), of both round and rectangular cross section, for fluids like water and refrigerants (R123, R113, R134a, and R245fa), 151 experimental data covering a range of mass velocities from 50 to 1600 kg/m^{2}s and pressure 101–888 kPa. An overall mean absolute error of 25.8% was noted. Better prediction was obtained for channels with thin walls and for flow without instability.

Kuan and Kandlikar [52] proposed an alternate mechanistic model of CHF based on the force balance at the interface of a vapor plug in a microchannel. Considering three

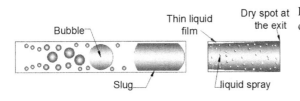

Figure 6.13 Flow regimes during saturated flow boiling.

different forces namely momentum force, inertia force, and surface force; they could obtain a closed form expression for CHF in teRMS of the contact angle θ and the channel height b:

$$q'' = h_{fg}\sqrt{\rho_g}\sqrt{\frac{2\sigma\cos\theta}{b} + \frac{G^2}{2\bar{\rho}}}$$ (6.14)

The model could predict the trend of CHF. It also compares very well with the experimental data of the authors.

6.7   Present State of Understanding and Prediction of CHF in Microchannels

In this section the present state of understanding and prediction capacity for CHF during flow boiling through mini and microchannels is briefly summarized.

The studies on this topic can be briefly classified into two categories. Early studies on single-channel narrow metallic tubes (1 mm $\leq$ D $\leq$ 3 mm) were made with high flow rates and relatively larger tube length due to the importance of high heat flux cooling particularly for nuclear reactors. Mainly water was considered as the coolant though some studies on liquid nitrogen, helium [42], and different refrigerants [51,56,92] as well as CO_2 [66] were reported. These investigations are important for nuclear reactors, particularly for the cooling of the first wall of fusion reactors. The studies in recent years are pertaining to heat sinks mainly targeted to the cooling of electronic components and other cutting edge applications. For this application, investigations have been made in both microtubes and microchannels of rectangular cross-sections. Copper, stainless steel as well as silicon have been taken as the channel material. A large number of studies have been made for parallel multichannel heat sinks. All the heat sink related studies are restricted to a low value of mass velocity and a relatively low value of length-to-diameter ratio of the tubes or channels.

From the experiments the following trends have emerged:

1. CHF increases with mass flow rate
2. CHF increases with the increase in channel hydraulic diameter. This is a common trend with saturated boiling. However, the trend is reverse for subcooled boiling.
3. CHF increases with the increase in L/D ratio.
4. CHF increases with the increase in saturation temperature.
5. There is an increase in CHF with the increase in degree of subcooling at the channel inlet.

All the above observations follow from the fundamental physics. They are intuitive. However, point 2 needs some explanation which has been provided elsewhere in the paper.

Over the years the experimental techniques for boiling in microchannel has matured as there has been a substantial improvement of the techniques for the fabrication of microchannels. With the help of micro machining and nonconventional manufacturing techniques [36], prototypes of microchannel heat sinks with suitable inlet and outlet manifolds are fabricated. In many of the experimental facilities, specially designed heating elements are embedded. Micro-thermocouples are also employed. High speed photography with a high resolution has been adopted in many studies. This has helped to reveal the flow regimes. As a consequence, now a reasonable understanding regarding the flow regimes during microchannel boiling exists.

Earlier, it was understood that CHF could be low in a multichannel heat sink due to flow oscillation and reverse flow. It was also suggested that the incorporation of some flow restriction at the inlet of the channels will improve the situation at the cost of extra pumping power [49]. Provision of artificial nucleation sites or variable channel cross section may also help.

Over the years a number of correlations have been used for the prediction of CHF in microchannels. It is interesting to note that some of the correlations which were not specifically developed for microchannels exhibit good agreement with microchannel data. The correlation by Katto [41] and Katto and Ohno [43] fall in that category. On the other hand, a host of new correlations have also been developed for CHF in microchannels. On the average, the correlations can predict the experimental data within ± 30% range.

Despite this insight effort for modeling CHF in microchannel is still in its infancy. Most of the postulations are based on dry-out of thin annular film. All the models are highly idealized and need experimental data for their closure. Many of the models are capable of predicting the correct trend of CHF variation with different parameters.

6.8   Gray Areas and Research Needs

Microchannel cooling has a great potential as miniaturization is the call of the day. Successful adoption of microchannel boiling in the future cooling technologies requires reliable methods of prediction not only for boiling heat transfer but also for CHF. The forward leap from single-phase liquid cooling to cooling with evaporation or boiling can be made possible only through a well-planned research program which addresses the gray areas in this field. Some of these areas are iterated below.

In a practical cooling system, a single microchannel is rarely expected to be used. However, the phenomenon of boiling in general and CHF in particular is to be studied through single microchannels. This is needed to understand the effect of confinement on CHF. Kandlikar [37] observed that for cooling with boiling heat transfer, normally channels less than 1 mm diameter have rarely been chosen. This is due to the dearth of confidence in using microchannels of narrower dimensions for this typical transport phenomenon. Further research is needed in this direction.

Conducting experimental study for boiling in single micron size channel possesses a number of practical difficulties. Correct assessment of temperature and heat flux is difficult due to the limitations of measurement. Special instrumentation (preferably non contact type) and measurement techniques need to be developed. Despite careful efforts, the strong conjugate effect cannot be neglected. Therefore, often the estimation of heat flux and surface temperature only through measurements is not reliable. Analysis based on inverse conduction is expected to enhance the accuracy of the derived data.

Flow visualization is extremely important. Though different boiling regimes have been reported in recent times from high speed photography, yet a lot more is to be done. Visualization is a real challenge for microchannel boiling. For microtubes, simultaneous visualization and parametric measurements are difficult to achieve. Innovative solutions are needed in this regard. It is not unrealistic to expect that micro PIV will be used in future to get more information during boiling in microchannels. Use of transparent thin film heating element over a quartz tube may be a viable solution and should be tried. In case of rectangular channels, visualization is possible if the top cover of the channel is made transparent and arrangements are made for adequate illumination. However, this violates the condition of uniformly heated channels. Further, the effect of inlet and outlet geometry on microchannel boiling needs to be assessed correctly.

Bergles and Kandlikar [7] have provided important suggestions regarding the measures to be taken for successful data reduction from microchannel experiments.

Though single-channel experiments are necessary, the need for further experimentations in multichannel heat sinks cannot be overlooked. The two instabilities mentioned by Bergles and Kandlikar [7], namely the instability due to the presence of upstream compressible volume and the excursive instability, reduce the CHF. These two instabilities should be studied thoroughly in connection with multi microchannel. The study should address two aspects. Firstly, one needs to know the extent by which these instabilities reduce CHF. Efforts should also be there to decrease margin of such instabilities by the proper design of the system.

A survey of literature on flow boiling through microchannels reveals two distinct classes of investigations. A large number of studies have been made to understand CHF at high mass flux [8,12,16,32,42,62]. These studies used d_h of $1-3$ mm for heated single channels, mostly tubes. Such applications are relevant for the cooling of the first wall of fusion reactors.

On the other hand, in recent times, most of the works on microchannel boiling are directed to heat sink design and involve low mass flux. Further, the range of d_h is sub-millimetric and investigation of multichannel with noncircular cross-section is a common trend. Obviously, there is a need for a comprehensive scheme of experiments which will encompass a large range of mass flux variation, variation of channel dimension and shape as well as single and multichannel keeping the rest of the test matrix fixed as far as possible.

So far, mostly two geometries of the microchannel have been considered in the study of flow boiling. They are round and rectangular. Very few trapezoidal and V-shaped channels have also been used. However, both for the applications of heat sink and high mass flux cooling, one cannot rule out the use of other geometries. Experimental studies are needed in these geometries to enhance the level of confidence in the prediction of CHF.

Surface roughness is expected to have a significant influence on CHF. In case of conventional channels, it has been shown that CHF can be enhanced by manipulating surface roughness. Quantification and manipulation of surface roughness in microchannels is a challenge. Experiments must be devised to explore the effect of surface roughness on the CHF in microchannels. Same systematic studies should be planned to explore the effect of dissolved gas in the liquid and its boiling characteristics.

A large number of studies on microchannels boiling have been done using a number of refrigerants. In all the cases pure refrigerant has been taken as the test fluid. If the refrigerant flow is driven by a compressor (as in a refrigeration cycle), the refrigerant mixes with the lubricants. Studies are pending on the CHF of oil-refrigerant mixture through narrow channels. Effects of other additives on the CHF of microchannel flow boiling are equally unknown. For large sized tubes different augmentation techniques have been adopted [14] to enhance CHF. This includes surface modification, use of additives in the base fluid etc. similar studies are needed for microchannels. In this regard the use of nanofluid [90] may be mentioned.

For tubes of larger dimensions, extensive efforts have been made to enhance the CHF. The techniques for such enhancements covers a wide range like surface modification, twisted tape insert, application of electric and rotational field, solid and liquid additives for the boiling liquid etc. In recent time use of nanofluids has been tried by a large number of researchers for this purpose in conventional systems. For the augmentation of CHF in microchannels rarely any investigation has been done. One effort in this direction is due to Vafaei and Wen [90]. They have used alumina nanofluids in a horizontal microchannel and observe around 51% increases in CHF for 0.1% addition of solid by volume. Obviously, a greater effort should be spent for the augmentation of CHF in microchannel.

A major initiative needs to be taken as far as the modeling of CHF through microchannels is concerned. So far, only a limited effort has been made to develop some mechanistic models of CHF in microchannels. These models are extensions of the existing models suitable for macro systems based on the assumed mechanism of dry-out. These models have shown some success in predicting the experimental data. The model of Revellin and Thome [71] may be mentioned in this regard. However, all these models are highly idealized and too specific. As they are low order models and need closures from experiment, standalone robust models critically verified by an extensive data bank of reliable experiments are the need of the day.

With the growing computational power and the better understanding of the physical phenomena, CFD has become a very powerful tool for prediction. In recent times reasonable success has been achieved in the modeling of boiling heat transfer in macro systems [20,54,60,61,82,87,94]. This has been achieved by direct numerical simulation where separate set of conservation equations are solved for each of the phases and a suitable algorithm is adopted for the modeling of the interface. In recent time computational techniques like molecular dynamics and Lattice Boltzmann method are also used to model different intricate processes of bubble growth, coalescence and moving contact line phenomena. A good survey of the available literature may be found in Chung et al. [15]. Unfortunately, such endeavors have not been taken to model CHF in microchannels. Researchers should pay attention to this. Through a combination of the bulk flow modeling and modeling for the micro layer, it is expected that the transport processes during CHF in microchannels will be better understood.

Finally, it should be appreciated that CHF is not totally isolated phenomenon. In case of flow boiling, it is regime dependent. It could be regarded as an extreme development of the nucleate boiling process or dry-out of a gradually thinning evaporating film, or a phenomenon triggered by the local dry patch formation over the heated surface. In case of microchannel, the process is further complicated by the wall confinement. In any case, an appropriate model of flow boiling should be able to capture CHF. For model development, this should be the ultimate goal of future research.

Nomenclature

a	Parameter
A	Cross-sectional area (m^2)
Bl	Boiling number $q/(Gh_{lv})$
Bo	Bond number $g(\rho_1 - \rho_g)d_h^2/\sigma$
c	Specific heat capacity (J/kg)
Ca	Capillary number $(\mu G/p\sigma)$
Co	Confinement number $(1/Bo^{1/2})$
d	Diameter (mm)
g	Acceleration due to gravity (m/s^2)
G	Mass flux $(kg/m^2 s)$
h	Enthalpy (J/kg)
k	Thermal conductivity (W/mK)
L	Length (m)
p	Pressure (Pa)
q	Heat flux (W/m^2)
Re	Reynolds number $(\rho u d/\mu)$
T	Temperature (K)
u	Velocity (m/s)
w	Mass flow rate (kg/s)
We	Weber number $(G^2 d_h/\rho\sigma)$
Y	Shah parameter

Greek letters

δ	Film thickness/wave thickness
ρ	Density (kg/m^3)
$\bar{\rho}$	Average density (kg/m^3)
μ	Dynamic viscosity (Pa-s)
σ	Surface tension (N/m)
χ	Vapor quality

Subscripts

ch	Channel
chf	Critical heat flux
cr	Critical
D	Diameter
e	Exit
$eq;o$	Equilibrium value at inlet
f	Saturated liquid
fg	Liquid to gas
h	Heated
hor	Horizontal
i	Interfacial
in	Inlet
l	Saturated liquid
lo	Liquid only
lv	Liquid to vapor
m	Mean
o	Outlet
r	Reduced w.r.t. critical quantity
sat	Saturated
sub	Subcooling
tab	Look-up table
v	Saturated vapor

Superscripts

''	Per unit area
'''	Per unit volume

References

[1]   B. Agostini, M. Fabbri, J.E. Park, L. Wojitan, J.R. Thome, B. Michel, State of the art of high heat flux cooling technology, Heat Transfer Eng. 28 (4) (2007) 258−281.

[2]   B. Agostini, R. Revellin, J.R. Thome, M. Fabbri, M. Michel, D. Calmi, U. Kloter, High heat flux flow boiling in silicon multi-microchannels—Part III: saturated critical heat flux of R236fa and two-phase pressure drops, Int. J. Heat Mass Transfer 51 (2008) 5426−5442.

[3]   S. Aoki, A. Inoue, M. Aritomi, Y. Sakamoto, Experimental study within on the boiling phenomena a narrow Gap, Int. J. Heat Mass Transfer 25 (7) (1982) 985−990.

[4]   A. Bar-Cohen, K. Geisler, E. Rahim, Pool and flow boiling in narrow gaps−application to 3D Chip Stacks, in: Proceedings of Fifth European Thermal-Sciences Conference, 2008.

[5]   A.E. Bergles, Subcooled burnout in tubes of small diameter, ASME Paper (1962). 63-WA-182.

[6]   A.E. Bergles, What is the real mechanism of CHF in pool boiling, in: V.K. Dhir, A.E. Bergles (Eds.), Pool and External Flow Boiling, ASME, New York, 1992, pp. 165−170.

[7]   A.E. Bergles, S.G. Kandlikar, On the nature of critical heat flux in microchannels, J. Heat Transfer 127 (2005) 101−107.

[8]   A.E. Bergles, V.J.H. Lienhard, G.E. Kendall, P. Griffith, Boiling and evaporation in small diameter channels, Heat Transfer Eng. 24 (1) (2010) 18−40.

[9]   S. Bhusan, S. Ghosh, G. Das, P. Das, Rise of Taylor bubbles through narrow rectangular channels, Chem. Eng. J. 155 (1−2) (2009) 326−332.

[10] J.A. Boure, A.E. Bergles, L.S. Tong, Review of two phase flow instability, Nucl. Eng. Des. 25 (1973) 165−192.

[11] M.B. Bower, I. Mudawar, High flux boiling in low flow rate, low pressure drop mini-channel and micro-channel heat sinks, Int. J. Heat Mass Transfer 37 (2) (1994) 321−332.

[12] G.P. Celeta, M. Cumo, A. Mariani, Burnout in highly subcooled water flow boiling in small diameter tubes, Int. J. Heat Mass Transfer 36 (1993) 1269−1285.

[13] G.P. Celata, A. Mariani, in: S.G. Kandlikar, M. Shoji, V.K. Dhir (Eds.), CHF and Post-CHF (Post-Dryout) Heat Transfer, Chapter 17, Handbook of Phase Change, Boiling and Condensation, Taylor and Francis, New York, 1999, pp. 443−493.

[14] S.H. Chang, W.P. Baek, Understanding, predicting and enhancing critical heat flux, in: The 10th International Topical Meeting on Nuclear Reactor Thermo-hydraulics (NURETH-10), Seoul, 2003.

[15] J.N. Chung, T. Chen, S.C. Maroo, A review of recent progress on nano/micro-scale Nucleate Boiling Fundamentals, Front. Heat Mass Transfer 2 (2011) 023004.

[16] R.S. Daleas, A.E. Bergles, Effects of Upstream Compressibility on Subcooled Critical Heat Flux, ASME, New York, 1965. Paper 65−HT−67.

[17] V.K. Dhir, S.P. Liaw, Framework for a unified model for nucleate and transition pool boiling, ASME J. Heat Transfer 111 (3) (1989) 739−746.

[18] P.K. Das, S. Chakraborty, S. Bhaduri, Critical heat flux during flow boiling in mini and microchannel—a state of the art review, Front. Heat Mass Transfer 3 (2012) 013008.

[19] Y. Fukuyama, M. Hirata, Boiling heat transfer characteristics with high mass flux and disappearance of CHF following to DNB, in: Proceedings of 7th International Heat and Mass Transfer Conference, vol. 4, 1982, pp. 273−278.

[20] P. Genske, K. Stephan, Numerical simulation of heat transfer during growth of single vapor bubbles in nucleate boiling, Int. J. Therm. Sci. 45 (3) (2006) 299−309.

[21] S.M. Ghiaasiaan, S.I. Abdel-Khalik, Two-phase flow in microchannels, Adv. Heat Transfer 34 (2001) 145−254.

[22] D.C. Groeneveld, S.C. Cheng, T. Doan, Aecl-uo critical heat flux Lookup table, Heat Transfer Eng. 7 (1986) 46−62.

[23] D.C. Groeneveld, The onset of dry sheath condition—a New Definition of Dry-out, Nuclear Eng. Des. 92 (1986) 135−140.

[24] S.J. Ha, H.C. No, A dry-spot model of critical heat flux in pool and forced convention boiling, Int. J. Heat Mass Transfer 41 (1998) 303−311.

[25] S.J. Ha, H.C. No, A dry-spot model of critical heat flux applicable to both pool boiling and subcooled forced convention boiling, Int. J. Heat Mass Transfer 43 (2000) 241−250.

[26] D.D. Hall, I. Mudawar, Critical heat flux (CHF) for water flow in tubes—II: subcooled CHF correlations, Int. J. Heat Mass Transfer 43 (2000) 2605−2640.

[27] Y. Haramura, Y. Katto, A new hydrodynamic model of critical heat flux applicable to both pool and forced convection boiling on submerged bodies in saturated liquids, Int. J. Heat Mass Transfer 26 (1983) 379−399.

[28] T. Harirchian, S.V. Garimella, The critical role of channel dimension, heat flux, and mass flux on flow boiling regimes in microchannel, Int. J. Multiphase Flow 35 (2009) 349−362.

[29] T. Harirchian, S.V. Garimella, The critical role of channel cross-sectional area in microchannel flow boiling heat transfer, Int. J. Multiphase Flow 35 (2009) 904−913.

[30] T. Harirchian, S.V. Garimella, Boiling heat transfer and flow regimes in microchannels—a comprehensive understanding, J. Electron Packag. 133 (1) (2011) 011001.

[31] G. Hetsroni, A. Mosyak, E. Pogrebnyak, Z. Segal, Periodic boiling in parallel microchannels at low vapor quality, Int. J. Multiphase Flow 32 (2006) 1141−1159.

[32] S. Hosaka, M. Hirata, N. Kasagi, Forced convective subcooled boiling heat transfer and CHF in small diameter tubes, in: Proceedings of 9th International Heat and Mass Transfer Conference, vol. 2, 1990, pp. 129−134.

[33] A.M. Jacobi, J.R. Thome, Heat transfer model for evaporation of elongated bubble flows in microchannels, J. Heat Transfer 124 (2002) 1131−1136.

[34] S.G. Kandlikar, A theoretical model to predict pool boiling CHF incorporating effects of contact angle and orientation, J. Heat Transfer 123 (6) (2001) 1071−1079.

[35] S.G. Kandlikar, Critical heat flux in subcooled flow boiling-an assessment of current understandings and future directions for research, Multiphase Sci. Technol. 13 (3) (2001) 207−232.

[36] S.G. Kandlikar, W.J. Grande, Evolution of microchannel flow passages—thermohydraulic performance and fabrication technology, Heat Transfer Eng. 24 (1) (2003) 3−17.

[37] S.G. Kandlikar, Effect of liquid−vapor phase distribution on the heat transfer mechanisms during flow boiling in minichannels and microchannels, Heat Transfer Eng. 27 (1) (2006) 4−13.

[38] S.G. Kandlikar, A scale analysis based theoretical force balance model for critical heat flux (CHF) during saturated flow boiling in microchannels and minichannels, in: Proceedings of ASME 2009 Second Micro/Nanoscale Heat and Mass Transfer International Conference, Shanghai, China, 2009.

[39] S.G. Kandlikar, Similarities and differences between flow boiling in microchannels and pool boiling, Heat Transfer Eng. 31 (3) (2010) 159−167.

[40] S.G. Kandlikar, A scale analysis based theoretical force balance model for critical heat flux (CHF) during saturated flow boiling in microchannels and minichannels, J. Heat Transfer 132 (2010) 081501.

[41] Y. Katto, A generalized correlation of critical heat flux for the forced convection boiling in vertical uniformly heated round tubes, Int. J. Heat Mass Transfer 21 (1978) 1527−1542.

[42] Y. Katto, S. Yokoya, Critical heat flux of liquid helium (I) in forced convection boiling, Int. J. Multiphase Flow 10 (1984) 401−403.

[43] Y. Katto, H. Ohno, An improved version of the generalized correlation of critical heat flux for the forced convective boiling in uniformly heated vertical tubes, Int. J. Heat Mass Transfer 27 (1984) 1641−1648.

[44] Y. Katto, Critical heat flux, Int. J. Multiphase Flow 20 (1994) 53−90.

[45] P.A. Kew, K. Cornwell, Correlations for prediction of flow boiling heat transfer in small-diameter channels, App. Therm. Eng. 17 (1997) 705−715.

[46] Y.H. Kim, S.J. Kim, S.W. Noh, Suh, et al., Critical heat flux in narrow gap in two-dimensional slices under uniform heating condition, in: Transactions of the 17th International Conference on Structural Mechanics in Reactor Technology (SMiRT 17), Prague, Czech Republic, 2003.

[47] J.J. Kim, Y.H. Kim, S.J. Kim, et al., Boiling visualization and critical heat flux phenomena in narrow rectangular Gap, Fourth Japan-Korea Symp. Nucl. Therm. Hydraul. Saf. 37 (5) (2004).

[48] S.M. Kim, I. Mudawar, Review of databases and predictive methods for heat transfer in condensing and boiling mini/micro-channel flows, Int. J. Heat Mass Transfer 77 (2014) 627−652.

[49] A. Kosar, A model to predict saturated critical heat flux in minichannels and microchannels, Int. J. Therm. Sci. 48 (2009) 261−270.

[50] A. Kosar, Y. Peles, Critical heat flux of R-123 in silicon-based microchannels, J. Heat Transfer 129 (2007) 844−851.

[51] W.K. Kuan, S.G. Kandlikar, Experimental study on saturated flow boiling critical heat flux in microchannels, in: Fourth International Conference on Nanochannels, Microchannels and Minichannels, Limerick, Ireland, 2006.

[52] W.K. Kuan, S.G. Kandlikar, Experimental study and model on critical heat flux of Refrigerants-123 and water in microchannels, J. Heat Transfer 130 (3) (2008) 1−5, 034503.

[53] J. Lee, I. Mudawar, Two-phase flow in high-heat-flux micro-channel heat sink for refrigeration cooling applications: Part II—Heat transfer characteristics, Int. J. Heat Mass Transfer 48 (2005) 941−955.

[54] J. Liao, R. Mei, J.F. Klausner, The influence of the bulk liquid thermal boundary layer on saturated nucleate boiling, Int. J. Heat Fluid Flow 25 (2) (2004) 196−208.

[55] S.P. Liaw, V.K. Dhir, Void fraction measurements during saturated pool boiling of water on partially wetted vertical surfaces, Trans. ASME J. Heat Transfer 111 (3) (1989) 731−738.

[56] S. Lin, P.A. Kew, K. Cornwell, Flow boiling of refrigerant R141b in small tubes, Trans. IChemE, Part A 79 (2001) 417−424.

[57] J.H. Lienhard, V.K. Dhir, Extended Hydrodynamic Theory of the Peak and Minimum Pool Boiling Heat Fluxes, 1973. NASA CR-2270, Contract No. NGL 18-001-035.

[58] A.W. Mauro, J.R. Thome, D. Toto, G.P. Vanoli, Saturated critical heat flux in a multi-microchannel heat sink fed by a split flow system, Exp. Therm. Fluid Sci. 34 (2010) 81−92.

[59] K. Moriyama, A. Inoue, The thermodynamic characteristics of the two-phase flow in extremely narrow channels, Heat Transfer Jpn. Res. 21 (8) (1992) 823–856.

[60] A. Mukherjee, V.K. Dhir, Study of lateral merger of vapor during nucleate pool boiling, J. Heat Transfer Trans. ASME 126 (6) (2004) 1023–1039.

[61] A. Mukherjee, S.G. Kandlikar, Numerical study of single bubbles with dynamic contact angle during nucleate pool boiling, Int. J. Heat Mass Transfer 50 (1–2) (2007) 127–138.

[62] H. Nariai, F. Inasaka, T. Shimuara, Critical heat flux of subcooled flow boiling in narrow tubes, ASME/JSME Therm. Eng. Joint Conf. 5 (1987) 45–462.

[63] S. Nukiyama, The maximum and minimum values of the heat Q transmitted from metal to boiling water under atmospheric pressure, Int. J. Heat Mass Transfer 9 (1966) 1419–1433.

[64] C.L. Ong, J.R. Thome, Macro-to-microchannel transition in two-phase flow: Part 2–flow boiling heat transfer and critical heat flux, Exp. Therm. Fluid Sci. 35 (2011) 873–886.

[65] J.E. Park, J.R. Thome, Critical heat flux in multi-microchannel copper elements with low pressure refrigerants, Int. J. Heat Mass Transfer 53 (2010) 110–122.

[66] J. Pettersen, Flow vaporization of CO_2 in microchannel tubes, Exp. Therm. Fluid Sci. 28 (2004) 111–121.

[67] S.L. Qi, P. Zhang, R.J. Wang, et al., Flow boiling of liquid nitrogen in microtubes: Part II–heat transfer characteristics and critical heat flux, Int. J. Heat Mass Transfer 50 (2007) 5017–5030.

[68] W. Qu, I. Mudawar, Measurement and correlation of critical heat flux in two-phase micro-channel heat sinks, Int. J. Heat Mass Transfer 47 (2003) 2045–2059.

[69] W. Qu, I. Mudawar, Measurement and correlation of critical heat flux in two-phase micro-channel heat sinks, Int. J. Heat Mass Transfer 47 (2004) 2045–2059.

[70] R. Revellin. Experimental Two-Phase Fluid Flow in Microchannels (Ph.D. thesis), Ecole Polytechnique Federale de Lausanne (EPFL), 2005.

[71] R. Revellin, J.R. Thome, A theoretical model for the prediction of the critical heat flux in heated microchannels, Int. J. Heat Mass Transfer 51 (2008) 1216–1225.

[72] R. Revellin, P. Haberschill, J. Bonjour, J. Thome, Conditions of liquid film dryout during saturated flow boiling in microchannels, Chem. Eng. Sci. 63 (2008) 5795–5801.

[73] R. Revellin, J.R. Thome, Critical heat flux during boiling in microchannels: a parametric study, Heat Transfer Eng. 30 (7) (2009) 556–563.

[74] R. Revellin, K. Mishima, J.R. Thome, Status of prediction methods for critical heat fluxes in mini and microchannels, Int. J. Heat Fluid Flow 30 (2009) 983–992.

[75] R. Revellin, J.,R. Thome, A. Bejan, J. Bonjour, Constructal tree – shaped microchannel networks for maximizing the saturated critical heat flux, Int. J. Therm. Sci. 48 (2009) 342–352.

[76] A.P. Roday, M.K. Jensen, Experimental investigation of the CHF condition during flow boiling of water in microtubes, in: ASME-JSME Thermal Engineering Summer Heat Transfer Conference, Vancouver, Canada, 2007.

[77] A.P. Roday, T.B. Tasciuc, M.K. Jensen, The critical heat flux condition with water in a uniformly heated microtube, J. Heat Transfer 130 (2008) 1–9.

[78] A.P. Roday, M.K. Jensen, Study of critical heat flux condition with water and R-123 during flow boiling in microtubes. Part I: experimental results and discussion of parametric effects, Int. J. Heat Mass Transfer 52 (2009) 3235–3249.

[79]   A.P. Roday, M.K. Jensen, Study of the critical heat flux condition with water and R-123 during flow boiling in microtubes. Part II–comparison of data with correlations and establishment of a new subcooled CHF correlation, Int. J. Heat Mass Transfer 52 (2009) 3250–3256.

[80]   S.K. Saha, G.P. Celata, Boiling and Evaporation, 2011.

[81]   M.M. Shah, Improved general correlation for critical heat flux during upflow in uniformly heated vertical tubes, Int. J. Heat Fluid Flow 8 (1987) 326–335.

[82]   G. Son, V.K. Dhir, Numerical simulation of nucleate boiling on a horizontal surface at high heat fluxes, Int. J. Heat Mass Transfer 51 (9–10) (2008) 2566–2582.

[83]   M.E. Steinke, S.G. Kandlikar, An experimental investigation of flow boiling characteristics of water in parallel microchannels, J. Heat Transfer 126 (2004) 518–526.

[84]   R.M. Stoddard, A.M. Blasick, S.M. Ghiaasiaan, S.I. Abdel-Khalik, S.M. Jeter, M.F. Dowling, Onset of flow instability and critical heat flux in thin horizontal annuli, Exp. Therm. Fluid Sci. 26 (2002) 1–14.

[85]   S.K. Saha, G. Zummo, G.P. Celata, Review of flow boiling in microchannels, Int. J. Microscale Nanoscale Therm. Fluid Transp. Phenomena. 1 (2) (2010) 1–67.

[86]   S.K. Saha, G.P. Celata, Thermofluid dynamics of boiling in MIcrochannels, Adv. Heat Transfer 43 (2011) 77–226.

[87]   G. Tomar, G. Biswas, A. Sharma, A. Agrawal, Numerical simulation of bubble growth in film boiling using a coupled level-set and volume-of-fluid method, Phys. Fluids 17 (11) (2005) 112103.

[88]   J.R. Thome, Engineering Data Book III, Wolverine Tube Inc, 2007.

[89]   L.S. Tong, Boundary-layer analysis of the flow boiling crisis, Int. J. Heat Mass Transfer 11 (1968) 1208–1211.

[90]   S. Vafaei, D. Wen, Critical heat flux of subcooled flow boiling of alumina nanofluids in a horizontal microchannel, J. Heat Transfer 132 (102404) (2010) 1–7.

[91]   C.L. Vandervort, A.E. Bergles, M.K. Jensen, An experimental study of critical heat flux in very high heat flux subcooled boiling, Int. J. Heat Mass Transfer 37 (suppl. 1) (1994) 161–173.

[92]   G.R. Warrier, V.K. Dhir, L.A. Momoda, Heat transfer and pressure drop in narrow rectangular channels, Exp. Therm. Fluid Sci. 26 (2002) 53–64.

[93]   L. Wojtan, R. Revelli, J.R. Thome, Investigation of saturated critical heat flux in a single, uniformly heated microchannel, Exp. Therm. Fluid Sci. 30 (2006) 765–774.

[94]   J.F. Wu, V.K. Dhir, Numerical simulations of the dynamics and heat transfer associated with a single bubble in subcooled pool boiling, J. Heat Transfer Trans. ASME 132 (11) (2010).

[95]   Z. Wu, W. Li, S. Ye, Correlations for saturated critical heat flux in microchannels, Int. J. Heat Mass Transfer 54 (2011) 379–389.

[96]   Z. Wu, W. Li, A new predictive tool for saturated critical heat flux in micro/mini-channels: effect of the heated length-to-diameter ratio, Int. J. Heat Mass Transfer 54 (2011) 2880–2889.

[97]   W. Yu, D.M. France, M.W. Wambsganss, J.R. Hull, Two-phase pressure drop, boiling heat transfer, and critical heat flux to water in a small-diameter horizontal tube, Int. J. Multiphase Flow 28 (2002) 927–941.

[98]   N. Zuber. Hydrodynamic Aspects of Boiling Heat Transfer (Ph.D. thesis), Research Laboratory, Los Angeles and Ramo-wooldridge Corporation, University of California, Los Angeles, CA, 1959.

[99]   W. Zhang, T. Hibiki, K. Mishima, Y. Mi, Correlation of critical heat flux for flow boiling of water in mini-channels, Int. J. Heat Mass Transfer 49 (2006) 1058–1072.

[100]   W. Qu, S.M. Soon, I. Mudawar, Two-phase flow and heat transfer in rectangular micro-channels, ASME Summer Heat Transfer Conference, Las Vegas, Nevada, USA, July 21−23, 2003, pp.1−14.

[101]   J. Lee, I. Mudawar, Fluid flow and heat transfer characteristics of low temperature two-phase micro-channel heat sinks − part 1: experimental methods and flow visualization results, Int. J. Heat Mass Transfer 51 (2008) 4315−4326.

[102]   F. Inasaka, H. Nariai, Critical heat flux and flow characteristics of subcooled flow boiling in narrow tubes, JSME Int. J. 30 (1987) 1595−1600.

[103]   G.P. Celata, M. Cumo, A. Mariani, M. Simoncini, G. Zummo, Rationalization of existing mechanistic models for the prediction of water subcooled flow boiling critical heat flux, Int. J. Heat Mass Transfer 37 (1994) 347−360.

[104]   L. Wojtan, R. Revellin, J.R. Thome, Investigation of saturated critical heat flux in a single, uniformly heated microchannel, Exp. Therm. Fluid Sci. 30 (2006) 765−774.

Instability in Flow Boiling through Microchannels

P.K. Das[1], A.K. Das[2]
[1]Department of Mechanical Engineering, Indian Institute of Technology Kharagpur, Kharagpur, West Bengal, India; [2]Department of Mechanical and Industrial Engineering, Indian Institute of Technology Roorkee, Roorkee, Uttarakhand, India

7.1 Introduction

Boiling is one of the most intriguing modes of heat transfer. It represents a very complex interplay among heat transfer, fluid dynamics, and phase change. As phase change is involved, a large amount of heat is transferred in the form of latent heat. This makes it possible to have a very high rate of heat transfer for a minimal temperature difference between the fluid and the heating surface. Therefore, it is no wonder that boiling heat transfer is a preferred mode of heat dissipation in a large number of industrial applications. However, the complexity of boiling phenomena often leads to unique flow behaviors like fluctuation, oscillation, excursion, flow instability, and even catastrophe. In the simplest situation of boiling, namely heat flux controlled pool boiling, the shift from nucleate boiling to film boiling at the critical heat flux (CHF) point with a sudden jump in surface temperature represents one such anomalous behavior. Such eventualities are more frequent in case of flow boiling. Instability is almost synonymous with two-phase flow. Much research interest has been exhibited in understanding and predicting instability in boiling systems for both forced flow [19] and flow driven by passive mechanisms [5,31].

Nevertheless, the issues of instability in connection to boiling systems are not yet well understood. Further, although a reasonable effort has been spent to investigate boiling instability in channels of conventional size, instability during boiling in mini- and microchannels is a grossly unexplored front.

Microchannel heat exchangers and heat sinks are uniquely preferred for their superior performance. As miniaturization is the call of the day in electronics, the requirement of the dissipation of higher and higher heat flux through a small substrate is ever increasing. Thermal management is very crucial for photonics, aviation, electronics, Microelectromechanical Systems (MEMS), medical electronics, data centers, etc. Microchannel heat sinks or heat exchangers are invariably used in such devices. In addition, conventional and hybrid automobiles, fusion reactors, fuel cells, gas turbine blades, rocket nozzles, etc. rely heavily on microchannel—based devices to fulfill their most stringent requirement of cooling.

Use of microchannels for cooling has gone through a clear shift of paradigm. From the use of gas as the coolant, preference has been given to cooling by single-phase liquids. Thereafter, microchannel heat exchangers and heat sinks are

Microchannel Phase Change Transport Phenomena. http://dx.doi.org/10.1016/B978-0-12-804318-9.00007-8

designed to have boiling heat transfer, which demonstrates some obvious advantages. Along with the capability of handling a very high heat flux at a low temperature difference, boiling heat transfer helps to maintain a uniformity of temperature reducing the risk of hot spot generation. One also requires a lesser inventory of the coolant and a smaller prime mover for circulation. However, the flow has to overcome a sufficiently large pressure drop and it is subjected to various instabilities [8]. Instabilities not only deteriorate the performance of the device; they may also cause potential damages to them.

The research endeavors on microchannel heat exchange devices are increasing, keeping pace with their applications in diverse fields. The growth of literature [4, 6, 9, 11, 20, 24, 33, 36–39] in this topic bears the testimony of this. Though various aspects of boiling through microchannels have been covered by the researchers to a relatively greater details, instability during flow boiling through microchannels have not been investigated that extensively. In this chapter, we present an overview of instabilities in flow boiling through microchannels based on the published literature.

7.2   Instability: A General Overview

Different instabilities have been encountered in two-phase flow even in adiabatic cases. Instabilities at the two phase interface even without a phase change is a topic of fundamental interest and has been investigated extensively [1, 12–14, 34]. In addition to the instabilities inherent to two phase flow, during boiling other instabilities are observed solely due to the change of phase. The inception, growth, and the manifestation of these instabilities are apparently not identical. It is always convenient to have a classification of physical phenomena so that their analysis can be made simpler. Similar efforts have also been made for instabilities observed during boiling, though the task is not simple. The task is challenging as the nature of all the instabilities observed are not clearly understood and more and more unique instabilities are revealed by recent investigations.

Though there have been different reviews on instabilities in two phase flow [3], for a basic overview of the phenomenon, one may refer to Nayak and Vijayan [31]. According to their review, a system is considered stable if, subjected to a perturbation, it returns to its original steady state. The system is neutrally stable, if once perturbed, it continues to oscillate with the same amplitude of the oscillation. If the system stabilizes to a new steady state or its amplitude continues to increase, the system is said to be in unstable equilibrium. It is obvious that the amplitude of oscillation cannot increase indefinitely. The amplitude will be limited by the nonlinearity inherent to the system and limit cycle oscillations, which can either be periodic or chaotic, will be finally established. It may be noted that in case of two-phase flow even for steady state flow, fluctuation with a small amplitude could be present in typical flow regime like slug flow. It is, therefore, important to quantify the magnitude of oscillation to discern the unsteady case. Values ranging from 10% to 30% of the mean value of the parameter are considered as an indication of instability.

Observation over the years have revealed that, there are various kinds of instabilities in boiling ranging from temperature excursions, flow excursions, boiling crisis, channel dry out, flow oscillations, and even reverse flow. Primarily, one can classify the instabilities into two categories, static instability and dynamic instability.

Static instabilities generally induce a shift of the existing operating point to a new operating point. Boiling crisis, bumping, chugging, geysering, Ledinegg instability, etc. fall in this category. Often, such types of anomalous behavior can be analyzed by stationary models. However, in some cases, there is hardly any reference regarding the origin for the instability or any well-accepted method for analyzing it [2].

Dynamic instability, on the other hand, is more complex and depends on the feedback mechanism inherent to the system. A system having a delayed feedback and with sufficient interaction between the inertia and compressibility, generally exhibits dynamic instability. There could be multiple feedbacks due to the complex interaction between pressure drop, flow rate, and change in density.

The most common dynamic instabilities are,

- Density-wave oscillation,
- Pressure-drop oscillation,
- Acoustic oscillation, and
- Thermal oscillation.

It may be noted that this discussion points out only the common type of instabilities. There are instabilities of coupled nature or there are cases where one instability leads to the other.

The research on boiling instability through microchannels has a main focus on the onset of flow instability (OFI). Most of the researchers have identified OFI to be a point where the system passes through the minimum pressure drop. Generally, it occurs at a point at which a small fluctuation can cause a large change in the flow rate. Figure 7.1 schematically depicts the demand curve (the variation of pressure drop with mass flow rate) of a uniformly heated channel.

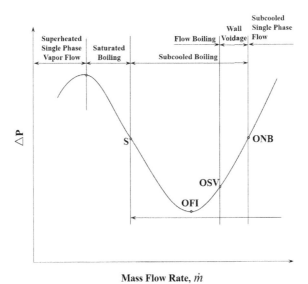

Figure 7.1 Demand curve for a heated channel with constant heat flux.

Subcooled single-phase liquid prevails in the entire channel at a high mass flux. As the flow rate through the channel is decreased, subcooled boiling, saturated boiling, and, ultimately, single-phase flow of vapor take place in succession. In Fig. 7.1, the terms ONB, OSV, OFI, and S refer to the onset of nucleate boiling, onset of significant void, onset of flow instability, and saturation conditions, respectively. At mass flow rates higher than the OSV point, some bubbles appear on the wall although the bulk liquid is subcooled. The condensation of these bubbles decreases the degree of subcooling and at one point, when sufficient bubble mass is generated, the flow becomes two phase. Further, the decrease of mass flow rate will ultimately result in an increase in the pressure drop, as the flow has to accelerate the generated vapor. As the mass flux through the heated channel decreases to the point of minimum pressure drop of the demand curve, a situation is created for flow excursion to take place. This is called OFI. One may refer to Fig. 7.2 to understand the possible flow excursion at OFI. If a positive displacement pump is used, it will give a situation denoted by case A. Case B represents the situation where a number of channels operate in parallel with the boiling channel. For this case, operating the systems steadily and continuously at position 1 is not possible. A small negative perturbation in the flow rate will shift it to point 2 and will lead to dry out of the channel. Case C represents the situation if a centrifugal pump with a relatively flat characteristic curve is used. Once again, continuous operation at position 1 is impossible. Finally, case D represents a situation where the circulation pump has steep characteristics and position 1 could be maintained steadily. The instability that has been described with the help of Figs 7.1 and 7.2 is a common instability for a boiling system. It is known as Ledinegg instability [26]. In the latter part of this chapter, this kind of instability will be referred to a number of times as it has been encountered by a number of researchers in different microchannel systems.

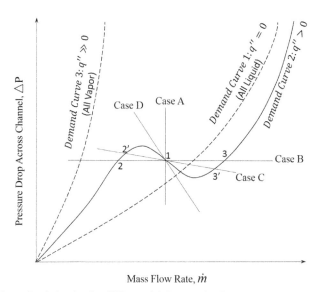

Mass Flow Rate, $\dot{m}$

Figure 7.2 Excursive behavior for different driving mechanisms.

A typical combination of demand curve and the circulation device is not the only cause of Ledinegg instability; it also depends on the operating conditions. Therefore, utmost care is needed to make the system free of this typical instability.

7.3   Experimental Investigations

As flow through microchannels in general and boiling in narrow geometries in particular is a less explored field, justifiably a comparatively large volume of experimental studies have been undertaken to understand the unique physics. The variety of experimental systems considered, the range of parameters explored, and the design of heat sinks and heat exchangers used for such studies are mind-boggling. Often, the information generated by such studies is confusing, though lately researchers are finding some common trends from seemingly different experiments.

A survey of the reported experiments on boiling through microchannels reveals that experiments have been conducted on single as well as multichannels, though recent studies mainly consider multichannel geometry due to the relevance in heat exchangers and heat sinks. Different geometries like rectangular, triangular, trapezoidal, diamond, and circular have been considered for the cross section of the microchannels, although the rectangular channels outnumber the others due to the ease of fabrication and a better use of the available envelope, which is generally small in a micro-system. Both metal and silicon have been used for channel construction, as well as channels fabricated from Polydimethylsiloxane (PDMS). Sometimes, Pyrex and quartz microtubes were used. Different fluids ranging from water, different hydrocarbons, and refrigerants to CO_2 have been tried. Visualization from high-speed image capturing and digital postprocessing has been used excessively to discern different types of instability. Additionally, temperature and pressure recordings have been used for the same purpose.

It may be noted that in many of the experiments, although the investigation of instability was not the primary goal, instability was observed as an inherent feature of the process in the typical system. Further, we report the results for channels with side dimensions of the order of 1 mm and below, as a universally accepted definition of microchannel is difficult to ascertain. The review by Saha and Celata [47] gives an important appraisal regarding the definition of microchannel in two-phase flow.

7.3.1   Experiments in Single Microchannel

Kennedy et al. [23] studied the onset of flow instability in copper microtubes of diameters 1.17 and 1.45 mm using degassed and deionized water as the fluid. To induce the flow instability, two different methods were adopted. Either flow rate was reduced from an initial high value, or, for a fixed flow rate, the heat flux was varied. In both the cases, the indication of the flow instability is taken as the pressure drop passes through the minimum value in the demand curve. The authors have generated a number of demand curves (pressure drop versus mass flux) for different values of heat flux. Each curve, obtained for a constant heat flux, denotes the onset of flow instability (OFI) for that particular condition. However, the existing correlations did not show good agreement with experimental data. For example, the Saha and Zuber [35] relationship for the onset of significant void significantly overpredicted the heat flux leading the OFI. A theoretical analysis of the stability of an evaporative meniscus has been provided by Hetsroni et al. [17]. The transition from stable to unstable flow has been obtained from geometrical and operating parameters.

Brutin and Tadrist [7] conducted a comprehensive study of flow boiling in a rectangular microchannel of 0.5×4 mm^2 cross section (hydraulic diameter: 889 μm). They have also used a model to calculate the pressure drop. Figure 7.3 shows the comparison between the predicted demand curve and the experimental results. To create different upstream conditions, they have incorporated a compliance at the upstream that may be disconnected if required. Figures 7.4 and 7.5 depict the variation of pressure drop with inlet Reynolds number for different heat flux values with the upstream buffer volume connected and disconnected, respectively. One may note the difference in the demand curve for the two cases and a shift in the boundary state. Further, one may note the fluctuations of the pressure at the inlet and exit for a typical case from Fig. 7.6, when the buffer tank (BT) is attached to the upstream of the test section.

Finally, the variation of the exit quality (X_{out}) against the inlet Reynolds number (Re_{IN}) has been plotted in Figs 7.7 and 7.8 for the two upstream conditions considered in the investigation. In these two plots, one can identify the segregation of the steady and unsteady state case. The investigation of Brutin and Tadrist [7] brings out that the behavior of flow boiling is different in microchannels and in large tubes. When a compliance source is incorporated, the unsteady state does not comply with Ledinegg criteria. Also, the removal of the compliance volume cannot remove the unsteady state completely. Finally, the researchers have commented regarding the need of further investigation.

Huh et al. [18] studied the instability related to flow pattern transition in a single microchannel of hydraulic diameter 103.5 μm, made of PDMS using deionized water for low values of mass fluxes (170 and 360 kg/m^2 s) and heat fluxes (200−530 kW/m^2).

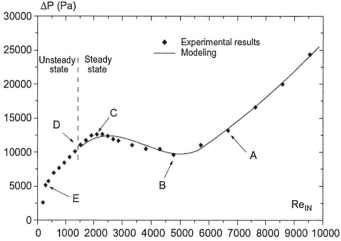

Figure 7.3 Average pressure loss versus inlet liquid Reynolds number for a heat flux of 89.9 kW/m^2, Brutin and Tadrist [7].

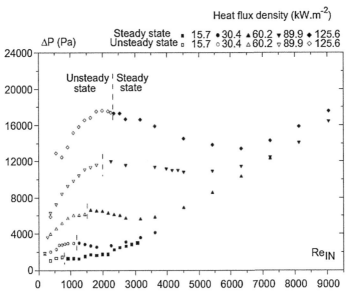

Figure 7.4 Average pressure loss versus inlet Reynolds number when the buffer is not connected to the loop for five heat fluxes, Brutin and Tadrist [7].

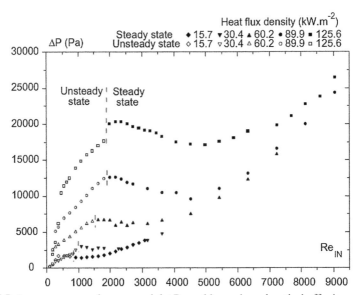

Figure 7.5 Average pressure loss versus inlet Reynolds number when the buffer is connected to the loop for five heat fluxes, Brutin and Tadrist [7].

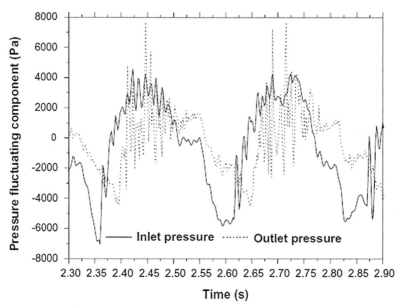

Figure 7.6 Inlet and outlet pressure evolution with BT connected, Brutin and Tadrist [7].

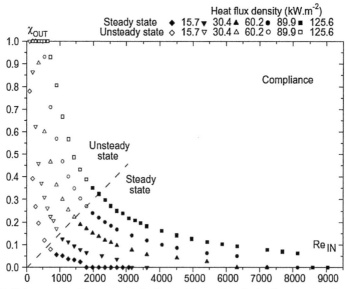

Figure 7.7 Exit vapor quality versus inlet Reynolds number for buffer connected to the loop case for five heat fluxes, Brutin and Tadrist [7].

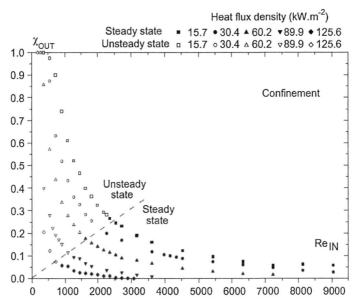

Figure 7.8 Exit vapor quality versus inlet Reynolds number for buffer not connected to the loop case for five heat fluxes, Brutin and Tadrist [7].

The test results showed that the heated wall temperature, pressure, and mass flux all fluctuated with a long time period and large amplitude. The periodic fluctuations indicated the transition from bubbly/slug flow to elongated slug/semiannular flow as revealed from the real-time visualization. Comparison of the fluctuations of different parameters at different flow rates are depicted in Figs 7.9 and 7.10. In test section 6, the number of thin film platinum micro-heaters denoted by $n = 1$ to $n = 6$ were attached for heating. They also served the purpose of temperature measurement. This study is a unique example of how flow regime transition can induce oscillations

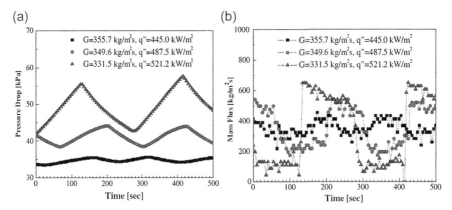

Figure 7.9 Comparison of fluctuations, Huh et al. [18]. (a) Pressure drop fluctuation. (b) Mass flux fluctuation.

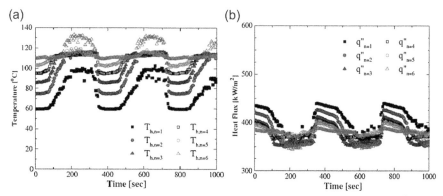

Figure 7.10 Fluctuations of (a) wall temperature and (b) heat flux at $G = 179.8$ kg/m^2 s, $q = 372.4$ kW/m^2, Huh et al. [18].

of very long periods (100−200 s) and large amplitudes in mass flow, temperature, and pressure. Wang and Cheng [40] studied stable and unstable flow boiling in a single microchannel of hydraulic diameter of 155 μm with deionized water as the coolant. Fifteen platinum serpentine micro-heaters were used at the bottom of the microchannel for both heating and temperature sensing. The phenomenon was investigated with simultaneous visualization and parametric measurements. Figure 7.11 shows the evolution of flow regimes at a typical operating condition, and Fig. 7.12 shows the wall temperature oscillations. It was reported that the periods of temperature and pressure drop oscillations during unstable flow boiling are in phase and increase with the decrease in heat flux. The oscillation period was larger than 100 s, as was also reported by Kennedy et al. [23]. Wang and Chen [41] also extended their study to partially heated microchannel and observed the emission of microbubble in flow boiling. From the discussions of this section it is obvious that instability and transition of flow regimes are closely related [32].

7.3.2 Experiments in Multi-Channel Configuration

Single-channel experiments, although important for understanding the basic mechanism of instabilities in confined geometries, are not adequate for appreciating the thermal hydraulics of multi-channel configuration, which is most common for heat exchangers and heat sinks. In multi-channel geometry, the hydrodynamics of a particular channel affects that of the neighboring channels in a unique way [15, 27]. Further, the connection of the channels (i.e., the inlet and exit plena or the headers) plays a very significant role in inducing or suppressing the instability. The research activities reported on the instability of boiling in multi-channels of millimetric and submillimetric side dimensions are large in volume and rich in variety.

Wu and Cheng [44] studied liquid-two phase vapor alternating flow of water boiling in silicon microchannels heated from below. They observed liquid/two-phase/vapor alternating flow (LTVAF) at high heat flux (22.6 W/cm^2) and a low

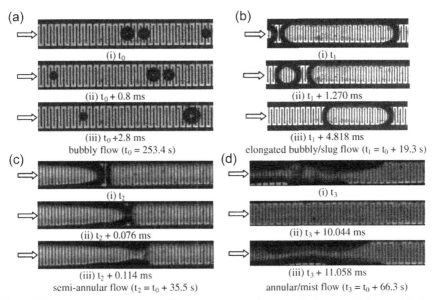

Figure 7.11 Images on evaluation of flow patterns near the outlet during a cycle in unstable flow boiling in a single microchannel with $D_h = 155$ μm at $q_{eff} = 567.4$ kW/m², $G = 117.8$ kg/m² s, and $T_{in} = 20$ °C, Wang and Cheng [40].

mass flux of 11.2 g/cm² s. Their temperature measurement and visualization indicated identical oscillation periods in all the phases. Typical oscillations of the fluid and solid wall temperatures are shown in Fig. 7.13.

Wu and Cheng [45] extended their study further. They have classified the observed instabilities into three categories: (1) liquid/two-phase alternating flow (LTAF) at low heat flux and high mass flux, (2) the continuous two-phase flow (CTF) at medium heat flux and medium mass flux, and (3) the liquid/two-phase/vapor alternating flow (LTVAF) at high heat flux and low mass flux. They have commented that as the generation of vapor bubble increases the pressure drop in the channel that decreases the mass flow rate.

Hetsroni et al. [16] studied explosive boiling in parallel microchannels of triangular cross-section with water and ethanol as the working fluids. They observed periodic boiling where the interval between the successive periodic decreased with the increase in boiling number. They also observed parallel channel instability, which was amplified before CHF. Figure 7.14 shows their observation in one of the microchannels. Vapor jet appears and penetrates through the bulk of liquid with flow in the reverse direction. A decreased mass flow rate also gives rise to a decrease in pressure drop, which, in turn, again increases the mass flow rate. During the increase of mass flow rate, if the wall heating is not sufficient to boil the liquid, single-phase liquid appears.

Xu et al. [46] investigated both static and dynamic flow instability in a parallel microchannel heat sink at a high heat fluxes. They observed the OFI (static flow instability) and the dynamic unsteady flow in a compact heat sink containing 26

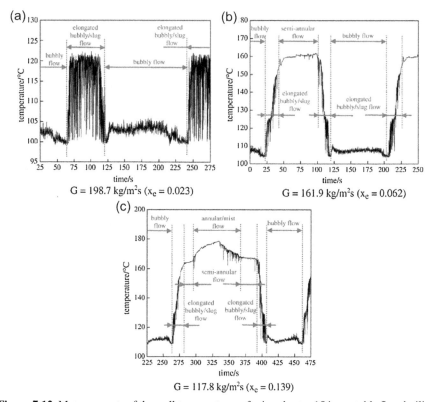

Figure 7.12 Measurements of the wall temperatures of micro-heater 15 in unstable flow boiling in a single microchannel with $D_h = 155$ μm at $q_{eff} = 570$ kW/m^2 and $T_{in} = 20$ °C, Wang and Cheng [40].

rectangular microchannels with 300-μm width and 800-μm depth. OFI was observed to occur at outlet temperature of 93—96 °C for water, which is lower than the saturation temperature. At a mass flux lower than that needed for OFI, three different types of oscillations were observed: large-amplitude/long-period oscillation (LALPO) superimposed with small-amplitude/short-period oscillation, small-amplitude/short-period oscillation (SASPO) and thermal oscillation (TO). Thermal oscillation accompanied the other two oscillations. Figure 7.15 shows an SASPO type oscillation, and Fig. 7.16 depicts pressure fluctuation in an LALPO type oscillation superimposed onto SASPO.

Chang and Pan [10] explored the boiling instability in a heat sink of 15 parallel microchannels with hydraulic diameter of 86.3 μm. Under the stable case they observed bubble nucleation, slug flow- and slug-annular flow in the forward direction. On the other hand, in the unstable case-forward or reversed slug/annular flow appear alternatively in all the channels. In the unstable case, oscillation of the length of the bubble/slug is also observed. The authors have provided a stability map based on two nondimensional numbers: N_{sub} subcooling number and N_{pch} phase change number

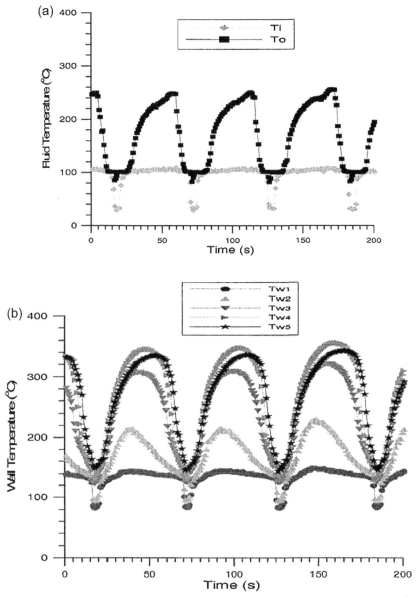

Figure 7.13 Oscillations of temperature measurements in LTVAF mode: $q = 22.6$ W/cm^2 and $m = 11.2$ g/cm^2 s, Wu and Cheng [44]. (a) Water temperature. (b) Wall temperature.

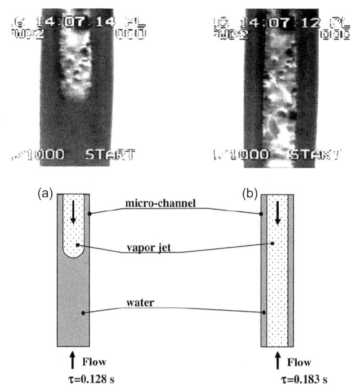

Figure 7.14 Flow pattern upstream of the ONB: (a) single-phase water flow, (b) jet that penetrates the bulk of the water, Hetsroni et al. [16].

(the details may be seen from the original article) as depicted in Fig. 7.17. This map is significantly different from similar maps for conventional channels.

Wang et al. [43] studied unstable and stable flow boiling both in single and parallel microchannels of identical trapezoidal cross sections with a hydraulic diameter of 186 μm. The boiling phenomenon shifted from unstable to stable zone depending on the mass flux. Boiling regime map has been constructed showing both stable and unstable regimes for the single channel as well as multichannel configuration. Water is used as the coolant at a subcooled condition at the inlet with a temperature of 35 °C. Both long period oscillation (time period more than 1 s) and short period oscillation (time period less than 1 s) for pressure and temperature were identified. Long period oscillation was due to expansion of the vapor bubble from the downstream, while the short period oscillation was due to the transition in the flow regime from annular to mist flow.

In case of multichannel configuration, there was significant interaction from the channels at the headers. For determining the instability, the authors have selected the heat to mass flux ratio q/G as the parameter. For stable boiling, the value of this parameter is below 0.09 kJ/kg and 0.96 kJ/kg, respectively, for single (Fig. 7.18)

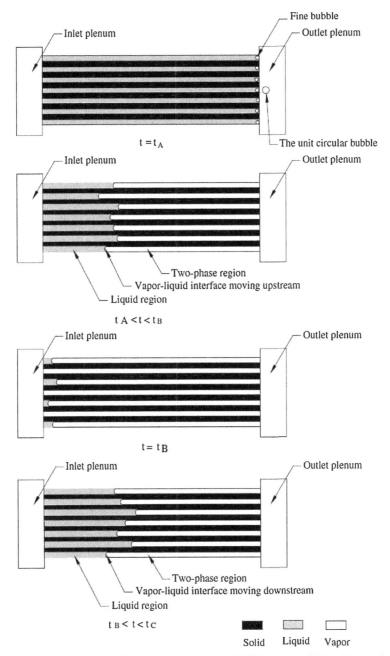

Figure 7.15 Dynamic vapor–liquid interface for the SASPO type oscillation, Xu et al. [46].

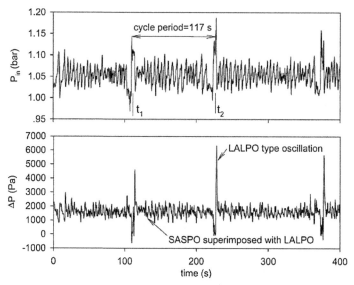

Figure 7.16 Recordings of T_{in}, T_{out}, P_{in}, and ΔP versus time for the LALPO type oscillation superimposed with SASPO type oscillation, Xu et al. [46].

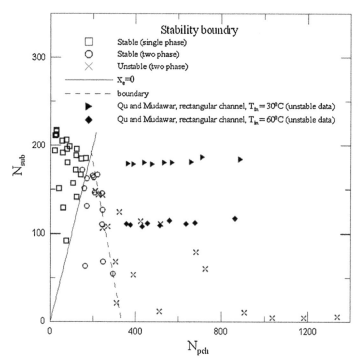

Figure 7.17 Stability boundary on the plane of subcooling number versus phase change number, Chang and Pan [10].

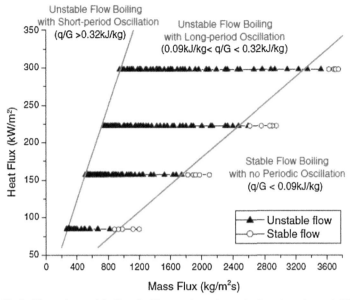

Figure 7.18 Stable and unstable flow boiling regimes in a single microchannel, Wang et al. [43].

and multichannel (Fig. 7.19) configurations. Unstable boiling oscillation mode existed for 0.96 kJ/kg < q/G < 2.14 kJ/kg in parallel microchannels, and for 0.09 kJ/kg < q/G < 0.32 kJ/kg in a single microchannel. Further, the oscillation period in a single channel was much greater (10–80 s) than that (2.5–9.8 s) in parallel channels with the same heat flux conditions. Short-period oscillation was observed with the expression of bubbles in the incoming subcooled liquid of the inlet plenum. The frequency increased with the increase in heat flux. Again, the inlet pressure oscillation was higher in the single channel. The short-period oscillation also generated some noise, which has not been reported earlier.

It goes without saying that instability during boiling in parallel channel configuration is important as such configurations are common in microchannel heat sinks and heat exchangers. In case of parallel channel configuration, the hydrodynamics of a particular channel is influenced by the flow and heat transfer of the neighboring channels. Therefore, the instability in parallel channel configuration cannot be analyzed properly with the knowledge of instability in a single channel. Further, in a parallel-channel configuration, the channels communicate with each other through the plenum. It is, therefore, extremely necessary to analyze the instability in parallel channel flow boiling, taking the geometry and hydrodynamics in the plenums or headers into consideration.

Wang et al. [42] made an investigation on the effects of inlet and outlet configurations on flow boiling instability in parallel microchannels. They have used simultaneous visualization and parametric measurements for the identification and analysis of flow boiling instability.

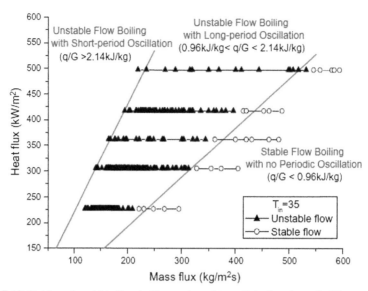

Figure 7.19 Stable and unstable flow boiling regimes in parallel microchannels, Wang et al. [43].

Wang et al. [42] considered three types of inlet-outlet connections for the microchannels: type A, type B, and type C, as shown in Fig. 7.20. Type A connection contains one inlet conduit and one outlet conduit perpendicular to the microchannels. This configuration creates restriction to both the inflow and outflow of the fluid. Type B connection facilitates free flow of liquid into and out of the microchannels. Type C represents a restrictive inlet for the fluid while the discharge from the outlet is free. A detail of the cross section restriction to each of the channels is shown in (c-ii) in the figure. The isosceles triangle at the inlet of each channels provide a 20% restriction of the cross-sectional area to the channel entry.

It has been observed that the amplitude of temperature and pressure oscillations are much smaller in type B configurations compared with those in type A connections for the same heat flux and mass flux condition, as shown in (c). On the other hand, type C configurations exhibit nearly steady flow boiling under the experimental conditions of Wang et al. [42]. Based on their observations, Wang et al. [42] suggested type C connection for high heat flux boiling in microchannels. They have used the minimum pressure drop in the demand curve as the criteria for instability.

7.4 Analysis of Instability in Flow Boiling through Microchannels

Instabilities in boiling are the outcome of a complex interplay between different mechanisms. The physics of all such instabilities are not yet fully understood. Flow boiling through microchannels is a relatively newly explored field. Though some experimental studies have been taken up, developing a reliable theory will take time. Nevertheless,

(a)

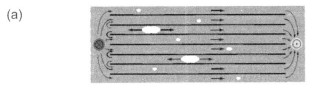

Type-A connections: flow entering and exiting from parallel
microchannels with restrictions because inlet/outlet conduits
perpendicular to microchannels

(b)

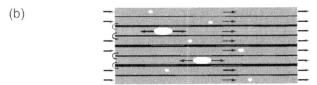

Type-B connection: flow entering to and exiting from
microchannels freely without restriction

(c)

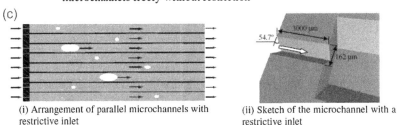

(i) Arrangement of parallel microchannels with   (ii) Sketch of the microchannel with a
restrictive inlet   restrictive inlet

Type-C connection: flow entering with restriction and exiting without restriction
in microchannels

Figure 7.20 Parallel microchannels with three different inlet/outlet connections, Wang et al. [42].

efforts have been made to quantify instability and to predict them from simplified
theory.

A first step toward the understanding of some physical phenomena, particularly
from experimental observations, is to identify some nondimensional numbers so
that some scaling can be done. It has been very logical that both heat flux q and
mass flux G will have unique influences on flow boiling instability. In some of the
work, (q/G) has been taken as a parameter for identifying the flow characteristics.
However, this is not a nondimensional number and it does not reflect the effect of
confinement offered by the microchannel geometry. It also does not take care of the
compliance of the system, which depends to a great extent on the design of the inlet
and outlet plena and influences the hydrodynamics considerably. Though this parameter gives some idea regarding the combined effect of these to parameters on boiling
instability in a particular microchannel design, it hardly gives any generalization.

Two nondimensional numbers, namely the subcooling number (N_{sub}) and the phase
change number (N_{pch}), are used to construct the stability maps for boiling even for conventional channels.

Researchers [7,10] have extended the use of these numbers even for microchannels. Generally, within a small value of N_{pch}, flow is stable for large values of N_{sub}. Beyond certain value of N_{pch} instability occurs for almost any value of N_{sub}. However, this picture is grossly qualitative and again is not generalized. However, a plot on the plane of these two numbers may be used for comparing the extent of the stable regions of flow boiling for different microchannel design.

Brutin and Tadrist [7] provided an analysis based on N_{pch} and N_{sub} considering the energy balance in the microchannel.

$$x_{out} = \frac{1}{L_V}\left(\frac{4q_w L_H}{\mu Re_L} - \Delta h_i\right) = N_{pch} - N_{sub} \tag{7.1}$$

Based on the used heat flux, separate curves are obtained. Here,

$$N_{pch} = \frac{4q_w}{\rho_0 U_0 L_V}\frac{L_H}{D_H} \tag{7.2}$$

$$N_{sub} = \frac{\Delta h_i}{L_V} \tag{7.3}$$

In Figs 7.8 and 7.9, the transition on boundary between unsteady and steady state has been denoted as a function of exit quality (x_{out}) and inlet Reynolds number (Re_{IN}) by a dashed line.

A linear relationship is reduced in the following form:

$$x_{out}^C = N_{pch}^C - N_{sub}^C = ARe^C \tag{7.4}$$

In this equation, the superscript C denotes the critical condition. The researchers have claimed a good agreement between the theory and the experimental results.

Kennedy et al. [23] used a heat transfer–based analysis (for their experimental data) to find the limiting mass flux for OFI. Using the correlation of Bergles and Rohsenow [48], they have predicted the heat flux at the ONB,

$$q_{ONB}'' = 5.30P^{1.156}\left[1.8(T_W - T_{sat})_{ONB}\right]^n \tag{7.5}$$

Further, the wall heat flux at ONB and the wall temperature T_w are also related by the following relation,

$$q_{ONB}'' = h(T_w - T_L)_{ONB} \tag{7.6}$$

Iterative solution is needed to find q_{ONB}'' and T_w. However, the reported agreement between the Bergles–Rohsenow [48] correlations is not convincing. Further, the authors have tried to compare the Saha and Zuber [35] correlation for net vapor generation with their experimental data. For calculating the Stanton number and Peclet number, they have used the following relationships.

$$St = \frac{q''}{\rho_L U_L C_{PL}(T_{sat} - T_L)} \tag{7.7}$$

$$Pe = \frac{GD_e C_{PL}}{k_L} \tag{7.8}$$

However, the agreement is not very encouraging in this case.

Finally, Kennedy et al. [23] tried simple correlation-based quantification for the onset of flow instability. They have calculated the heat flux value at the point of saturated liquid

$$q''_{sat} = \frac{GA\left(\hat{h}_f - \hat{h}_I\right)}{P_H L_H} \tag{7.9}$$

They have also assumed that

$$q''_{OFI} = 0.9 q''_{sat} \tag{7.10}$$

A good agreement has been obtained for the experimental results as can be seen from the figure provided by the researchers.

A similar relationship can be found for the mass flux at OFI and the mass flux (G_{sat}) needed to get the saturation condition at a given heat flux.

$$G_{sat} = \frac{q'' P_H L_H}{A\left(\hat{h}_f - \hat{h}_I\right)} \tag{7.11}$$

From the experimental data, the following relationship was obtained,

$$G_{OFI} = 1.11 G_{sat} \tag{7.12}$$

However, Eqs (7.10) and (7.12) are for typical datasets. Their validity beyond the range of experimental result is doubtful. For the details of the mathematical expressions the reader may refer the original articles.

7.5 Efforts to Suppress the Instability in Flow Boiling through Microchannels

As can be appreciated from the previous sections, flow boiling through microchannels, particularly through parallel microchannels, is very susceptible to different kinds of instabilities. Though one cannot claim the understanding of boiling instability through microchannels is complete, some efforts have already been made to suppress instability through different approaches.

The endeavor of Wang et al. [42] has been mentioned in this regard. They have proposed the suppression of boiling instability through the proper design of the flow inlet and outlet. They have also showed that providing some restriction at the inlet and keeping the outlet restriction free will reduce the instability to a greater extent. The three types of inlet outlet connections considered by them along with some details are shown in Fig. 7.20. Figure 7.21 shows the pressure fluctuation for different connections of the microchannels comparable experimental conditions. One may appreciate the effect of channel connections in supporting instability.

There have been efforts to reduce the instability by controlling the bubble nucleation characteristics of the microchannel. In this regard, the proposition of Kandlikar [21] is worth mentioning. Kandlikar and Balasubramanian [49] observed that nucleation is succeeded by rapid bubble growth and ultimately filling up of the entire channel by vapor plug. This in turn produces a back flow of the incoming fluid. This report may be referred to for a photographic record of the phenomenon. From the consideration of heat transfer, it can be shown that high rate of heat transfer in a microchannel may reduce the degree of subcooling near the nucleation site, even the local liquid could be in the superheated state. This gives rise to a very rapid growth of the bubble and eventually back flow of the liquid.

As a bubble grows over a cavity, the maximum pressure inside the bubble is given by the saturation pressure corresponding to the heater temperature. This introduces a pressure spike as shown in Fig. 7.22. This may cause a back flow.

Kandlikar [21] postulated two ways to reduce the instabilities:

1. Reducing the liquid superheating at ONB, and
2. By introducing a pressure drop element at the entrance of each channel.

Kandlikar [21] has further argued that if nucleation occurs near the exit end of the channel, the chance of back flow of liquid is relatively less. On the other hand, if the

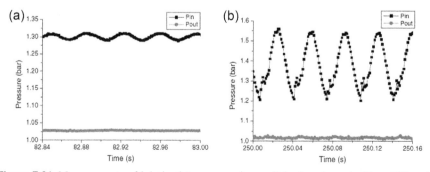

Figure 7.21 Measurements of inlet/outlet pressures in parallel microchannels ($D_h = 186\ \mu m$) in annular/mist alternating flow boiling regime: (a) type B connection with $q = 485.52\ kW/m^2$, $G = 91.43\ kg/m^2s$, and $T_{in} = 35\ °C$ ($x_e = 0.745$) and (b) type A connection with $q = 484.70\ kW/m^2$, $G = 89.53\ kg/m^2s$, and $T_{in} = 35\ °C$ ($x_e = 0.762$), Wang et al. [42].

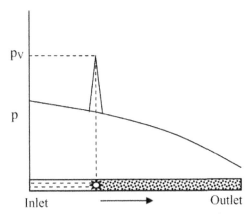

Figure 7.22 Schematic representation of pressure variation following nucleation during flow boiling in a microchannel under stable operation, Kandlikar [21].

nucleation occurs very near the inlet end, liquid may flow in the reserve direction into the inlet plenum as shown in Fig. 7.22. Kandlikar et al. [22] experimentally demonstrated the stabilization of flow boiling in microchannels using pressure drop elements and fabricated nucleation sites. Flow restrictions were introduced after the inlet manifold and artificial sites were introduced by laser drilling. Incorporation of both methods shows a remarkable improvement in the suppression of flow instability. Figure 7.23 schematically shows the effect of the position of the nucleation site on back flow.

Kuo and Peles [25] suggested the use of reentrant cavities to suppress flow boiling oscillations and instabilities. They have done experiments in three different types of channels: one with reentrant cavities, the second one with interconnected reentrant cavities, and the third one with plane wall. As shown in Table 7.1, it was observed that the structured reentrant cavities can mitigate the flow instability by reducing the superheat and pressure at the initial stages of bubble nucleation. They delayed and moderated the boiling instability, thereby extending the stable boiling region and increasing CHF. Reentrant cavities could mitigate the rapid bubble growth and parallel channel type as well as pressure drop/compressible volume type instabilities.

Cosar et al. [50] suggested the use of inlet flow restriction in suppressing flow instability. Several efforts have also been made to suppress boiling instability by modifying the channel cross section along the channel length. Lu and Pan [28] used channels with diverging cross-sectional area along the direction of flow. They have plotted the stability boundary on a plane of subcooling number (N_{sub}) against the phase change

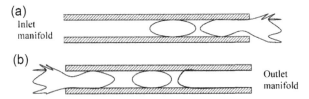

Figure 7.23 Effect of nucleation location on flow boiling characteristics, Kandlikar [21].

Table 7.1 **Microchannels with Different Wall Configurations, Kuo and Peles [25]**

Device	Description	SEM Images
1	Nonconnected reentrant cavity microchannel	
2	Interconnected reentrant cavity microchannel	
3	Plain wall microchannel	

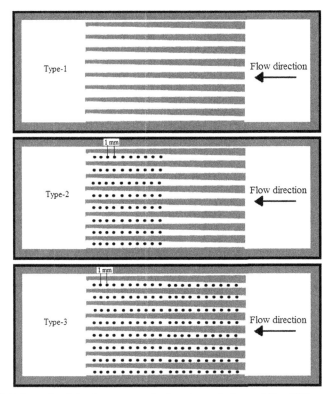

Figure 7.24 Schematic of the three types of microchannels, Lu and Pan [28].

number (N_{pch}) and demonstrated the effectiveness of the variable area channels in suppressing the boiling instability compared with parallel channels with uniform cross-sectional area.

Lu and Pan [29,30] proposed the use of a combination of diverging channel area and artificial nucleation sites. They have used three different channel designs as shown in Fig. 7.24. Flow visualization demonstrates that the combination of diverging channels with artificial nucleation sites at the floor of the channel demonstrates remarkable suppression of flow boiling. Further, the fluctuation of pressure and temperature at the inlet of the channels could also be reduced by this arrangement. It has also been observed that the diverging channel with half of its length provided with artificial nucleation sites exhibits the best result in diminishing the boiling instability.

7.6 Reduction of Instability in Flow Boiling through Microchannels—Achievements and Challenges

From the overview presented earlier regarding the instability in flow boiling through microchannels, certain trends can be identified. First, we like to discuss the achievements so far. A reasonable number of experiments have been conducted to investigate the issue of boiling instability in microchannels. Most of the studies have been done in multi-channel geometries barring a few in single channels. Water has been used as the coolant in maximum number of investigations. Further, most of the studies report basically one kind of boiling instability. The phenomenon of rapid bubble growth and the back flow of the coolant into the inlet plenum have been observed in most of the investigations. Such fluctuation in flow and the reverse flow also generates the fluctuations in pressure and temperature at the inlet, as has been reported by many of the researchers. Efforts have been made to quantify the instability using correlations available for channels of conventional size.

The researchers have rightly felt the need for devising techniques to mitigate the boiling instability and to enhance the range of stable boiling. Kandlikar [21] identified two reasons for the OFI, observed by most of the researchers during the flow boiling of in parallel microchannels: rapid increase in local superheat and rapid rise of fluid pressure due to fast growth of vapor bubble. His group has demonstrated that with the combination of inlet flow restriction and artificial nucleation sites, the flow boiling instability could be suppressed by a very significant amount. Instead of a local flow restriction at the inlet, the use of flow channels with area increasing in the direction of flow has also been suggested [28]. A combination of such area along with artificial nucleation sites at the floor of the microchannels exhibited a good potential in delaying and reducing the flow instability. However, until currently, efforts made to observe, identify, and understand instabilities in flow boiling are limited in scope. The scope of experimental studies should be extended to explore the following aspects.

1. Only a few studies have been done to explore flow boiling instability in single microchannels. Though most of the industrial applications use parallel microchannels, the understanding of boiling stability in the presence of extreme confinement and in the absence of the influence of the neighboring channels is an important requirement. Increasing number of experimental investigations in single microchannels over a larger range of parameters is solicited. Apart from the effects of the operating parameters like pressure, flow rate, inlet temperature, heat flux, and confinement (dimensions of the cross section), ratio of channel length to hydraulic diameter should be clear from the single-channel experiments.
2. Though most of the studies have been made on multichannel geometries, hardly any investigation has been done on the effect of number of channels on the instability. Further flow maldistribution and the nonuniform distribution of the heat flux surely affect the stability of boiling phenomenon. Well-planned experimental investigations should be taken up to investigate these issues.
3. Studies have been made mainly on one type of boiling instability. Other types of boiling instabilities need to be investigated through rigorous experiments.
4. The range of experiments including the type of channel cross section and type of fluids needs to be extended.
5. The parallel microchannels are connected to each other through the inlet and exit plenums. The volume of the inlet plenum acts as storage and provides some compressibility to the system. The design of the inlet plenum has therefore a profound effect on the instability of the system. A systematic study is necessary to understand the effect of the geometry and volume of the plenum on the instability of flow boiling. It is also important to arrive at the designs of the plenum that will reduce the flow oscillations even if there is a substantial generation of vapor.

Until currently, little has been achieved as far as theoretical analysis of the flow boiling instability through microchannels is concerned. It need not be exaggerated that only through a consorted effort of both experimental investigations and theoretical analysis can this critical problem be addressed. A conjugate analysis of fluid dynamics, heat transfer, and phase change considering the microchannels, substrate, and the inlet and exit plenums needs to be done to understand the phenomenon with its fullest rigor. This will enable the designers to design robust microchannel heat sinks and heat exchangers, which will guarantee stability within the specified range of operations.

Nomenclature

A	Area (m^2)
C_P	Specific heat (J/kg K)
D_H, D_e	Hydraulic diameter (m)
G	Mass flux (kg/m^2 s)
h	Heat transfer coefficient (W/m K)
$\widehat{h}$	Enthalpy (J/kg)
k	Thermal conductivity (W/m K)
L_H	Heated length (m)
L_V	Latent heat of vaporization (J/kg)
N_{pch}	Phase change number ($-$)
N_{sub}	Subcooling number ($-$)
P	Pressure (P)
P_H	Heated perimeter (m)
Pe	Peclet number
q, q''	Heat flux (W/m^2)
Re	Reynolds number
St	Stanton number
T	Temperature (K)
U	Velocity (m/s)
Greek Symbols	
χ	Vapor quality
Δ	Difference
μ	Dynamic viscosity (Pa s)
ρ	Density (kg/m^3)

Continued

Subscripts and Superscripts	
C	Critical
I, i	Inlet
out	Outlet
L	Liquid
OFI	Onset of flow instability
Sat	Saturation

References

[1]   V.S. Ajaev, G.M. Homsy, S.J.S. Morris, Dynamic response of geometrically constrained vapor bubbles, J. Colloid Interface Sci. 254 (2) (2002) 346–354.

[2]   J.C. Barber, Hydrodynamics, Heat Transfer and Flow Boiling Instabilities in Microchannels, 2010.

[3]   A.E. Bergles, Review of instabilities in two-phase systems, in: Two-phase Flows and heat Transfer, NATO Advanced Study Institute, Istanbul, 1976, pp. 383–423.

[4]   A.E. Bergles, Critical heat flux in microchannels: experimental issues and guidelines for measurements, in: 1st International Conference on Microchannels and Minichannels, Rochester, 2003.

[5]   S. Bhattacharyya, D.N. Basu, P.K. Das, Two-phase natural circulation loops: a review of the recent advances, Heat Transfer Eng. 33 (4–5) (2012) 461–482.

[6]   D. Brutin, Flow boiling instability, in: Encyclopedia of Microfluidics and Nanofluidics, Springer US, 2008, pp. 687–695.

[7]   D. Brutin, L. Tadrist, Pressure drop and heat transfer analysis of flow boiling in a minichannel: influence of the inlet condition on two-phase flow stability, Int. J. Heat Mass Transfer 47 (10) (2004) 2365–2377.

[8]   D. Brutin, L. Tadrist, Destabilization mechanisms and scaling laws of convective boiling in a minichannel, J. Thermophys. Heat Transfer 20 (4) (2006) 850–855.

[9]   D. Brutin, F. Topin, L. Tadrist, Experimental study of unsteady convective boiling in heated minichannels, Int. J. Heat Mass Transfer 46 (16) (2003) 2957–2965.

[10] K.H. Chang, C. Pan, Two-phase flow instability for boiling in a microchannel heat sink, Int. J. Heat Mass Transfer 50 (11) (2007) 2078–2088.

[11] P. Das, S. Chakraborty, S. Bhaduri, Critical heat flux during flow boiling in mini and microchannel-a state of the art review, Front. Heat Mass Transfer (FHMT) 3 (1) (2012).

[12] V.E.B. Dussan, S.H. Davis, On the motion of a fluid-fluid interface along a solid surface, J. Fluid Mech. 65 (01) (1974) 71–95.

[13] V. Dussan, Moving contact line in waves on fluid interfaces, in: Proceedings of a Symposium. Madison, WI, USA, Academic Press Inc., New York, NY, USA, 1983.

[14] W. Fritz, Maximum volume of vapor bubbles, Phys. Z. 36 (1935) 379.

[15] G. Hetsroni, A. Mosyak, Z. Segal, E. Pogrebnyak, Two-phase flow patterns in parallel micro-channels, Int. J. Multiphase Flow 29 (3) (2003) 341–360.

[16]   G. Hetsroni, A. Mosyak, E. Pogrebnyak, Z. Segal, Periodic boiling in parallel micro-
        channels at low vapour quality, Int. J. Multiphase Flow 32 (2006) 1141−1159.

[17]   G. Hetsroni, L.P. Yarin, E. Pogrebnyak, Onset of flow instability in a heated capillary
        tube, Int. J. Multiphase Flow 30 (12) (2004) 1421−1449.

[18]   C. Huh, J. Kim, M.H. Kim, Flow pattern transition instability during flow boiling in a
        single microchannel, Int. J. Heat Mass Transfer 50 (5) (2007) 1049−1060.

[19]   S. Kakac, T.N. Veziroglu, A review of two phase instabilities, in: S. Kakac, M. Ishii
        (Eds.), Advances in Two Phase Heat Transfer, vol. 2, MartunusNijihoff, Boston, 1983.

[20]   S.G. Kandlikar, Fundamental issues related to flow boiling in minichannels and micro-
        channels, Exp. Therm. Fluid Sci. 26 (2) (2002) 389−407.

[21]   S.G. Kandlikar, Nucleation characteristics and stability considerations during flow
        boiling in microchannels, Exp. Therm. Fluid Sci. 30 (5) (2006) 441−447.

[22]   S.G. Kandlikar, W.K. Kuan, D.A. Willistein, J. Borrelli, Stabilization of flow boiling in
        microchannels using pressure drop elements and fabricated nucleation sites, J. Heat
        Transfer 128 (4) (2006) 389−396.

[23]   J.E. Kennedy, G.M. Roach, M.F. Dowling, S.I. Abdel-Khalik, S.M. Ghiaasiaan,
        S.M. Jeter, Z.H. Quershi, The onset of flow instability in uniformly heated horizontal
        microchannels, J. Heat Transfer 122 (1) (2000) 118−125.

[24]   S.M. Kim, I. Mudawar, Review of databases and predictive methods for heat transfer in
        condensing and boiling mini/micro-channel flows, Int. J. Heat Mass Transfer 77 (2014)
        627−652.

[25]   C.J. Kuo, Y. Peles, Flow boiling instabilities in microchannels and means for mitigation
        by reentrant cavities, J. Heat Transfer 130 (7) (2008) 072402.

[26]   M. Ledinegg, Instability flow during natural forced circulation, Die Wärme 61 (8) (1938)
        891−898.

[27]   H.Y. Li, P.C. Lee, F.G. Tseng, C. Pan, Two-phase flow instability of boiling in a double
        microchannel system at high heating powers, in: ASME 2003 1st International Confer-
        ence on Microchannels and Minichannels, American Society of Mechanical Engineers,
        2003, pp. 615−621).

[28]   C.T. Lu, C. Pan, Stabilization of flow boiling in microchannel heat sinks with a diverging
        cross-section design, Journal of Micromechanics and Microengineering, 18 (2008) 1−13.

[29]   C.T. Lu, C. Pan, Convective boiling in a parallel microchannel heat sink with a diverging
        cross-section design and artificial nucleation sites, in: ECI International Conference on
        Boiling Heat Transfer, May 2009, pp. 3−7.

[30]   C.T. Lu, C. Pan, A highly stable microchannel heat sink for convective boiling, Journal
        of Micromechanics and Microengineering, 19 (2009) 1−13.

[31]   A.K. Nayak, P.K. Vijayan, Flow instabilities in boiling two-phase natural circulation
        systems: a review, Sci. Technol. Nucl. Install. 2008 (2008).

[32]   Y. Peles, Two-phase boiling flow in microchannels: instabilities issues and flow regime
        mapping, in: ASME 2003 1st International Conference on Microchannels and Mini-
        channels, American Society of Mechanical Engineers, January 2003, pp. 559−566.

[33]   W. Qu, I. Mudawar, Measurement and prediction of pressure drop in two-phase micro-
        channel heat sinks, Int. J. Heat Mass Transfer 46 (15) (2003) 2737−2753.

[34]   F. Renk, P.C. Wayner, G.M. Homsy, On the transition between a wetting film and a
        capillary meniscus, J. Colloid Interface Sci. 67 (3) (1978) 408−414.

[35]   P. Saha, N. Zuber, Point of net vapour generation and vapour void fraction in subcooled
        boiling, in: Proceedings of the 5th International Heat Transfer Conference, Tokyo, Japan,
        1974, pp. 175−179.

[36] S.K. Saha, G. Zummo, G.P. Celata, Review on flow boiling in microchannels, IJMNTFTP 1 (2) (2010) 111−178.

[37] S.K. Saha, G.P. Celata, Boiling and evaporation. Heat Transfer and Fluid Flow in Microchannels, in: Heat Transfer Design Handbook, Begell House, USA, 2011, 2.13.4.

[38] S.K. Saha, G.P. Celata, S.G. Kandlikar, Thermofluid dynamics of boiling in micro-channels, in: Y.I. Cho, G.A. Greene (Eds.), Advances in Heat Transfer, vol. 43, Elsevier, 2011, pp. 77−226.

[39] A. Serizawa, Z. Feng, Z. Kawara, Two-phase flow in microchannels, Exp. Therm. Fluid Sci. 26 (6) (2002) 703−714.

[40] G. Wang, P. Cheng, An experimental study of flow boiling instability in a single microchannel, Int. Commun. Heat Mass Transfer 35 (10) (2008) 1229−1234.

[41] G. Wang, P. Cheng, Subcooled flow boiling and microbubble emission boiling phe-nomena in a partially heated microchannel, Int. J. Heat Mass Transfer 52 (1) (2009) 79−91.

[42] G. Wang, P. Cheng, A.E. Bergles, Effects of inlet/outlet configurations on flow boiling instability in parallel microchannels, Int. J. Heat Mass Transfer 51 (9) (2008) 2267−2281.

[43] G. Wang, P. Cheng, H. Wu, Unstable and stable flow boiling in parallel microchannels and in a single microchannel, Int. J. Heat Mass Transfer 50 (21) (2007) 4297−4310.

[44] H.Y. Wu, P. Cheng, Liquid/two-phase/vapour alternating flow during boiling in microchannels at high heat flux, Int. Commun. Heat Mass Transfer 30 (3) (2003) 295−302.

[45] H.Y. Wu, P. Cheng, Boiling instability in parallel silicon microchannels at different heat flux, Int. J. Heat Mass Transfer 47 (17) (2004) 3631−3641.

[46] J. Xu, J. Zhou, Y. Gan, Static and dynamic flow instability of a parallel microchannel heat sink at high heat fluxes, Energy Convers. Manage. 46 (2) (2005) 313−334.

[47] S.K. Saha, G.P. Celata, Thermofluid dynamics of boiling in microchannels, in: Y.I. Cho, G.,A. Greene (Eds.), Part I, Advances in Heat Transfer, vol. 43, 2011, pp. 77−159.

[48] A.E. Bergles, W.M. Rohsenow, The determination of forced-convection surface-boiling heat transfer, J. Heat Transfer 86 (3) (1964) 365−372.

[49] S.G. Kandlikar, P. Balasubramanian, An experimental study on the effect of gravitational orientation on flow boiling of water in $1054 \times 197\mu m$ parallel minichannels, J. Heat Transfer 127 (8) (2005) 820−829.

[50] A. Koşar, K. Chih-Jung, Y. Peles, Suppression of boiling flow oscillations in parallel microchannels by inlet restrictors, J. Heat Transfer 128 (3) (2006) 251−260.

Condensation in Microchannels

Gherhardt Ribatski, Jaqueline D. Da Silva
Heat Transfer Research Group, Department of Mechanical Engineering, Escola de
Engenharia de São Carlos (EESC), University of São Paulo (USP), São Carlos, São Paulo,
Brazil

8.1   Introduction

Microchannels are increasingly used in compact heat exchangers to achieve extremely
high heat transfer rates under confined conditions. In this context, condensation inside
small-diameter channels can be found in heat pipes and compact heat exchangers for
electronic equipment and spacecraft thermal control, automotive and residential air
conditioning systems, and refrigeration systems. A significant portion of the studies
concerning condensation inside small-diameter tubes involves refrigerants characterized by high ozone depletion potential (ODP) and/or high global warming potential
(GWP). Recently, considerable effort has been exercised to replace these refrigerants
by natural refrigerants such as ammonia (R717), hydrocarbons such as propane
(R290), butane (R600), carbon dioxide (R744), and low-GWP refrigerants as
R1234ze and R1234yf. In general, these fluids are capable of minimizing environmental impacts without penalizing the system efficiency. However, hydrocarbons
are considered hazardous, as they are flammable fluids. On the other hand, ammonia,
a highly toxic fluid, is found mainly in industrial applications. Likewise, the HFOs
(R1234ze, R1234yf) are still expensive and are not used extensively. In this context,
heat exchangers based on microchannels allow reduction of the refrigerant inventory
and, consequently, the total refrigerant cost of the system and the risk of accidents
inherent to refrigerant leaks. This statement is corroborate by the study of Hrnjak
and Litch [47] that concerns an evaluation of an ammonia chiller (13 kW cooling capacity) with an air-cooled condenser containing minichannels and a plate-type evaporator. By using low-charge heat exchangers, Hrnjak and Litch [47] were able to reduce
the refrigerant charge to 20 g/kW.

This chapter presents an overview on the literature concerning condensation inside
small-diameter tubes. Initially, an overview on the mechanisms involved during
condensation inside conventional channels and the differences relative to small-
diameter tubes are presented. Then, experimental studies from literature concerning
condensation inside small-diameter channels are schematically described. Finally,
comparisons among prediction methods for pressure drop and heat transfer coefficient
from the literature and condensation experimental data inside small-diameter tubes are
presented. Based on this analysis, predictive methods for void fraction, pressure drop,
and heat transfer coefficient during condensation inside small-diameter tubes are
suggested.

Microchannel Phase Change Transport Phenomena. http://dx.doi.org/10.1016/B978-0-12-804318-9.00008-X

8.2 Convective Condensation

Heat transfer and pressure drop behaviors during convective condensation inside tubes, as in flow boiling, are intrinsically related to the flow pattern occurring during the condensation process. Figure 8.1 illustrates a schematic of the flow pattern evolution with the variation of vapor quality during condensation inside horizontal channels. This figure also depicts the corresponding behavior of the heat transfer coefficient according to the flow patterns. This schematic illustration is based on a constant and uniform wall temperature along the cross section and tubing length.

As shown in Fig. 8.1 for conventional channels (internal diameter > 3 mm), initially the superheated vapor flows at a dry wall desuperheating zone. In this region, the main heat transfer mechanism is single-phase forced convection, and the frictional pressure drop can be estimated through single-phase correlations. With further increasing of the thermodynamic vapor quality, a liquid film surrounding the tube internal wall is established under a condition of superheated vapor. For liquid flowing in an annulus with vapor flowing within its core, the condensation heat transfer process can be characterized as follow: (1) Gravitational dominated condensation, also denominated as stratified flow. Gravitational effects are dominant for low mass velocities. Under this condition, the contribution of the shear effects on the liquid film to the heat transfer coefficient is minimum and the condensate formed on the upper region of the tube flows down along the tube perimeter and is agglomerated on the tube lower region, then, flows according to the flow main direction. Usually, for gravitational dominated condensation, the heat transfer coefficient on the tube upper part is modeled according to the Nusselt [65] theory for falling film condensation. (2) Shear stress

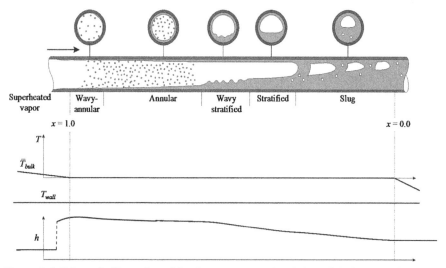

Figure 8.1 Schematic illustration of the flow patterns and variation of the heat transfer coefficient during the condensation process.

dominated condensation, also denominated as annular flow. Under high mass velocity conditions, the liquid film tends to present a uniform thickness along the tube perimeter and the average velocity of the liquid within the film follows the flow main direction. For shear stress dominated condensation, the thermal resistance is a function of the liquid film thickness and convective effects dominate the heat transfer process. It is important to emphasize that the transition from gravitational to shear stress dominated condensation is not abrupt; instead, is a progressive process, characterized by the relative decrease of gravitational effects and increase of shear stress effects with increasing mass velocity.

At the beginning, the condensing superheated vapor region is followed by wavy-annular and annular flow patterns. Along the wavy-annular flow region, the heat transfer coefficient increases compared to the previous region due to the improvement of the interfacial heat transfer area as result of the wavy structure and reduction of the thermal resistance at the wave valleys. Additionally, the vapor, at higher velocity than the liquid, may promote the detachment of droplets from the film, inducing additional increments to the heat transfer coefficient. Due to the vapor condensation along the channel, the two-phase velocity decreases inducing the reduction of the shear effects on the liquid–vapor interface. Both the decreasing of two-phase velocity and the reduction of superficial void fraction with decreasing vapor quality are responsible for increasing the liquid film thickness.

At the transition from annular to stratified-wavy, as the liquid film thickness increases and the void fraction decreases due to the vapor condensation, the flow topology changes progressively from elongated bubble, to bubbly flow, and finally, after the collapse of all bubbles, to single-phase liquid flow. Convective effects are dominant for elongated bubble and bubbly flows; therefore, the heat transfer decreases as the two-phase flow velocity decreases until achieve a single-phase condition. The heat transfer coefficient for single-phase liquid flow can be estimated by a classic correlation from the literature Gnielinski [42]. The annular flow pattern is observed close to the channel inlet section; however, for low mass velocities, the flow patterns evolve rapidly to intermittent flow characterized by waves with high amplitude that achieve the channel upper part. The direct transition from annular to stratified-wavy is also observed, and under this situation, the interface is characterized by waves of small amplitude, evolving to smooth-stratified flow pattern with decreasing the vapor quality.

8.3 Condensation Inside Small Diameter Channels

Contrary to flow boiling heat transfer, for which the transition between micro- and macro-scale has been exhaustively analyzed, without achieving a common criterion among authors so far; studies concerning the transition for condensation are still rare. Figure 8.2 by Ribatski [73] shows a comparison of the micro- to macro-scale transitional criteria proposed in the literature for convective boiling. According to Ribatski [73], the following aspects were taken into account in the literature when proposing criteria for the transition between micro- and macro-scale flow boiling: (1) channel

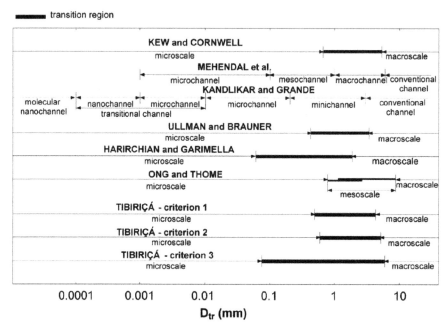

Figure 8.2 Micro- to macro-transitional criteria for in-tube convective boiling, Ribatski [73].

manufacturing process and the heat exchanger application; (2) the degree of confinement of a bubble nucleating within a channel; (3) the degree of uniformity of the film thickness along the channel perimeter; and (4) the possible occurrence of stratified flows. The first criterion does not take into account the two-phase flow characteristics; therefore, its adoption as a transitional criterion in a heat transfer predictive method is not recommendable for condensation or for convective boiling heat transfer. The second criterion is not suitable to have its application extended to condensation due to the absence of bubble nucleation during condensation. On the other hand, exploiting the possibility of applying the third and fourth criteria to in-tube condensation seems promising. This approach would take into account the competitive forces acting on the two-phase flow topology and establishing the predominant heat transfer mechanisms during condensation.

Studies available in the literature concerning condensation inside small-diameter channels highlighted the fact that flow patterns should be investigated as they strongly influence the heat and momentum transfer processes. Figure 8.3 illustrates the flow pattern transition during condensation inside small-diameter channels as proposed by Kim and Mudawar [55]. This figure also shows the pressure profile along the channel and its gradient as the flow pattern changes from single-phase vapor to single-phase liquid flow. The main difference between the flow patterns displayed in Fig. 8.1 for conventional channels and Fig. 8.3 for small-diameter channels is the fact that stratification effects are negligible for the last one. In fact, during condensation inside small-diameter channels and under conditions of low mass velocities, surface tension effects rather than gravitational effects are predominant. The opposite is observed for

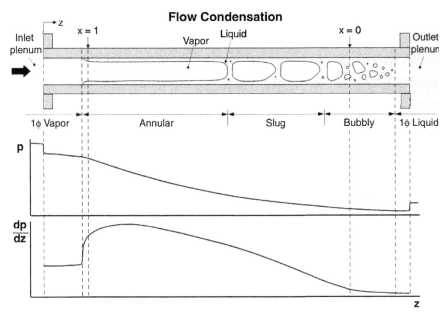

Figure 8.3 Schematic illustration of pressure drop gradient during condensation inside microchannels by Kim and Mudawar [55].

conventional channel (hydraulic diameters larger than 3 mm). On the other hand, under conditions of high mass velocities, shear stress effects are dominant for both micro- and macro-scale flow. Based on a theoretical model solved numerically, Wang and Rose [89] pointed out that surface tension and viscosity control the film thickness along the tube perimeter and, consequently, the local average heat transfer coefficient during condensation in small-diameter channels.

Coleman and Garimella [17,19] and Garimella et al. [39] observed air–water adiabatic flow and R134a condensation into square and circular horizontal tubes with hydraulic diameters ranging from 1 to 4.91 mm. They found that tube diameter has a substantial effect on two-phase flow patterns and their transition. Several studies on pressure drop and heat transfer coefficient during condensation in small-diameter channels have been published recently; however, most of them only presented experimental results for pressure drop and heat transfer coefficient without taking into account the flow pattern effects.

Table 8.1 presents a summary of the studies available in the literature concerning pressure drop during condensation and adiabatic two-phase flows inside small-diameter tubes. In general, these studies cover channel diameters down to 100 μm, and most of them were performed for channels with circular cross sections, for single and multi-channels configurations, and were made of copper, aluminum, glass, silicon, acrylic, or stainless steel. Triangular, rectangular, square, and flat tubes and trapezoidal geometries were also evaluated. The majority of the studies were performed for halocarbon refrigerants (R11, R12, R113, R123, R134a, R152a, R22, R236ea, R245fa,

Table 8.1 Summary of Frictional Pressure Drop Experimental Studies

Authors	Fluid/Test Conditions	Hydraulic Diameter (mm)/Material of Channel	Flow Orientation/Saturation Temperature (°C)	Geometry/Configuration	Mass Velocity (kg/m² s)	Test Section Length (mm)	Reduced Pressure (−)/Pressure Drop Gradient (kPa/m)	Vapor Quality (−)/Heat Flux[a] (kW/m²)	Average Roughness (μm)
Lin et al. [58]	R12/adiabatic	0.66; 1.17/Copper	Horizontal/17–53	Circular/single channel	1440–5090	150	0.12–0.31/200–730	0–0.36/–	–
Mishima and Hibiki [62]	Air–water/adiabatic	1–4/Aluminum; glass	Vertical upward/—	Circular/single channel	71–2374	210–1000	–	–	–
Tipplet et al. [87]	Air–water/adiabatic	1.1; 1.45/Glass 1.09; 1.49/–	Horizontal/–	Circular; semitriangular/single and multiple channels	43–6000	200; 195; 192; 193	–/0–186.2		–
Yan and Lin [91]	R134a/Condensation	2.0/Copper	Horizontal/25–50	Circular/multiple channels	100; 150; 200	200	0.16–0.32/5–30	0.1–0.9/10; 20	–
Coleman and Garimella [18]	R134a/adiabatic; condensation	1.52; 0.76; 0.51, 0.76	Horizontal/55	Circular; square/multiple channels	150–750	–	0.37/2–150	0.1–0.9/–	–
Chen et al. [12]	Air–water/adiabatic R410A/adiabatic	1.02–7.02/– 3.17/–	Horizontal/– Horizontal/5; 25	Circular/single Circular/single channel	50–3000 50–600	150; 995 700	–/1.3–13 –	0–0.9/– –	– –
Lee and Lee [57]	Air–water/adiabatic	0.78–6.67/Acrylic	Horizontal/–	Rectangular/single channel	72–2050	–	–	–	–
Zhang and Webb [94]	R134a; R22; R404A/adiabatic	3.25/Aluminum	Horizontal/20–65	Circular/single channel	400–1000	914	0.14–0.54/0–48	0.2–0.88/–	–
Kawahara et al. [51]	N_2 + water/adiabatic	0.1/Silicon	Horizontal/–	Circular/single channel	238–3983	64.5	–/0–150	–	–
Baird et al. [3]	R123; R11/adiabatic	0.92; 1.95/Copper	Horizontal/20–72	Circular/single channel	70–600	–	0.02–0.1/0–100	0–1/15–110	–

Koyama et al. [56]	R134a/Condensation	1.11; 0.8/Aluminum	Horizontal/47–65	Circular/multiple channel	100–700	600	0.3–0.46/	0–1.0	—
Garimella et al. [39]	R134a	0.52–0.75/–	Horizontal/52.3	Circular; square; triangular; rectangular/multiple channels	150–750	375	0.34/0–60	0.1–0.9/–	0.35–4
Chung and Kawaji [14]	N_2 + water/adiabatic	0.53/Silicon + polyamide	Horizontal/22.9; 29.9	Circular/single channel	22–3575	46; 277	—	—	—
Shin and Kim [80]	R134a/Condensation	0.691/Copper	Horizontal/40	Circular/single channel	100–600	171	0.25/5–200	0.15–0.85/5–20	—
Garimella et al. [40]	R134a/Condensation	0.506; 1.52; 0.761; 3.05/Aluminum	Horizontal/52.3	Circular; flat/multiple channels	150–750	—	0.34/0–100	0.05–0.9/–	—
Cavallini et al. [8]	R236ea; R134a; R410A/adiabatic	1.4/Aluminum	Horizontal/40	Rectangular/multiple channels	200–1400	—	0.1–0.5/0–200	0.25–0.75/–	—
Pehlivan et al. [69]	Air–water/adiabatic	0.8; 1.0; 3.0/Borosilicate glass	Horizontal/–	Circular/single channel	236–2252	200	—	—	—
Revellin and Thome [72]	R134a; R245fa/adiabatic	0.509; 0.709/Stainless steel	Horizontal/26; 30; 35	Circular/single channel	200–2000	110	0.04–0.21/100–1350	0–0.95/preheater: 3.1–415	—
Field and Hrnjak [36]	R134a; R410a; R290; R717/adiabatic	0.148/Aluminum	Horizontal/–	Rectangular/single channel	290–590	—	0.09–0.36/0–40	0–0.98/–	—
Quan et al. [70]	Water/Condensation	0.109; 0.142; 0.151; 0.259/Silicon	Horizontal/–	Trapezoidal/multiple channel	90–288	60	—	0.15–0.85	—
Saisorn and Wongwises [74]	Air–water/adiabatic	0.15; 0.53/Silicon	Horizontal/–	Circular/single channel	—	104; 320	–/0–720	—	—
Park and Hrnjak [67]	R744/adiabatic	0.89/Aluminum	Horizontal/–25; –15	Circular/multiple channel	200–800	500	0.22–0.31/4–70	0.05–0.85/–	—

Continued

Table 8.1 Summary of Frictional Pressure Drop Experimental Studies—cont'd

Authors	Fluid/Test Conditions	Hydraulic Diameter (mm)/Material of Channel	Flow Orientation/Saturation Temperature (°C)	Geometry/Configuration	Mass Velocity (kg/m² s)	Test Section Length (mm)	Reduced Pressure (−)/Pressure Drop Gradient (kPa/m)	Vapor Quality (−)/Heat Flux[a] (kW/m²)	Average Roughness (μm)
Dutkowski [34]	Air–water/adiabatic	1.05–2.3/Stainless steel	Horizontal/10–30	Circular/single channel	170–7350	300	−/0–590	0.0003–0.22/–	–
Bohdal et al. [5]	R134a; R404A/Condensation	0.31–3.30/Stainless steel	Horizontal/20–40	Circular/single channel	100–1300	100	0.14–0.25/10–1300	1–0/–	–
Da Silva et al. [20]	R245fa/adiabatic	1.1/Stainless steel	Horizontal/31; 41	Circular/single channel	100–700	143	0.05–0.07/20–160	0.05–0.98/Preheater: 0–95	0.375
Ducoulombier [33]	CO_2/adiabatic	0.529/Stainless steel	Horizontal/−10; −5; 0; 5	Circular/single channel	200–1400	1690	0.350.5/–	0–1/–	0.8
Tibiriçá et al. [83]	R134a/adiabatic	2.32/Stainless steel	Horizontal/31; 41	Circular/single channel	100–700	464	0.05–0.07/10–70	0.1–0.99/Preheater: 0–55	0.33
Charun [10]	R404A/Condensation	1.4; 1.60; 1.94; 2.3; 3.3/Stainless steel	Horizontal/20–40	Circular/single channel	100–1000	100	−/5–40	1–0/	–
Del Col et al. [24]	R290/adiabatic	0.96/Stainless steel	Horizontal/40	Circular/single channel	200–800	40; 42	0.32/0–350	0.1–0.95/–	1.3
Garimella and Fronk [41]	Ammonia/Condensation	1.435/Stainless steel	Horizontal/–	Circular/single channel	75–150	–	0.10–0.23	0–1/–	–
Kim et al. [54]	FC72/Condensation	1.0/Copper	Horizontal/57.2–62.3	Square/multiple channels	68–367	299	0.05–0.07/0–30	0–1/4.3–32.1	–
Zhang et al. [95]	R22; R410A; R407C/Condensation	1.088; 1.289/Stainless steel	Horizontal/30; 40	Circular/single channel	300–600	200	0.24–0.30/0–40	0.1–0.9	–

Reference	Refrigerant/condition	Diameter (mm)/Material	Orientation/angle	Shape/channel	Mass flux (kg/m²s)		Range	Range	
Da Silva and Ribastki [21]	R245fa; R134a/adiabatic	1.1/Stainless steel	Horizontal/22; 31; 41	Circular/single channel	100–1300	155	0.03–0.25/10–200	0.05–0.98/0–95	0.375
Del Col et al. [25]	R134a/adiabatic R134a/adiabatic	0.96/Copper 2.0/Copper	Horizontal/30; 40; 50 Horizontal/40; 50	Circular/single channel Circular/single channel	200–800 200–500	22 44	0.18–0.32/0–180	0.09–0.97/– 0.02–0.95/–	1.3 1.7
Heo et al. [46]	CO_2/Condensation	1.5/Aluminum	Horizontal/–5; 0; 5	Rectangular/multiple channels	400–1000	450	0.41–0.53/20–170	0.1–0.9/–	–
Liu et al. [59]	R152a/Condensation	1.152/Stainless steel 0.952/Stainless steel	Horizontal/40; 50	Circular/single channel Rectangular/single channel	200–800 200–800	336 352	0.2–0.26/0–180	0.1–0.9/–	2.0 3.2
Del Col et al. [26]	R290/adiabatic	0.96/Copper	Horizontal/40–42	Circular/single channel	200–800	230	0.33/0–360	0.1–0.95/–	1.3
López-Belchí et al. [60]	R1234yf; R134a; R32/Condensation	1.16/Aluminum	Horizontal/20–55	Square/multiple channels	350–940	259	0.2–037/10–120	0.13–094/–	0.226
Sakamatapan and Wongwises [76]	R134a/Condensation	1.1; 1.2/Aluminum	Horizontal/35–45	Rectangular/multiple channels	345–685	250	0.21–0.28	0.1–0.85/15–25	–
Del Col et al. [29]	R1234ze/adiabatic	0.96/Copper	Horizontal/40	Circular/single channel	100–800	230	0.21/0–320	0.2–1.0/–	1.3

[a]The indication "preheater" means that a preheater was used to achieve the experimental condition at the inlet of an adiabatic test section.

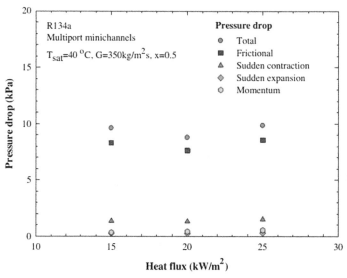

Figure 8.4 Pressure drop parcels during condensation inside a channel with hydraulic diameter of 1.1 mm according to Sakamatapan and Wongwises [76].

R32, R404A, R407C, R410A) and mixtures of noncondensable gases (air, N_2) and water. Pressure drop gradients were also experimentally evaluated for hydrocarbons (R290), ammonia, carbon dioxide, water, and, most recently, for HFOs (R1234yf, R1234ze). In general, experiments for mixtures of noncondensable gases and water were performed for wider mass velocity ranges, not typical of practical applications, covering values from 22 to 7350 kg/m² s, while for typical fluids of refrigeration and air conditioning systems, the experiments were performed for lower mass velocities, covering values from 50 to 1400 kg/m² s. Experiments were performed for adiabatic and cooling conditions.

Figure 8.4 from Sakamatapan and Wongwises [76] displays the magnitude of the pressure drop parcels. This figure reveals that the frictional pressure drop is the main contribution to the total pressure drop, corresponding to approximately 79.4% of total pressure gradient. Based on their experimental results for the total pressure drop and on estimative of its components for a multiple microchannel condenser, Sakamatapan and Wongwises [76] indicated that the sudden contraction pressure drop, the sudden expansion pressure drop, and the momentum pressure drop amount were 22.35%, 4.21%, and 5.98% of total pressure drop, respectively.

Figure 8.5 illustrates the effect of the mass velocity on the pressure drop gradient with increasing vapor quality. As expected and commonly observed in the literature, the pressure drop increases dramatically with increasing mass velocity. Moreover, the frictional pressure gradient increases with augmentation of the vapor quality. Usually, under similar experimental conditions, low-pressure refrigerants present higher pressure drop gradient than high-pressure refrigerants, due to the higher vapor specific volume of the low-pressure refrigerants and, consequently, higher two-phase flow velocities. The pressure drop decreases with increasing saturation temperature because

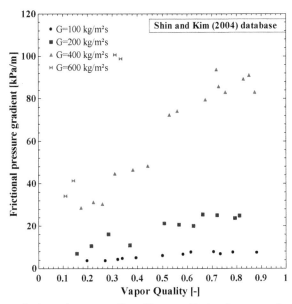

Figure 8.5 Mass velocity and vapor quality effects on frictional pressure drop during condensation inside small-diameter channels according to the experimental data of Shin and Kim [80]: R134a, $D = 0.691$ mm, $T_{sat} = 40\,^{\circ}$C.

the vapor specific volume decreases with increasing the saturation temperature. Decreasing the tube diameter also causes an increase of the pressure drop related to the augmentation of shear stress effects.

Table 8.2 displays a summary of the studies concerning experimental investigations on the heat transfer coefficient during convective condensation inside small-diameter tubes. According to this table, studies were performed only for horizontal flows and hydraulic diameters down to 127 µm. As for pressure drop, most of studies were performed for circular channels; however, experimental results for square, rectangular, trapezoidal, triangular, and semicircular transversal cross sections and flat tubes are also available. Channels within W- and N-shaped inserts were also evaluated. It is important to emphasize that for small channels with noncircular shapes, the effect of surface tension on the liquid film thickness along the channel perimeter may become relevant due to the corners. At high vapor qualities and mass velocities of 75 and 150 kg/m²s, Derby et al. [30], observed higher heat transfer coefficients for square and triangular geometries compared to the semicircle channels. However, for a mass velocity of 300 kg/m²s, the triangular channel provided lower heat transfer coefficients than the other geometries, even for vapor qualities greater than 0.5. The authors justified this behavior based on the fact that triangular channels experience flooding earlier than the other geometries. Moreover, Agarwal et al. [2] observed the highest heat transfer coefficients for the rectangular geometry followed by square and circular channels. Agarwal et al. [2] performed experiments for single and aluminum multiple channels. The single channels, evaluated by them, were composed of the following materials: silicon, copper, and stainless steel.

Table 8.2 Summary of Studies Concerning the Heat Transfer Coefficient During Convective Condensation

Authors	Fluid	Hydraulic Diameter (mm)/ Material of Channel	Flow Orientation	Geometry/ Configuration	Saturation Temperature (°C)	Mass Velocity (kg/m²)	Test Section Length (mm)/ Reduced Pressure (−)	Cooling Method	Vapor Quality (−)/ Heat Flux (kW/m²)	Heat Transfer Coefficient (kW/m² K)
Yang and Webb [92]	R12	2.63; 1.56/ Aluminum	Horizontal	Flat tube/multiple channels	65	400−1400	508/0.41	Modified Wilson plot/ Cooling water flowing externally to an annular region	0.12−0.97/ 4−12	1−5
Yan and Lin [91]	R134a	2.0/Copper	Horizontal	Circular/multiple channels	25−50	100; 150; 200	2000/0.16−0.32	Modified Wilson plot/ Cooling water external cross sections	0.1−0.9/10; 20	1−5
Garimella and Bandhauer [38]	R134a	0.76/Aluminum	Horizontal	Square/multiple channels	−	150−750	304/−	Thermal amplification technique	0.15−0.8	2−10
Wang et al. [88]	R134a	1.46/Aluminum	Horizontal	Rectangular/multiple channels	45−63	75−750	610/0.28−0.44	Cooling water cross section	0−0.95	0−8
Baird et al. [3]	R123; R11	0.92; 1.95/Copper	Horizontal	Circular/single channel	20−72	70−600	30/0.02−0.1	Thermoelectric coolers (TEC) −Peltier	0−1.0/15−110	0.5−10
Koyama et al. [56]	R134a	1.11; 0.8/ Aluminum	Horizontal	Circular/multiple channels	47−65	100−700	600/0.3−0.46	Thermoelectric sensors of 75 mm/Cooling water jackets	0.0−1.0/−	−
Kim et al. [52]	R410A; R22	1.41/Aluminum	Horizontal	Flat tube/multiple channels	45	200−600	455/0.34	Modified Wilson Plot/ Cooling water flowing externally to an annular region	0.1−0.9/5−15	2−7
Shin and Kim [80]	R134a	0.691/Copper	Horizontal	Circular/single channel	40	100−600	171/0.25	Air cooling	0.1−0.9/5−20	2−12

Reference	Refrigerant	Diameter/Material	Orientation	Geometry/Channel		Mass flux		Cooling method		
Bandhauer et al. [4]	R134a	0.506–1.524/Aluminum	Horizontal	Circular/multiple channels	—	150–750	304/–	Thermal amplification technique	0.15–0.85	1–14
Matkovic et al. [61]	R134a; R32	0.96/Copper	Horizontal	Circular/single channel	40	100–1200	230/0.25–0.42	Coolant temperature profile/Cooling water flowing externally to an annular region	0.05–0.95/–	1–25
Park and Hrnjak [67]	R744	0.89/Aluminum	Horizontal	Circular/multiple channels	−15;−25	200–800	150/0.22–0.31	Secondary cooling fluid by conduction through bass pieces in the test section	0.1–0.9	1–10
Del Col et al. [22]	R1234yf	0.96/Copper	Horizontal	Circular/single channel	24–40	200–1000	230/0.19– 0.3	Coolant temperature profile/Cooling water flowing externally to an annular region	0.1–0.85/–	1–12
Agarwal et al. [2]	R134a	0.424–0.839/–	Horizontal	Square; barrel; triangular; rectangular; W shape insert; N shape insert/ multiple channels	55	150–750	0.3048/0.36	Thermal amplification technique	0.15–0.85/–	2–18
Quan et al. [71]	Water	0.127; 0.173/ Silicon	Horizontal	Trapezoidal/single channel	—	54–559	60/–	Cooling air	0.25–0.8/–	50–200
Bohdal et al. [5]	R134a; R404A	0.31–3.30/ Stainless Steel	Horizontal	Circular/single channel	20–40	100–1300	100/0.14–0.25	Coolant temperature profile/Cooling water flowing externally to an annular region	0.0–1.0/–	0–15
Del Col et al. [23]	R134a	1.23/Copper	Horizontal	Square/single channel	40	200–800	224/0.25	Coolant temperature profile/Cooling water flowing externally to an annular region	0.05–0.98/ 30–84	2–16
Oh and Son [66]	R22; R134a; R410A	1.77/Copper	Horizontal	Circular/single channel	40	450–1050	160/0.25; 0.3	Coolant temperature profile/Cooling water flowing externally to an annular region	0.07–0.95/–	2–20

Continued

Table 8.2 Summary of Studies Concerning the Heat Transfer Coefficient During Convective Condensation—cont'd

Authors	Fluid	Hydraulic Diameter (mm)/Material of Channel	Flow Orientation	Geometry/Configuration	Saturation Temperature (°C)	Mass Velocity (kg/m²s)	Test Section Length (mm)/Reduced Pressure (—)	Cooling Method	Vapor Quality (—)/Heat Flux (kW/m²)	Heat Transfer Coefficient (kW/m² K)
Park et al. [68]	R1234ze; R134a; R236fa	1.45/Aluminum	Vertical	Rectangular/single channel	25–70	50–260	260/0.08–0.52	Cooling water flowing externally to an annular region	0–0.92/1–62	1–3.5
Charun [10]	R404A	1.4; 1.60; 1.94; 2.3; 3.3/Stainless steel	Horizontal	Circular/single channel	20–40	100–1000	100/–	Coolant temperature profile/Cooling water flowing externally to an annular region	0.0–1.0/2.5–22.5	0–30
Del Col et al. [24]	R290	0.96/Copper	Horizontal	Circular/single channel	40	100–800	230/0.32	Coolant temperature profile/Cooling water flowing externally to an annular region	0–0.9/–	1–18
Derby et al. [30]	R134a	1.0/Copper	Horizontal	Square; triangular; semicircle/multiple channels	35; 45	75–450	100/0.21; 0.28	Heat flux sensor	0.05–0.9/23.5–46	1–7
Zhang et al. [95]	R22; R410A; R407C	1.088; 1.28/Stainless steel	Horizontal	Circular/single channel	30; 40	300–600	200/0.23–0.3	Cooling water flowing externally to an annular region	0.1–0.9/–	0–8
Kim and Mudawar [53]	FC72	1.0/Copper	Horizontal	Square/multiple channels	55	68–367	299/0.51	Heat flux sensor	1–0/–	1–10
Goss and Passos [43]	R134a	0.77/Copper	Horizontal	Circular/multiple channels	29; 33; 38	230–445	105/0.18–0.24	Thermoelectric coolers (TEC) —Peltier	0.55–1.0/17–53	10–50

Reference	Fluid	Diameter/Material	Orientation	Channel	Temperature	Mass flux		Technique		
Heo et al. [46]	CO_2	1.5/Aluminum	Horizontal	Rectangular/multiple channels	−5; 0; 5	400–1000	400/0.41–0.53	Modified Wilson plot/Cooling mixture (water + ethylene glycol) flowing externally to an annular region	0.0–1.0/–	2–7
Liu et al. [59]	R152a	1.152; 0.952/Stainless steel	Horizontal	Circular; square/single channel	40; 50	200–800	336; 352/0.2; 0.26	Cooling water flowing externally to an annular region	0.1–0.9/–	70
Sakamatapan et al. [75]	R134a	1.1; 1.2/Aluminum	Horizontal	Rectangular/multiple channels	35–45	340–690	250/0.21–0.29	Cooling water flowing externally to an annular region	0.1–0.9/15–25	1.3–2.4
Del Col et al. [26]	R290	0.96/Copper	Horizontal	Circular/single channel	40–41	100–1000	230/0.33	Coolant temperature profile/Cooling water flowing externally to an annular region	0.05–0.9/10–240	1–21
Del Col et al. [27]	R134a; R32	1.23/Copper	Horizontal; inclined from 15° to 90°	Square/single channel	40	100–390	224/0.25	Coolant temperature profile/Cooling water flowing externally to an annular region	0.05–0.9/–	1–12
Del Col et al. [29]	R1234ze(E)	0.96/Copper	Horizontal	Circular/single channel	40	100–800	230/0.21	Coolant temperature profile/Cooling water flowing externally to an annular region	0.05–0.85/–	1–13

According to Table 8.2, most authors have obtained heat transfer coefficient results for condensation of R134a, but studies for other halocarbon refrigerants (R11, R12, R123, R152a, R22, R236fa, R32, R407C, R410A) and HFOs (R1234yf, R1234ze) as well as hydrocarbons (R290), water, and CO_2 were also performed. However, it is important to highlight the absence of studies involving condensation of ammonia. For halocarbon refrigerants and HFOs, the saturation temperature effect on the heat transfer coefficient was investigated for temperatures from 20 to 70 °C, ranges typical of air-cooled condensers. Due to its peculiar characteristics, characterized by a lower critical temperature than most of halocarbon refrigerants, condensation tests with CO_2 were run for lower saturation temperatures, covering conditions from −15 to 25 °C. Heat transfer coefficients up to 50 kW/m² K were observed for halocarbon refrigerants, while values up to 200 kW/m² K were achieved for water. As pointed out by Del Col and Bortolin [28], a reduced number of authors have investigated heat transfer coefficient for low mass velocities during condensation, which surface tension effects are predominant.

Contrary to studies concerning flow boiling, for which the heating effect can be easily obtained through the Joule effect by applying an electrical current directly on the tube surface and the wall temperature is measured directly through thermocouples fixed on the tube surface, for condensation, the evaluation of heat transfer coefficient is a much more difficult task. According to Table 8.2, the following procedures were used to evaluate the heat transfer coefficient and the vapor quality during condensation experiments: (1) heat is rejected to water at low temperature flowing, generally, in an annulus involving the test section for single-channels, and through parallel plates sandwiching the test section containing multiple channels. Then, local heat flux and vapor quality are estimated based on energy balance and the temperature profile along the test section length. The wall temperature is measured through thermocouples within tube wall. Averaged heat transfer coefficients are also determined using water as cooling source by using the modified Wilson plot method. Both methods are deeply dependent on the uncertainty of the temperature measurements that should be at least lower than 0.1 °C. Moreover, for these methods the uncertainty in the heat transfer coefficient increases substantially with increasing its value. Cooling water combined with heat flux sensors and thermocouples for wall temperature measurements were also employed; (2) cooling air with the heat transfer coefficient given as the ratio of the heat transferred to the air and the average wall subcooling; (3) thermoelectric coolers (Peltier) with local heat transfer coefficients estimated based on the wall and saturation local temperatures and on the power removed by the Peltier coolers. In this context, Wang and Rose [89] highlighted the fact that local temperatures of the vapor along the microchannels cannot be directly measured. Instead, they are estimated from measured temperatures and pressures at the test section inlet and outlet, and, so, their uncertainties are difficult to quantify.

Figure 8.6 presented by Sakamatapan et al. [75] illustrates the effect of channel diameter on the heat transfer coefficient for condensation of R134a. As expected, the heat transfer coefficient increases with decreasing the channel diameter. Agarwal et al. [2] pointed out similar behavior to Figure 8.6 for the heat transfer coefficient with channel diameter.

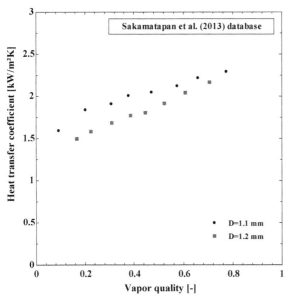

Figure 8.6 Diameter effect on condensation heat transfer coefficient according to Sakamatapan et al. [75]; R134a, $G = 500$ kg/m^2 s, $T_{sat} = 40\,°C$, $\dot{q} = 25$ kW/m^2.

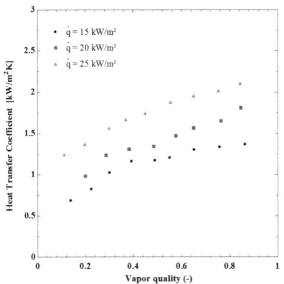

Figure 8.7 Heat flux effect on local heat transfer coefficient according to the data of Sakamatapen et al. [75]; R134a, $G = 350$ kg/m^2 s at $T_{sat} = 40\,°C$.

According to Figure 8.7 from Sakamatapan et al. [75], the heat transfer coefficient increases with increasing heat flux. Such a result is somewhat surprising because it is expected that condensation heat transfer coefficient is dominated by convective effects, therefore, negligible effects of heat flux on the heat transfer coefficient are expected. A different behavior was observed by Derby et al. [30] and Bortolato [6], with an

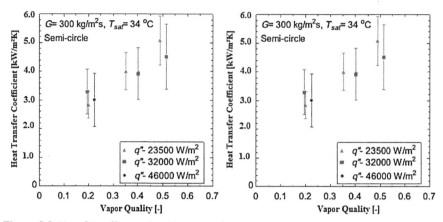

Figure 8.8 Heat flux effect on local heat transfer coefficient according to the data of Derby et al. [30].

almost negligible effect of the heat flux on the heat transfer coefficient as shown in Figs 8.8 and 8.9, respectively. For the study of Derby et al. [30], the differences in the heat transfer coefficient with varying heat flux are within the uncertainty of their measurements. In the study of Bortolato [6], the augmentation of the heat flux was obtained by decreasing the temperature of the inlet water responsible for the cooling effect.

Figure 8.10 illustrates the effect of the mass velocity on the heat transfer coefficient during condensation. As expected for a heat transfer mechanism dominated by convective effects, the heat transfer coefficient increases with increasing mass velocity. Figures 8.6−8.11 have depicted also that the heat transfer coefficient increases with increasing vapor quality. This behavior agrees with the schematic shown in Fig. 8.1

Figure 8.9 Comparison among experimental local heat transfer coefficient for different inlet water temperature from Bortolato [6]; R1234ze(E), $G = 200$ kg/m^2 s and $T_{sat} = 40$ °C.

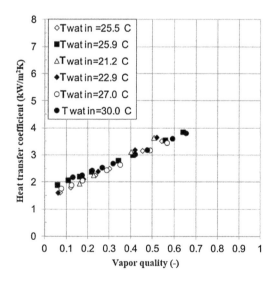

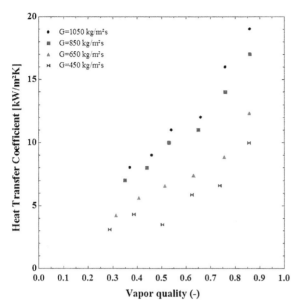

Figure 8.10 Illustration of the mass velocity effect on the heat transfer coefficient according to the data of Oh and Son [66]; R22, $D = 1.77$ mm, $T_{sat} = 40\ °C$.

and is result of the increment of the two-phase flow velocity, and consequently, of the convective effects with increasing the vapor quality.

In general and as shown in Fig. 8.11, the heat transfer coefficient increases with decreasing the saturation temperature. According to Fig. 8.11, this effect is enhanced

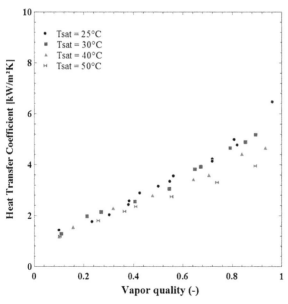

Figure 8.11 Saturation temperature effect on condensation heat transfer coefficient according to the data of Yan and Lin [91]; $G = 200$ kg/m^2 s, $D = 2$ mm, $\dot{q} = 10$ kW/m^2.

under conditions of high vapor qualities. Convective effects and, consequently, the heat transfer coefficient are enhanced by increasing the two-phase velocity as results of augmenting the vapor specific volume, due to the saturation temperature decreasing.

8.4 Methods for Prediction of Heat Transfer Coefficient and Pressure Drop for Condensation inside Small-Diameter Channels

8.4.1 Pressure Drop

For two-phase flows inside small-diameter channels, the liquid phase is laminar for most of applications, what is rare for conventional channels. In addition, the effect of surface tension becomes progressively more pronounced as the channel dimension decreases. As consequence, the influence of gravity may become less important and stratified flows are not verified. On the other hand, the annular flow becomes predominant, and shear effects are intensified due to higher velocity gradients compared to conventional channels for a similar two-phase flow velocity. Bubbly flow pattern also is rare, because the bubbles lifespan is very short, as the bubbles collapses very quickly.

The total pressure drop is given by the sum of frictional, gravitational and accelerational parcels. The parcel corresponding to the gravitational pressure drop is estimated as follows:

$$\Delta p_g = \int_0^L \rho_{2\phi} g \, dz \tag{8.1}$$

where $\rho_{2\phi}$ is the two-phase density given by the following equation:

$$\rho_{2\phi} = \rho_L(1 - \alpha) + \rho_V \alpha \tag{8.2}$$

The accelerational pressure drop for two-phase flows inside tubes with constant transversal areas is given as:

$$\Delta p_a = G^2 \left\{ \left[\frac{(1-x)^2}{\rho_L(1-\alpha)} + \frac{x^2}{\rho_V \alpha} \right]_{out} - \left[\frac{(1-x)^2}{\rho_L(1-\alpha)} + \frac{x^2}{\rho_V \alpha} \right]_{in} \right\} \tag{8.3}$$

The subscripts *in* and *out* refer to the two-phase flow characteristics at the channel inlet and outlet sections. Several predictive methods are found in the literature concerning the estimative of the superficial void fraction, α, defined as the time average of instantaneous ratio between the superficial area occupied by the vapor and the total cross sectional area. According to Tibiriçá and Ribatski [84], few authors have published void fraction results for small-diameter tubes and most of them for adiabatic condition. Shedd [79] and Winkler et al. [90] have performed void fraction measurements during condensation of halocarbon refrigerants, while Yashar [93] has measured volumetric void fraction of evaporating halocarbon refrigerant through the quick-closing valves method. It should be highlighted that, on contrary to conventional channels, for small-diameter tubes, the accelerational parcel could become relevant even under adiabatic conditions due to the flashing effect.

Recently, Kanizawa and Ribastki [49] proposed a new predictive method for void fraction based on the principle of minimum energy dissipation. They performed a regression analysis to correlate the momentum coefficients ratio using a database obtained from the literature comprising more than 3400 data points. The proposed method has proven more accurate than the current methods available in the literature, as it predicts 92% and 85% of the experimental data for horizontal and vertical saturated flows, respectively, within an error band of ±10%. The experimental database includes the experimental results of Shedd [79] for condensation, and involves saturated flow conditions and covers mass velocities ranging from 37 to 4500 kg/m² s and tube internal diameters from 0.5 to 89 mm. Their method for horizontal two-phase flow is recommended for predicting void fraction inside small-diameter tubes. Suggesting the use of only the method for horizontal tubes, which is based on the data for smaller channels, seems reasonable because as the tube diameter decreases, surface tension effects progressively overcome gravitational effects. So, similar results are expected independently of the channel orientation. Their method is given as follows:

$$\alpha = \left[1 + 1.021 \cdot Fr^{-0.092} \left(\frac{\mu_L}{\mu_V} \right)^{-0.368} \left(\frac{\rho_V}{\rho_L} \right)^{1/3} \left(\frac{1-x}{x} \right)^{2/3} \right]^{-1} \qquad (8.4)$$

where Fr is the mixture Froude number defined as follows:

$$Fr = \frac{G^2}{(\rho_L - \rho_V)^2 gD} \qquad (8.5)$$

The method proposed by Kanizawa and Ribastki [49] can also be used to evaluate the amount of refrigerant in a channel by integrating the amount of refrigerant contained in a differential element along the channel length, as follows.

$$m = \int_0^L [\rho_L(1-\alpha) + \rho_V \alpha] A dz \qquad (8.6)$$

The relation between dz and the variation of vapor quality along the differential element can be obtained from an energy balance. For a constant heat flux boundary condition, this relationship is given as follows.

$$dz = \left(\frac{i_{LV} GA}{P\dot{q}} \right) dx \qquad (8.7)$$

where A and P are the cross-sectional area and the perimeter of the channels, respectively.

Felcar and Ribatski [35] performed a comprehensive comparison involving 17 predictive methods for the frictional pressure drop and a broad database from literature comprising experimental data from 15 laboratories. The database includes experimental results obtained under adiabatic, evaporation and condensation conditions. Based on their comparison [35], the homogeneous model with the dynamic viscosity given by Cicchitti et al. [15] provides the best predictions of the database. However, they pointed out that despite of being the best, this method is still unsatisfactory because predicts only 58% of the data within and error band of ±30%.

The discrepancies among experimental databases contribute for the low performance of models and correlations. The differences among experimental results for similar conditions can be related to the following aspects: (1), comparisons between databases are performed and trends are suggested without a complete knowledge of the experimental conditions; (2) flow boiling data were obtained in the presence of flow oscillations; (3) absence of procedures to validate the experimental results as single-phase flow energy balances and single-phase flow pressure drop measurements; (4) flashing effects are neglected and inappropriate data regression procedures are used; (5) as pointed out by Tibiriçá and Ribatski [84], even when using state-of-the-art instrumentation and calibration procedures, experimental errors, most of the times higher than ±20%, still play an important role in the pressure drop data analyses.

Table 8.3 displays the results of comparisons among frictional pressure drop predictive methods from literature and databases from Table 8.1 including only results for condensation inside small-diameter tubes. The methods present in Table 8.3 were selected because they are frequently mentioned in the literature as the most accurate for predicting pressure drop inside small-diameter tubes. The experimental data considered in this analysis were gathered from graphs and tables available in the original publications and comprise only the frictional pressure drop parcel estimated from measurements under condensation conditions.

According to Table 8.3, by adopting as the evaluation parameter the fraction of data predicted within ±30%, the homogeneous model with the two-phase viscosity proposed by Cicchitti et al. [15] provided the best predictions of the overall database, followed by the method of Friedel [37]; ranked as the second best. In general, these two methods, as well as the methods of Müller-Steinhagen and Heck [64], Cioncolini et al. [16] and Da Silva and Ribatski [21], predicted approximately 50% of the database within an error band of ±30%. The methods proposed by Friedel [37] and Da Silva and Ribatski [21], which is a modified version of Müller-Steinhagen and Heck [64], provided the best prediction of more than one independent database.

Figures 8.12 and 8.13 display the comparisons between experimental data and the corresponding predictions according to the homogeneous model, with the two-phase viscosity defined by Cicchitti et al. [15]; and the method of Friedel [37]; respectively. Based on Fig. 8.13, it can be concluded that the method proposed by Friedel [37] provides worst predictions for low pressure drops.

Generally, according to Fig. 8.12, the homogeneous model underpredicted a significant parcel of the experimental results. These data correspond to the results of

Table 8.3 Statistical Parameter from Comparison of Experimental Data from the Literature and Frictional Pressure Drop Predicted Methods

	Homogeneous Model-Cicchitti et al. [15]		Chisholm [13]		Friedel [37]		Muller-Steinhagen and Heck [64]		Mishima and Hibiki [62]		Cioncolini et al. [16]		Da Silva and Ribatski [21]	
	ε %	λ %	ε %	λ %	ε %	λ %	ε %	λ %	ε %	λ %	ε %	λ %	ε %	λ %
Yan and Lin [91]	65.12	22.45	138.55	10.96	122.87	8.22	80.60	12.3	**36.76**	**43.84**	249.14	2.74	50.52	23.29
Garimella et al. [39]	43.00	44.16	67.82	32.49	**29.34**	**58.88**	43.50	35.03	72.69	5.08	34.68	41.12	44.18	35.03
Shin and Kim [80]	71.70	12.33	190.70	20.41	136.12	28.57	**94.13**	**29.93**	47.23	0.00	154.28	12.24	96.44	17.69
Garimella et al. [40]	**14.99**	**95.67**	89.35	29.33	24.73	68.27	16.03	87.98	54.20	10.10	73.60	7.69	16.72	90.87
Heo et al. [46]	56.82	0.00	60.97	27.66	39.53	28.72	46.37	6.38	76.22	0.00	**25.00**	**68.09**	55.96	0.00
Liu et al. [59]	31.27	37.50	64.21	37.50	**24.61**	**85.00**	23.30	80.00	71.99	0.00	31.28	70.00	34.25	32.50
Lopez Belchi et al. [60]	10.18	97.44	129.24	15.38	29.59	61.54	18.51	74.36	55.58	2.56	90.47	0.00	**9.78**	**100.00**
Sakamatapan and Wongwises [76]	30.89	54.35	154.21	13.04	69.23	2.17	41.09	19.57	51.11	2.17	129.67	0.00	**24.99**	**69.57**
Overall database	**47.44**	**54.57**	121.67	29.03	**65.78**	**52.69**	53.13	51.21	66.91	8.74	102.29	28.09	51.35	51.75

ε, mean absolute error; λ, Fraction of data predicted within $\pm30\%$. The bold values indicate the method that provided the best prediction of the corresponding database.

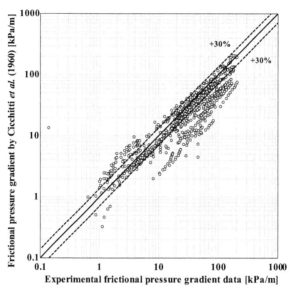

Figure 8.12 Comparison of the experimental data for frictional pressure gradient and the corresponding predictions from the homogeneous model with two-phase viscosity defined by Cicchitti et al. [15].

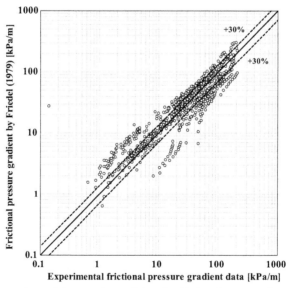

Figure 8.13 Comparison of the experimental data for frictional pressure drop gradient and the corresponding predictions by Friedel [37].

Sakamatapan and Wongwises [76], Shin and Kim [80], and Yan and Lin [91]. So, from a comparison of Figs 8.12 and 8.13 and taking into account that the method proposed by Friedel [37], statistical parameters indicate that his method is only marginally less accurate than the homogeneous model. Thus, in the present chapter, the method of Friedel [37] is recommended for predicting the frictional pressure drop parcel during condensation inside small-diameter tubes. However, it should be emphasized that additional experimental studies are still necessary in order to build up a wider database with reliable results, allowing the development of an accurate predictive method for the frictional pressure drop parcel during condensation inside small-diameter tubes.

The method of Friedel [37] is given as follows:

$$\phi_{LO}^2 = \frac{\Delta p_{2\phi}}{\Delta p_{LO}} = E + \frac{3.24F\,H}{Fr_H^{0.045}\,We_H^{0.035}} \tag{8.8}$$

where ϕ_{LO}^2 is the two-phase multiplier, $\Delta p_{2\phi}$ is the frictional pressure drop for condensation, and Δp_{LO} is the frictional pressure drop assuming the two-phase mixture flowing as only liquid. The dimensionless numbers in Eq. (8.8) are defined as follows:

$$E = (1-x)^2 + x^2\,\frac{(dp/dz)_{VO}}{(dp/dz)_{LO}} \tag{8.9}$$

$$H = \left(\frac{\rho_L}{\rho_V}\right)^{0.91} \left(\frac{\mu_V}{\mu_L}\right)^{0.19} \left(1 - \frac{\mu_V}{\mu_L}\right)^{0.7} \tag{8.10}$$

$$F = x^{0.78}(1-x)^{0.224} \tag{8.11}$$

where $(dp/dz)_{VO}$ and $(dp/dz)_{LO}$ are the frictional pressure gradients that would result if the two-phase mixture flowed through the tube as only vapor and only liquid, respectively.

The homogeneous Froude and Weber numbers are estimated according to the following equations:

$$Fr_H = \frac{G^2}{gD\,\rho_H^2} \tag{8.12}$$

$$We_H = \frac{G^2D}{\sigma\,\rho_H} \tag{8.13}$$

where ρ_H is the homogeneous density given by:

$$\rho_H = \left(\frac{x}{\rho_V} + \frac{1-x}{\rho_L}\right)^{-1} \tag{8.14}$$

8.4.2   Heat Transfer Coefficient

In general, predictive methods for heat transfer condensation inside tubes can be grouped according to the following classification:

1. Strictly empirical: these methods can be subdivided into two subgroups. The first one, using an approach similar to the one used for pressure drop, obtaining a two-phase multiplier function for the ratio between heat transfer coefficients for condensation and for single-phase forced convection, given by a correlation from literature. The second group consists on a simple correlation of dimensionless numbers, usually considering Dittus–Boelter as initial step, through the adjustment of empirical constants and exponents based on a broad database.
2. Flow pattern based methods: usually these methods are based on flow pattern transitional criteria which can be based on heat transfer behaviors and flow pattern visualizations. Once characterized the flow pattern, then the heat transfer process is modeled according to the two-phase flow topology.

Among the strictly empirical methods, the correlations proposed by Akers et al. [1], Cavallini and Zechin [7], and Shah [77] are commonly mentioned in the literature for condensation inside conventional channels. Recently, Shah [78] modified his previous predictive method in order to extend its application to a wider range of conditions. The method proposed by Shah [78] was adjusted using a database that covers 22 fluids, hydraulic diameters from 2 to 49 mm, reduced pressures from 0.0008 to 0.9, and mass velocity from 4 to 820 kg/m^2 s. These data are from 39 independent sources. According to Shah [78], his method has predicted the database used on its development with an average absolute error of 14.4%.

Generally, flow pattern based methods consider stratified and annular flow patterns. Significant parcel of the heat transfer predictive methods characterized within this group was developed only for stratified flow or for annular flow. In stratified flows, the heat transfer process is mainly related to gravitational effects, and the heat transfer coefficient on the upper part of the tube is given accordingly to the theory proposed by Nusselt [65] for film condensation on horizontal tubes. The condensate liquid on the tube upper part is drained down to the tube lower part. Then, the heat transfer coefficient of bottom region of the tube is modeled as single-phase forced convection. The main aspect of this approach is estimating the perimeter over which each mechanism is predominant. Based on the fact that the heat transfer coefficient for forced convection is much lower than for falling film condensation, authors such as Chato [11] and Jaster and Kosky [48] have neglected the perimeter where the prevailing mechanism is forced convection.

Moser et al. [63] proposed an analytical model for annular flow, corresponding to shear-controlled condensation inside smooth tubes, based on an equivalent Reynolds number that requires predictive methods for the single-phase heat transfer coefficient, and for the two-phase frictional pressure gradient. From a comparison of the predictions given by their model and a database comprising 1197 data points, covering seven halocarbon refrigerants, tube diameters between 3.14 and 20 mm, mass velocities from 87 to 862 kg/m^2 s and saturation temperatures from 22 to 52 °C. Moser et al. [63] found a mean absolute deviation for the predictions of 13.6%. Comparisons of the same database and the methods of Shah [77], Traviss and Rohsenow [85] and Traviss et al. [86], revealed mean absolute deviations of 14.3% and 20.0%, respectively.

Flow pattern—based methods developed for predicting the heat transfer coefficient along all the condensation process, instead of methods developed for only one flow pattern, were proposed by Dobson and Chato [32] and Thome et al. [82]. Dobson and Chato [32] developed their method based on the flow pattern transition criteria proposed by Soliman [81], modified in order to predict their transitions. The method of Dobson and Chato [32] includes only stratified wavy and annular flow patterns. For shear-controlled condensation, corresponding to annular flow, the heat transfer coefficient for condensation is given as the product of the two-phase multiplayer and the single-phase heat transfer coefficient for only liquid flow, calculated according to Dittus and Boelter [31]. For stratified flow, the method of Dobson and Chato [32] takes into account the film of condensate flowing from the top of tube toward its bottom and the forced convection in the stratified liquid. The empirical constants and exponents of the Dobson and Chato [32] method were adjusted based on experimental data points for R134a, R22, R32, R125, mass velocity from 25 to 800 kg/m^2 s, saturation temperature from 35 to 45 °C, and heat flux from 5 to 15 kW/m^2.

The method proposed by Thome and coworkers considers the following flow patterns: stratified, stratified-wavy, intermittent, annular, and mist flow patterns. First, Thome et al. [82] developed their method by modifying the flow pattern map of Kattan et al. [50] in order of predicting flow patterns under condensation conditions. In their heat transfer model for stratified flow, falling film condensation is considered on the upper parcel of the tube, corresponding to the region where the condensate flows down to the portion of stratified liquid, while convective heat transfer is considered in the lower region flooded with liquid. For annular flow, convective condensation is assumed over all tube perimeter, and the liquid film is the main thermal resistance. The axial liquid film flow is assumed as turbulent and waviness on the film surface enhancing the heat transfer coefficient is also taken into account. The method for annular flow has its application extended to intermittent and mist flow patterns. The empirical constants and exponents of the method proposed by Thome et al. [82] were adjusted based on experimental data from nine independent laboratories, including 15 fluid and covering mass velocities ranging from 24 to 1022 kg/m^2 s, vapor quality from 0.03 to 0.97, reduced pressures from 0.02 to 0.80, and tube internal diameters from 3.1 to 21.4 mm.

Koyama et al. [56] has proposed a modification on the predictive method of Haraguchi et al.[44,45] by replacing in their method, the correlation for the two-phase multiplier proposed by Mishima and Hibiki [62]. The method proposed by Koyama et al. [56] consists of an asymptotic sum of terms related to gravity controlled condensation and convective condensation, adopting an asymptotic coefficient equal to two. The modified method predicted almost all the database of Koyama et al. [56] described in Table 8.2 within an error margin of +50 to −20%.

Table 8.4 presents the statistical parameters obtained from a comparison among experimental data from literature described in Table 8.2 and the predictive methods

Table 8.4 Statistical Parameters from the Comparison of Experimental Data Described in Table 8.1 and Predictive Methods from the Literature

	Cavallini and Zecchin [7]		Shah [77]		Dobson and Chato [32]		Moser et al. [63]		Thome et al. [82]		Koyama et al. [56]		Cavallini et al. [9]		Shah [78]	
	ε %	λ %	ε %	λ %	ε %	λ %	ε %	λ %	ε %	λ %	ε %	λ %	ε %	λ %	ε %	λ %
Yan and Lin [91]	22.2	70.9	12.5	98.2	18.9	87.2	17.8	81.8	54.9	36.3	15.2	87.2	**9.1**	**100**	39.3	50.9
Kim et al. [52]	61.9	5.2	45.5	20	54.8	10	**21.8**	**84.2**	20.9	73.6	11.1	100		–	25.6	50
Shin and Kim [80]	22.8	63.3	**18.3**	**80**	20.7	70	24.9	56.6	25.4	60	35.7	31.6	17.5	78.5	31.5	38.3
Park and Hrnjak [67]	62.5	17.6	43.3	49	52.2	25.4	29.7	49	27.3	58.8	**17.7**	**88.2**	25.4	54.9	50.6	19.6
Matkovic et al. [61]	28.9	55	18	75.3	25	61.3	13.9	94.9	48.3	21.5	24.7	62	**7.5**	**98.1**	59.8	12
Agarwal et al. [2]	39.1	54.8	30.4	62.2	32.2	62.2	26.9	62.2	34.4	45.1	30.3	48.9	**22.8**	**74**	50.4	14.8
Park et al. [68]	29.1	52.1	35.9	43.4	32.3	39.1	11.3	100	12.5	91.3	12.6	82.6	**9.7**	**100**	47.8	30.4
Oh and Son [66]	37.3	57.5	29	69.4	36.9	56.9	**22.3**	**75.4**	45.4	24.5	24.1	72.6	21.7	74	44.9	22.2
Derby et al. [30]	28.5	54.4	22.1	75.2	23.9	69.1	21.2	73.7	21.5	74.1	27.2	57.1	**14.9**	**88.8**	30.6	52.5
Goss and Passos [43]	43.4	12.5	52.8	0	**42**	**12.5**	67.1	0	70.9	0	63.1	0	–	–	63.2	0
Heo et al. [46]	232	0	209	0	217	0	140	0	106	0	114	1.1	126	1.1	34.5	50
Liu et al. [59]	37.2	31.4	21.5	75.7	34.6	31.4	17.2	85.7	22.1	67.1	16.1	85.7	**12.6**	**98.5**	41.7	24.2
Sakamatapan et al. [75]	254	0	214	0	243	0	173	0	106	0	124	0	–	–	**82.6**	**20**
Del Col et al. [26]	54.4	3.1	66.1	9.5	45.4	15.8	16.4	92	44.3	22.2	**9**	**100**	12.1	100	57	20.6
Del Col et al. [29]	16.9	90	11.9	100	18.2	86.3	10.4	95.4	28.8	54.5	25.3	72.2	**6.4**	**100**	42.5	31.8
Overall database	56	44.7	47.1	60	51.3	51.2	35.3	67.5	37.1	47.2	34.9	57.8	**24.2**	**80**	45.1	30.9

ε, mean absolute error; λ, Fraction of data predicted within ±30%.

proposed by Cavallini and Zecchin [7], Shah [77], Dobson and Chato [32], Moser et al. [63], Thome et al. [82], Koyama et al. [56], Cavallini et al. [9] and Shah [78]. Some experimental databases listed in Table 8.2 were not included in the comparison due to the lack of information in the original article concerning the experimental conditions for which the data were obtained.

In Table 8.4, the methods found as the best for each database and for the overall database, comprising 1877 data points, are highlighted with bold letters. According to this table, the method proposed by Cavallini et al. [9] provides the best predictions of the overall database, followed by the methods proposed by Moser et al. [63], Koyama et al. [56] and Shah [77]. Moreover, the Cavallini et al. [9] method was ranked between the two best for seven databases while the Moser et al. [63] and Koyama et al. [56] methods were ranked between the two best for only two databases. The experimental results of Goss and Passos [43] and Sakamatapan et al. [75] were poorly predicted by all the methods evaluated in the present study. The reason for this is not clear from an analysis of their database, as they have performed experiments for R134a, as most of studies, and for experimental conditions within the range of majority of the other studies in the table.

Figures 8.14 and 8.15 display comparisons of the overall database listed in Table 8.4 and the predictive methods proposed by Cavallini et al. [9] and Moser et al. [63], respectively.

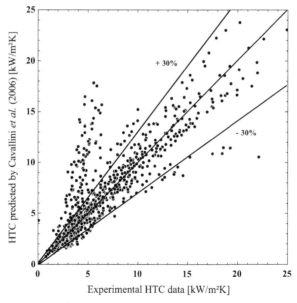

Figure 8.14 Comparison of the heat transfer coefficient experimental (HTC) data and the corresponding predictions according to the method of Cavallini et al. [9].

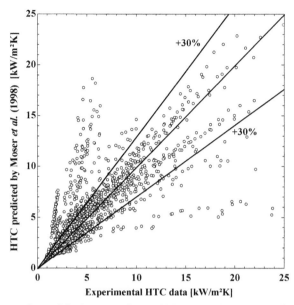

Figure 8.15 Comparison of the heat transfer coefficient experimental (HTC) data and the corresponding predictions according to the method of Moser et al. [63].

According to Fig. 8.14 the method proposed by Cavallini et al. [9] presents an almost equal distribution of the data around the 45° line. In Figs 8.14 and 8.15, sets of data with large deviation from the predictions are clearly identified. These data correspond to the databases of Heo et al. [46] and Goss and Passos [43].

The fact that the method of Moser et al. [63] has provided reasonable predictions of the experimental database listed in Table 8.4 is not surprising, since this method was developed for annular flow, which is predominant for small-diameter channels. Based on the fact that the method proposed by Cavallini et al. [9] has provided the best predictions of the experimental data for the heat transfer coefficient, this method is recommended as design tool for heat exchanger designers. Cavallini et al. [9] developed their method based on previous data obtained by them and also based on experimental results gathered in the literature. The database considered in the development of their method covers tube diameters from 3.1 to 17 mm, mass velocities from 24 to 2240 kg/m^{2}s and the following fluids: R22, R134a, R404A, R410A, R125, R32, R236ea, R407C, R502, R507A, R123, R142b, R32, R290, R600, R600a, R1270, R125/R236ea, R32/R125, R290/R600a, R290/R600, water, CO$_2$ and ammonia.

According to the method of Cavallini et al. [9], the transition from gravitational to shear stress dominated condensation is given as a function of the vapor superficial velocity j_V and the Martinelli parameter X_{tt} as follows:

$$j_V^T = \left\{ \left[7,5 \Big/ 4,3X_{tt}^{1,111} + 1 \right]^{-3} + C_T^{-3} \right\}^{\frac{-1}{3}} \tag{8.15}$$

where C_T assumes the value of 1.6 for hydrocarbons and 2.6 for fluids distinct than hydrocarbons.

For vapor superficial velocities higher than the value of j_V^T given by Eq. (8.15), corresponding to shear stress dominated condensation (annular flow), the heat transfer coefficient is given as follows:

$$h_{annular} = h_{LO} \left[1 + 1.128 x^{0,817} \left(\frac{\rho_L}{\rho_V} \right)^{0,3685} \left(\frac{\mu_L}{\mu_V} \right)^{0,2363} \left(\frac{1 - \mu_V}{\mu_L} \right)^{2,144} Pr_L^{-0,1} \right] \tag{8.16}$$

For vapor superficial velocities lower than j_V^T given by Eq. (8.15), corresponding to gravitational dominated condensation (stratified-smooth and stratified-wavy) the heat transfer coefficient is given as follows:

$$h_{SW} = \left[h_{annular} \left(j_V^T / j_V \right)^{0,8} - h_{strat} \right] \left(j_V / j_V^T \right) + h_{strat} \tag{8.17}$$

$$h_{strat} = 0.725 \left\{ 1 + 0.741[(1 - x)/x]^{0,3321} \right\}^{-1}$$
$$\left[k_L^3 \rho_L (\rho_L - \rho_G) g i_{LV} \Big/ \mu_L D (T_{sat} - T_{wall}) \right]^{0,25} + \left(1 - x^{0,087} \right) h_{LO} \tag{8.18}$$

The single-phase heat transfer coefficient h_{LO} in Eq. (8.16) and (8.18), assuming the two-phase mixture as liquid, is given by the Dittus-Boelter correlation:

$$h_{LO} = 0.023 Re_{LO}^{0.8} Pr_L^{0,4} \frac{k_L}{D} \tag{8.19}$$

where Re_{LO} is the Reynolds number for the two-phase mixture flowing as liquid.

For zeotropic mixtures with two and three components, Cavallini et al. [9] recommend using their method for single fluids with the Bell and Ghaly [96] correction to account the relative heat transfer penalization.

The average heat transfer coefficient is obtained by integrating numerically the local heat transfer coefficient along the tube length, calculated through Eqs. (8.15) to (8.17), as follows:

$$\overline{h} = \frac{1}{L} \int_0^L h \, dz \tag{8.20}$$

Nomenclature

A	Cross-sectional area (m^2)
D	Diameter (m)
g	Gravitational acceleration (m/s^2)
G	Mass velocity (kg/m^2 s)
h	Heat transfer coefficient (kW/m^2 K)
i	Enthalpy (kJ/kg)
j	Superficial velocity
k	Thermal conductivity (W/mK)
L	Channel length (m)
P	Channel perimeter (m)
p	Pressure (kPa)
p_r	Reduced pressure (−)
$\dot{q}$	Heat flux (kW/m^2)
Ra	Average roughness (μm)
T	Temperature (K)
z	Distance from the tube inlet (m)
Greek Symbols	
α	Void fraction
Δp	Pressure drop (kPa)
ε	Mean absolute error (%)
ϕ^2	Two-phase multiplier
λ	Parcel of data predicted within an error band of ±30% (%)
μ	Dynamic viscosity (Pa s)
ρ	Density (kg/m^3)
σ	Surface tension (N/m)
χ	Vapor quality (−)
Dimensionless Number	
E	Defined by Eq. (8.9)
F	Defined by Eq. (8.11)
Fr	Froude number
H	Defined by Eq. (8.10)

Pr	Prandtl number
Re	Reynolds number
We	Weber number
Subscripts	
2ϕ	Two-phase
a	Accelerational
annular	Annular flow
eq	Equivalent
g	Gravitational
h	Hydraulic
L	Liquid
L0	Mixture flowing as only liquid
LV	Vapor–liquid difference
sat	Saturation
strat	Stratified flow
SW	Stratified-wavy flow
V	Vapor
V0	Mixture flowing as only vapor
wall	Tube wall
wat in	Inlet water

References

[1] W.W. Akers, H.A. Deans, O.K. Crosser, Condensing heat transfer within horizontal tubes, Chem. Eng. Progr. Symp. Ser. 55 (1959) 171–176.

[2] A. Agarwal, T.M. Bandhauer, S. Garimella, Measurement and modeling of condensation heat transfer in non-circular microchannels, Int. J. Refrig. 33 (2010) 1169–1179.

[3] J.R. Baird, D.F. Fletcher, B.S. Haynes, Local condensation heat transfer rates in fine passages, Int. J. Heat Mass Transfer 46 (2003) 4453–4466.

[4] T.M. Bandhauer, A. Agarwal, S. Garimella, Measurement and modeling of condensation heat transfer coefficients in circular microchannels, J. Heat Transfer 128 (2006) 1050–1059.

[5] T. Bohdal, H. Charun, M. Sikora, Comparative investigations of the condensation of R134a and R404A refrigerants in pipe minichannels, Int. J. Heat Mass Transfer 54 (2011) 1963–1974.

[6] M. Bortolato, Two-phase Heat Transfer inside Microchannels: Fundamental and Applications in Refrigeration and Solar Technology (Ph.D. thesis), Università degli Studi di Padova, Padova, Italy, 2014.

[7] A. Cavallini, R. Zecchin, Dimensionless correlation for heat transfer in forced convection condensation, in: Proceedings of the 5th International Heat Transfer Conference, JSME, 1974, pp. 309−313.

[8] A. Cavallini, D. Del Col, L. Doretti, M. Matkovic, L. Rosseto, C. Zilio, Condensation heat transfer inside multiport minichannels, Heat Transfer Eng. 26 (2005) 45−55.

[9] A. Cavallini, D. Del Col, L. Doretti, M. Matkovic, L. Rosseto, C. Zilio, Condensation in horizontal smooth tubes: a new heat transfer model for heat exchanger design, Heat Transfer Eng. 27 (8) (2006) 31−38.

[10] H. Charun, Thermal and flow characteristics of the condensation of R404A refrigerant in pipe minichannels, Int. J. Heat Mass Transfer 55 (2012) 2692−2701.

[11] J.C. Chato, Laminar condensation inside horizontal and inclined tubes, ASHRAE J. 4 (4) (1962) 36.

[12] I.Y. Chen, K. Yang, Y. Chang, C. Yang, Two-phase pressure drop of air-water and R410A in small horizontal tubes, Int. J. Multiphase Flow 27 (2001) 1293−1299.

[13] D. Chisholm, A theoretical basis for the Lockhart-Martinelli correlation for two-phase flow, Int. J. Heat Mass Transfer 10 (1967) 1767−1778.

[14] P.M. Chung, M. Kawaji, The effect of channel diameter on adiabatic two-phase flow characteristics in microchannels, Int. J. Multiphase Flow 30 (2004) 735−761.

[15] A. Cicchitti, C. Lombardi, M. Silvestri, G. Soldaini, R. Zavattarelli, Two-phase cooling experiments-pressure drop, heat transfer and burnout measurements, Energ. Nucl. 7 (1960) 407−425.

[16] A. Cioncolini, J.R. Thome, C. Lombardi, Unified macro-to-microscale method to predict two-phase frictional pressure drops of annular flows, Int. J. Multiphase Flow 35 (2009) 1138−1148.

[17] J.W. Coleman, S. Garimella, Characterization of two-phase flow patterns in small diameter round and rectangular tubes, Int. J. Heat Mass Transfer 42 (15) (1999) 2869−2881.

[18] J.W. Coleman, S. Garimella, Two-phase Flow Regime Transitions in Microchannel Tubes: The Effect of Hydraulic Diameter, American Society of Mechanical Engineers, Heat Transfer Division, Orlando, FL, 2000, pp. 71−83.

[19] J.W. Coleman, S. Garimella, Two-phase flow regimes in round, square and rectangular tubes during condensation of refrigerant R134a, Int. J. Refrig. 26 (1) (2003) 117−128.

[20] J.D. Da Silva, C.B. Tibiriçá, G. Ribatski, Two-phase frictional pressure drop and flow boiling heat transfer for R245fa in a 1.1 mm Tube, in: 23rd. IIR International Congress of Refrigeration, Prague, Czech Republic, 2011.

[21] J.D. Da Silva, G. Ribatski, Two-phase frictional pressure drop of halocarbon refrigerants inside small diameter tubes. Data analysis and preposition of a new frictional pressure drop correlation, in: 8th International Conference on Multiphase Flow, Jeju − South Korea, 2013.

[22] D. Del Col, D. Torresin, A. Cavallini, Heat transfer and pressure drop during condensation of the low GWP refrigerant R1234yf, Int. J. Refrig. 33 (2010) 1307−1318.

[23] D. Del Col, S. Bortolin, A. Cavallini, M. Matkovic, Effect of cross sectional shape during condensation in a single square minichannel, Int. J. Heat Mass Transfer 54 (2011) 3909−3920.

[24] D. Del Col, S. Bortolin, M. Bortolato, L. Rosseto, Condensation heat transfer and pressure drop with propane in a minichannel, in: International Refrigeration and Air-conditioning Conference, 2012.

[25] D. Del Col, A. Bissseto, M. Bortolato, D. Torresin, L. Rosseto, Experiments and updated model for two phase frictional pressure drop inside minichannels, Int. J. Heat Mass Transfer 67 (2013) 326−337.

[26] D. Del Col, M. Bortolato, S. Bortolin, Comprehensive experimental investigation of two-phase heat transfer and pressure drop with propane in a minichannel, Int. J. Refrig. 47 (2014) 66−84.

[27] D. Del Col, M. Bortolato, M. Azzolin, S. Bortolin, Effect of inclination during condensation inside a square cross section minichannel, Int. J. Heat Mass Transfer 78 (2014) 760−777.

[28] D. Del Col, S. Bortolin, Condensation in minichannels: experimental investigation and numerical modelling, in: 101 Eurotherm Seminar-Heat, 2014.

[29] D. Del Col, M. Bortolato, M. Azzolin, S. Bortolin, Condensation heat transfer and two-phase frictional pressure drop in a single minichannel with R1234ze(E) and other refrigerants, Int. J. Refrig. 50 (2015) 87−103.

[30] M. Derby, H.J. Lee, Y. Peles, M.K. Jesen, Condensation heat transfer in square, triangular, and semi-circular mini-channels, Int. J. Heat Mass Transfer 55 (1−3) (2012) 187−197.

[31] F.W. Dittus, L.M.K. Boelter, Heat transfer in Automobile radiators of the tubular type, in: Publications in Engineering, vol. 2, University of California, Berkeley, 1930, p. 443.

[32] M.K. Dobson, J.C. Chato, Condensation in smooth horizontal tubes, J. Heat Trans., Transac. ASME 120 (1) (1998) 193−213.

[33] M. Ducoulombier, Carbon dioxide flow boiling in a single microchannel−part II: heat transfer, Exp. Therm. Fluid Sci. 35 (4) (2011) 597−611.

[34] K. Dutkowski, Air−water two-phase frictional pressure drop in minichannels, Heat Transfer Eng. 31 (4) (2010) 321−330.

[35] H.O.M. Felcar, G. Ribatski, Avaliação de métodos preditivos para perda de carga durante o escoamento bifásico e a ebulição convectiva em micro-canais, in: Proceedings of the 1st Brazillan Meeting on Boiling, Condensation and Multiphase Flow, Florianópolis, Brazil, Paper MF-104, 2008.

[36] B.S. Field, P.S. Hrnjak, Two-phase pressure drop and flow regime of refrigerants and refrigerant-oil mixtures in small channel, Technical Report TR-261, in: Air Conditioning and Refrigeration Center, University of Illinois, Urbana-Champaign, IL, 2007.

[37] L. Friedel, Improved friction pressure drop correlations for horizontal and vertical two-phase pipe flow, in: European Two-phase Flow Group Meeting, Paper E2, Ispra, Italy, 1979.

[38] S. Garimella, T.M. Bandhauer, Measurement of condensation heat transfer coefficients in microchannel tubes, in: ASME International Mechanical Engineering Congress and Exposition, American Society of Mechanical Engineers, New York, NY, United states, 2001, pp. 243−249.

[39] S. Garimella, J.D. Killion, J.W. Colleman, An experimental validated model for two-phase pressure drop in intermittent flow non circular microchannels, J. Fluid Eng. 125 (2003) 887−894.

[40] S. Garimella, A. Agawal, J.D. Killion, Condensation pressure drop in circular micro-channels, Heat Transfer Eng. 26 (2005) 28−35.

[41] S. Garimella, B.M. Fronk, Single and multi-constituent condensation of fluids and mixtures with varying properties in microchannels, in: ECI 8 Th International Conference on Boiling and Condensation Heat Transfer, 2012.

[42] V. Gnielinski, New equations for heat and mass transfer in turbulent pipe and channel flow, Int. Chem. Eng. 16 (2) (1976) 359−368.

[43] G. Goss, J.C. Passos, Heat transfer during the condensation of R134a inside eight parallel microchannels, Int. J. Heat Mass Transfer 59 (2013) 9−19.

[44]   H. Haraguchi, S. Koyama, T. Fujii, Condensation of refrigerants HCFC22, HFC134a
        and HCFC123 in a horizontal smooth tube (1st report, proposal of empirical expressions
        for the local frictional pressure drop), Trans. JSME (B) 60 (1994) 239–244
        (in Japanese).
[45]   H. Haraguchi, S. Koyama, T. Fujii, Condensation of refrigerants HCFC22, HFC134a and
        HCFC123 in a horizontal smooth tube (2nd report, proposal of empirical expressions for
        the local heat transfer coefficient, Trans. JSME 60 (1994) 245–252 (in Japanese].
[46]   J. Heo, H. Park, R. Yun, Condensation heat transfer and pressure drop characteristics of
        CO2 in a microchannel, Int. J. Refrig. 36 (2013) 1657–1668.
[47]   P. Hrnjak, A. Litch, Microchannel heat exchangers for charge minimization in air-cooled
        ammonia condensers and chillers, Int. J. Refrig. 31 (2008) 658–668.
[48]   H. Jaster, P.G. Kosky, Condensation heat transfer in a mixed flow regime, Int. J. Heat
        Mass Tranfer 19 (1976) 95–99.
[49]   F.T. Kanizawa, G. Ribastki, A new void fraction predictive method based on the mini-
        mum energy dissipation, J. Braz. Soc. Mech. Sci. Eng. (under review) (2015).
[50]   N. Kattan, J.R. Thome, D. Favrat, Flow boiling in horizontal tubes: part 1—development
        of a diabatic two phase flow pattern map, J. Heat Transfer 120 (1998) 140–147.
[51]   A. Kawahara, P.M.Y. Chung, M. Kawaji, Investigation of two-phase flow pattern, void
        fraction and pressure drop in a microchannel, Int. J. Multiphase Flow 28 (2002)
        1411–1435.
[52]   N.H. Kim, J.P. Cho, J.O. Kim, B. Youn, Condensation heat transfer of R-22 and R-410A
        in flat aluminum multi-channel tubes with or without micro-fins, Int. J. Refrig. 26 (2003)
        830–839.
[53]   S.M. Kim, I. Mudawar, Flow condensation in parallel micro-channels - part 2: heat
        transfer results and correlation technique, Int. J. Heat Mass Transfer 55 (2012) 984–994.
[54]   S.M. Kim, J. Kim, I. Mudawar, Flow condensation in parallel micro-channels – part 1:
        experimental results and assessment of pressure drop correlations, Int. J. Heat Mass
        Transfer 55 (2012) 971–983.
[55]   A.M. Kim, I. Mudawar, Review of databases and predictive methods for pressure drop in
        adiabatic, condensing and boiling mini/micro-channel flows, Int. Journal Heat Mass
        Transfer 77 (2014) 74–97.
[56]   S. Koyama, K. Kuwara, K. Nakashita, S. Kudo, K. Yamamoto, An experimental study on
        condensation of refrigerant R134a in a multi-port extruded tube, Int. J. Refrig. 24 (2003)
        425–432.
[57]   H.J. Lee, S.Y. Lee, Pressure drop correlations for two-phase flow within horizontal
        rectangular channels with small heights, Int. J. Multiphase Flow 27 (2001) 783–796.
[58]   S. Lin, C.C.K. Kwok, R.Y. Li, Z.H. Chen, Z.Y. Chen, Local frictional pressure drop
        during vaporization for R-12 through capillary tubes, Int. J. Multiphase Flow 17 (1991)
        95–102.
[59]   N. Liu, J.M. Li, J. Sun, H.S. Wang, Heat transfer and pressure drop during condensation
        of R152a in circular and square microchannels, Exp. Therm. Fluid Sci. 47 (2013) 60–67.
[60]   A. López Belchí, F. Illan-Gomez, F. Vera-Garcia, J.R. Garcia-Cascales, Experimental
        condensing two-phase frictional pressure drop inside mini-channels. Comparisons and
        new model development, Int. J. Heat Mass Transfer 75 (2014) 581–591.
[61]   M. Matkovic, A. Cavallini, D. Del Col, L. Rosseto, Experimental study on condensation
        heat transfer inside a single circular minichannel, Int. J. Heat Mass Transfer 52 (2009)
        2311–2323.
[62]   K. Mishima, T. Hibiki, Some characteristics of air-water two-phase flow in small
        diameter vertical tubes, Int. J. Multiphase Flow 22 (1996) 703–712.

[63] K.W. Moser, R.L. Webb, B. Na, A new equivalent Reynolds number model for a condensation in smooth tubes, Trans. ASME 120 (1998) 410−417.

[64] H. Müller-Steinhagen, K. Heck, A simple friction pressure drop correlation for two-phase flow in pipes, Chem. Eng. Process 20 (1986) 297−308.

[65] M. Nusselt, Die Oberflächen Kondensation des wasserdampfes, Z. Ver. Dtsch. Ing. 60 (1916) 541.

[66] H.K. Oh, C.H. Son, Condensation heat transfer characteristics of R-22, R-134a and R-410Ain a single circular microtube, Exp. Therm. Fluid Sci. 35 (2011) 706−716.

[67] C.Y. Park, P. Hrnjak, CO_2 flow condensation heat transfer and pressure drop in multi-port microchannels at low temperatures, Int. J. Refrig. 32 (2009) 1129−1139.

[68] C. Park, F. Vakili-Farahani, L. Consolini, J.R. Thome, Experimental study on condensation heat transfer in vertical minichannels for new refrigerant R1234ze(E) versus R134a and R236fa, Exp. Therm. Fluid Sci. 35 (2011) 442−454.

[69] K. Pehlivan, I. Hassan, M. Vaillancourt, Experimental study on two-phase flow and pressure drop in millimeter-size channels, Appl. Therm. Eng. 26 (2006) 1506−1514.

[70] X. Quan, P. Cheng, H. Wu, An experimental investigation on pressure drop of steam condensing in silicon microchannels, Int. J. Heat Mass Transfer 51 (2008) 5454−5458.

[71] X. Quan, L. Dong, P. Cheng, Determination of annular condensation heat transfer coefficient of steam in microchannels with trapezoidal cross sections, Int. J. Heat Mass Transfer 53 (2010) 3670−3676.

[72] R. Revellin, J.R. Thome, Adiabatic two-phase frictional pressure drop in microchannel, Exp. Therm. Fluid Sci. 31 (2007) 673−685.

[73] G. Ribatski, A critical overview on the recent literature concerning flow boiling and two-phase flows inside micro-scale channels, Exp. Heat Transfer 26 (2013) 198−246.

[74] S. Saisorn, S. Wongwises, Flow pattern, void fraction and pressure drop of two-phase air−water flow in a horizontal circular micro-channel, Exp. Therm. Fluid Sci. 32 (2008) 748−760.

[75] K. Sakamatapan, J. Kaew-On, A.S. Dalkilic, O. Mahian, S. Wongwises, Condensation heat transfer characteristics of R-134a flowing inside the multiport minichannels, Int. J. Heat Mass Transfer 64 (2013) 976−985.

[76] K. Sakamatapan, S. Wongwises, Pressure drop during condensation of R134a flowing inside a multiport minichannel, Int. J. Heat Mass Transfer 75 (2014) 31−39.

[77] M.M. Shah, A general correlation for heat transfer during film condensation inside pipes, Int. Heat Transfer 22 (1979) 547−556.

[78] M.M. Shah, An improved and extended general correlation for heat transfer during condensation in Plain tube, HVAC&R Res. 15 (2009) 889−913.

[79] T.A. Shedd, Void Fraction and Pressure Drop Measurements for Refrigerant R410a Flows in Small Diameter Tubes, 2010. Preliminary AHRTI Report No. 20110−20201.

[80] J.S. Shin, M.H. Kim, An experimental study of condensation heat transfer inside a mini-channel with a new measurement technique, Int. J. Multiphase Flow 30 (2004) 311−325.

[81] H.M. Soliman, Mist−annular transition during condensation and its influence on the heat transfer mechanism, Int. J. Multiphase Flow 12 (2) (1986) 277−288.

[82] J.R. Thome, J. El Hajal, A. Cavallini, Condensation in horizontal tubes, part 2: new heat transfer model based on flow regimes, Int. J. Heat Mass Transfer 46 (2003) 3365−3387.

[83] C.B. Tibiriçá, J.D. Da Silva, G. Ribatski, Experimental investigation of flow boiling pressure drop of R134a in a microscale horizontal smooth tube, J. Therm. Sci. Eng. Appl. 3 (2011) 0110061−0110068.

[84]   C.B. Tibiriçá, G. Ribatski, Flow boiling in micro-scale channels - synthesized literature review, Int. J. Refrig. 36 (2013) 301—324.

[85]   D.P. Traviss, W.M. Rohsenow, Flow regimes in horizontal two-phase flow with condensation, ASHRAE Trans. 79 (Part 2) (1973) 31—39.

[86]   D.P. Traviss, W.M. Rohsenow, A.B. Baron, Forced-convection condensation inside tubes: a heat transfer equation for condenser design, ASHRAE Trans. 79 (Part 1) (1973) 157—165.

[87]   K.A. Triplett, S.M. Ghiaasiaan, S.I. Abdel-Khalik, D.L. Sadowski, Gas—liquid two-phase flow in microchannels part I: two-phase flow patterns, Int. J. Multiphase Flow 25 (1999) 377—394.

[88]   W.W. Wang, T.D. Radcliff, R.N. Christensen, A condensation heat transfer correlation for millimeter-scale tubing with flow regime transition, Exp. Therm. Fluid Sci. 26 (2002) 473—485.

[89]   H.S. Wang, J.W. Rose, Theory of heat transfer during condensation in microchannels, Int. J. Heat Mass Transfer 54 (2011) 2525—2534.

[90]   J.M. Winkler, J. Killion, S. Garimella, Void fractions for condensing refrigerant flow in small channels. Part II: void fraction measurement and modeling, Int. J. Refrig. 35 (2012) 246—262.

[91]   Y.Y. Yan, T.F. Lin, Condensation heat transfer and pressure drop of refrigerant R134a in a small pipe, Int. J. Heat Mass Transfer 42 (1999) 697—708.

[92]   C.Y. Yang, R.L. Webb, Condensation of R-12 in small hydraulic diameter extruded aluminum tubes with and without micro-fins, Int. J. Heat Mass Transfer 39 (1996) 791—800.

[93]   D.A. Yashar, Experimental Investigation of Void Fraction during Horizontal Flow in Smaller Diameter Refrigeration Applications (M.S. Thesis), University of Illinois, Urbana Champaign, IL, 1998.

[94]   M. Zhang, R.L. Webb, Correlation of two-phase friction for refrigerants in small-diameter tubes, Exp. Therm. Fluid Sci. 25 (2001) 131—139.

[95]   H. Zhang, J.M. Li, B.X. Wang, Experimental investigation of condensation heat transfer and pressure drop of R22, R410A and R407C in mini-tubes, Int. J. Heat Mass Transfer 55 (2012) 3522—3532.

[96]   K.J. Bell, M.A. Ghaly, An approximate generalized method for multicomponent/partial condenser, AIChE Symp. Ser. vol. 69 (1973), pp. 72—79.

Conclusions

Sujoy K. Saha[1], *Gian P. Celata*[2]
[1]Department of Mechanical Engineering, Indian Institute of Engineering Science and Technology, Shibpur, Howrah, West Bengal, India; [2]Energy Technology Department, ENEA Casaccia Research Centre, S. M. Galeria, Rome, Italy

With the growing interest in microchannels, specifically in electronic cooling and surgical field, many research findings pertaining to fluid flow and heat transfer have been reported but no general/concrete conclusion seems to have emerged. Some investigators have reported increased heat transfer and/or pressure drop, while others reported the opposite. To some degree, these deviations may be due to the difficulties in measuring the parameters necessary for the theoretical calculation. The pressure drop in laminar flow is inversely proportional to the diameter to the power of 4, meaning that a very precise value of the diameter is necessary for determining the friction factor. Further, although an explanation for discrepancy could come from experimental uncertainties; many authors have not considered the entrance and exit losses or the flow developing length in their discussion and reported a great discrepancy. About 76% of the investigators studied heat transfer and pressure drop of single-phase microchannels in the laminar region. This shows that the behavior of fluids in the turbulent region remained unexplored and this is the main reason that the experimental results of many researchers were overpredicted or underpredicted.

Because boiling phenomenon is in general considered to be more efficient than the single-phase flow due to the large latent heat transfer at a constant temperature in a liquid, the advantages have not yet been realized in microchannels. Two-phase electronic cooling is a better alternative than single-phase cooling because substrate material will be subjected to uniform thermal stress rather than varying thermal stress (as occurs with single-phase cooling). The understanding of two-phase flow through microchannels is at its infancy. Flow boiling instabilities are well recognized as a hindrance to the realization of evaporative micro-scale heat exchangers. They introduce temperature oscillations, thermal stress cycling, and vibration, and they can compromise the structural integrity of the system. More importantly, they can lead to premature transition to the limiting critical heat flux condition, and this is perhaps one of the most difficult situations that can occur in flow boiling system. It marks the transition from a very effective heat transfer process to a very ineffective process. Since the heat transfer coefficient drops very rapidly following the critical heat flux condition, the surface can reach unacceptable temperatures causing complete burnout of the channel. Methods developed so far to suppress flow boiling instabilities have either increased pressure drop or lacked sufficient capabilities to suppress the flow instabilities. New techniques need to be developed to overcome the drawbacks associated with the

Microchannel Phase Change Transport Phenomena. http://dx.doi.org/10.1016/B978-0-12-804318-9.00009-1

current techniques. Regardless of the effort to develop new techniques, fundamental studies need to be continued to map flow boiling instabilities and identify their modes. This is desirable because instabilities are multidimensional problems with many variables controlling their occurrence. In short, more fundamental and applied research is needed on the present topic of this book.

Index

Note: Page numbers followed by "f" and "t" indicate figures and tables, respectively.